T0132648

INTRODUCTION TO
Computational Models with Python

Chapman & Hall/CRC
Computational Science Series

SERIES EDITOR

Horst Simon
Deputy Director
Lawrence Berkeley National Laboratory
Berkeley, California, U.S.A.

PUBLISHED TITLES

PUBLISHED TITLES CONTINUED

INTRODUCTION TO
Computational
Models with
Python

José M. Garrido

Kennesaw State University,
Kennesaw, Georgia, USA

CRC Press
Taylor & Francis Group
Boca Raton London New York

CRC Press is an imprint of the
Taylor & Francis Group, an **informa** business

A CHAPMAN & HALL BOOK

CRC Press
Taylor & Francis Group
6000 Broken Sound Parkway NW, Suite 300
Boca Raton, FL 33487-2742

© 2016 by Taylor & Francis Group, LLC
CRC Press is an imprint of Taylor & Francis Group, an Informa business

No claim to original U.S. Government works

Printed on acid-free paper
Version Date: 20150623

International Standard Book Number-13: 978-1-4987-1203-3 (Hardback)

Visit the Taylor & Francis Web site at
http://www.taylorandfrancis.com

and the CRC Press Web site at
http://www.crcpress.com

Contents

SECTION III **Data Structures, Object Orientation, and Recursion**

SECTION V Linear Optimization Models

CHAPTER 20 ▪ Linear Optimization Modeling   325

List of Figures

List of Tables

Preface

A *computational model* is a computer implementation of the solution to a (scientific) problem for which a mathematical representation has been formulated. These models are applied in various areas of science and engineering to solve large-scale and complex scientific problems. Developing a computational model involves studying the problem, formulating a mathematical representation, implementing the solution, and validating the model by applying computer science concepts, principles, and methods, and usually includes applying techniques of high-performance computing (HPC).

Computational modeling focuses on reasoning about problems using computational thinking and multidisciplinary/interdisciplinary computing for developing computational models to solve complex problems. It is the foundation component of computational science, which is an emerging discipline that includes concepts, principles, and methods from applied mathematics and computer science.

This book presents an introduction to computational models and their implementation using the Python programming language. This is one of the most popular programming languages, especially among scientists of a wide variety of disciplines. One advantage of Python is the higher level of abstraction at the programming level. Although Fortran and C programming languages are the ones most suitable for high-performance computing (HPC), Python provides several packages to improve performance of numerical computing, comparable to C and Fortran.

Python is a high-level interpreted language that is much slower than C in general; however, Python provides many ways to optimize critical parts of code by rewriting them in C. Numpy and all standard libraries are already optimized in this way, and custom application code can also be optimized, for instance with PyPy or Cython.

People who think Python is slow for serious number crunching haven't used the Numpy and Scipy modules available with Python. Developing software with Python on relatively small projects provides one an appreciation of the dynamically typed nature of this language, which often shortens the development time. For larger projects this may be a hindrance, as the code would run slower than, say, its equivalent in C++. Using Numpy and/or Scipy modules with Python, the code would run as fast as a native C++ program (where the code in C++ would sometimes take longer to develop). The speed is achieved from using modules written in C or libraries written in Fortran.

An interesting Python/Numpy/C++ related performance question appears on *Benchmarking (python vs. c++ using BLAS) and (numpy)* by J.F. Sebastian. He writes "There is no difference between C++ and Numpy on my machine."

Some researchers have commented on the Web that Python has been found to be a comprehensive, flexible, and easy-to-use language, much more convenient for data science than C/C++ or Java, or even R and MATLAB® in data mining and big data analysis (using Python modules such as Panda).

The primary goal of this book is to present fundamental and introductory principles for developing computational models for a wide variety of applications. The prerequisites are knowledge of programming and Calculus I. Emphasis is on reasoning about problems, conceptualizing the problem, mathematical formulation, and the computational solution that involves computing results and visualization.

The book emphasizes analytical skill development and problem solving. The main software tools for implementing computational models are: the Python programming language interpreter, the several packages available from the huge Python Library, and an Integrated Development Environment. These tools are open-source and platform independent (Linux, MacOS, and Windows).

The material in this book is aimed at beginners to advanced undergraduate science (and engineering) students. However, the vision in the book is to promote and introduce the principles of computational modeling as early as possible in the undergraduate curricula and to introduce the approaches of multidisciplinary and interdisciplinary computing.

This book provides a foundation for more advanced courses in scientific computing, including parallel computing using MPI, grid computing and other methods and techniques used in high-performance computing. Additional applied mathematical concepts outside the scope of this book are non-linear equations, partial differential equations, non-linear optimization, and other techniques.

Please note that there are many sites on the Web that maintain tutorials on various additional aspects of Python that fall outside the scope of this book. One such aspect is the installation of the Python interpreter and the necessary library modules in various operating systems (MacOS, Linux, Windows). Another topic is the use of several IDEs for developing Python programs. Another related aspect is the use of graphics for developing GUIs, which may improve the use of Python programs, by end users.

The material in the book is presented in five parts or sections. The first part is an overview of problem solving and introduction to simple Python programs. This part introduces the basic modeling and techniques for designing and implementing problem solutions, independent of software and hardware tools.

The second part presents an overview of programming principles with the Python programming language. The relevant topics are basic programming concepts, data definitions, programming structures with flowcharts and pseudo-code, solving problems, and algorithms.

The third part introduces Python lists, arrays, basic data structures, object orientation, linked lists and recursion, and running programs under Linux.

The fourth part applies programming principles and fundamental techniques to implement the relatively simple computational models. It gradually introduces numerical methods and mathematical modeling principles. Simple case studies of problems that apply mathematical models are presented. Case studies are of simple linear, quadratic, geometric, polynomial, and linear systems using the NumPy package. Computational models that use polynomial evaluation, computing roots of polynomials, interpolation, regression, and systems of linear equations are discussed. Examples and case studies demonstrate the computation and visualization of data produced by computational models.

The fifth part introduces the modeling of linear optimization problems and several case studies are presented. The problem formulation to implementation of computational models with linear optimization is shown.

All the Python programs in source code, the data files used, and several output files (text mode) are posted on the following website:

`cs.kennesaw.edu/~jgarrido/comp_models/python_mod`

José M. Garrido
Kennesaw, Georgia

I

Problem Solving

## CHAPTER 1

Problem Solving and Computing

1.1 INTRODUCTION

Computer problem solving attempts to derive a computer solution to a real-world problem, and a computer *program* is the implementation of the solution to the problem. A *computational model* is a computer implementation of the solution to a (scientific) problem for which a mathematical representation has been formulated. These models are applied in various areas of science and engineering to solve large-scale and complex scientific problems.

A computer program consists of data definitions and a sequence of instructions. The instructions allow the computer to manipulate the input data to carry out computations and produce desired results when the program executes; an appropriate programming language is used.

This chapter discusses problem solving principles and presents elementary concepts and principles of problem solving, computational models, and programs.

1.2 COMPUTER PROBLEM SOLVING

Problem solving is the process of developing a computer solution to a given real-world problem. The most challenging aspect of this process is discovering the method to solve the problem. This method of solution is described by an algorithm. A general process of problem solving involves the following steps:

1. Understand the problem.

2. Describe the problem in a clear, complete, and unambiguous form.

3. Design a solution to the problem (algorithm).

4. Develop a computer solution to the problem.

An *algorithm* is a description of the sequence of steps performed to produce the desired results, in a clear, detailed, precise, and complete manner. It describes the computations on the given data and involves a sequence of instructions or operations that are to be performed on the input data in order to produce the desired results (output data).

A program is a computer implementation of an algorithm and consists of a set of data definitions and sequences of instructions. The program is written in an appropriate programming language and it tells the computer how to transform the given data into correct results by performing a sequence of computations on the data. An algorithm is described in a semi-formal notation such as pseudo-code and flowcharts.

1.3 ELEMENTARY CONCEPTS

A *model* is a representation of a system, a problem, or part of it. The model is simpler than, and should be equivalent to, the real system in all relevant aspects. In this sense, a model is an abstract representation of a problem. *Modeling* is the activity of building a model.

Every model has a specific purpose and goal. A model only includes the aspects of the real problem that were decided as being important, according to the initial requirements of the model. This implies that the limitations of the model have to be clearly understood and documented.

An essential modeling method is to use mathematical entities such as numbers, functions, and sets to describe properties and their relationships to problems and real-world systems. Such models are known as *mathematical models*.

A *computational model* is an implementation in a computer system of a mathematical model and usually requires high-performance computational resources to execute. The computational model is used to study the behavior of a large and complex system. Developing a computational model consists of:

- applying a formal software development process; and

- applying *Computer Science* concepts, principles, and methods, such as:

  - abstraction and decomposition
  - programming principles
  - data structures
  - algorithm structures
  - concurrency and synchronization
  - Modeling and simulation
  - multi-threading, parallel, and distributed computing for high performance (HPC)

Abstraction is a very important principle in developing computational models. This is extremely useful in dealing with large and complex problems or systems. Abstraction is the hiding of the details and leaving visible only the essential features of a particular system.

One of the critical tasks in modeling is representing the various aspects of a system at different levels of abstraction. A good abstraction captures the essential elements of a system, and purposely leaves out the rest.

Computational thinking is the ability of reasoning about a problem and formulating a computer solution. Computational thinking consists of the following elements:

- Reasoning about computer problem solving.

- The ability to describe the requirements of a problem and, if possible, design a mathematical solution that can be implemented in a computer.

- The solution usually requires *multi-disciplinary* and *inter-disciplinary* approaches to problem solving.

- The solution normally leads to the construction of a *computational model*.

Computational Science integrates concepts and principles from applied mathematics and computer science and applies them to the various scientific and engineering disciplines. Computational science is:

- An emerging multidisciplinary area.

- The intersection of the more traditional sciences, engineering, applied mathematics, and computer science, and focuses on the integration of knowledge for the development of problem-solving methodologies and tools that help advance the sciences and engineering areas. This is illustrated in Figure 1.1.

- An area that has as a general goal the development of high-performance computer models.

- An area that mostly involves multi-disciplinary computational models including simulation.

When a mathematical analytical solution of the model is not possible, a numerical and graphical solution is sought and experimentation with the model is carried out by changing the parameters of the model in the computer, and studying the differences in the outcome of the experiments. Further analysis and predictions of the operation of the model can be derived or deduced from these computational experiments.

One of the goals of the general approach to problem solving is modeling the problem at hand, building or implementing the resulting solution using an interactive tool environment (such as MATLAB® or Octave), Python with an IDE, or with some appropriate programming language, such as C, C++, Fortran, or Ada.

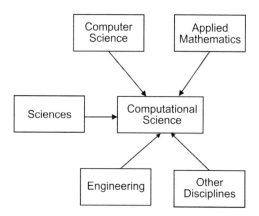

Figure 1.1 Computational science as an integration of several disciplines.

1.4 DEVELOPING COMPUTATIONAL MODELS

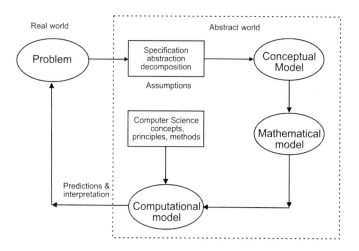

Figure 1.2 Development of computational models.

The process of developing computational models consists of a sequence of activities or stages that starts with the definition of modeling goals and is carried out in a possibly iterative manner. Because models are simplifications of reality, there is a trade-off as to what level of detail is included in the model. If too little detail is included in the model, one runs the risk of missing relevant interactions and the resultant model does not promote understanding. If too much detail is included in the model, the model may become overly complicated and actually preclude the development of understanding. Figure 1.2 illustrates a simplified process for developing computational models.

Computational models are generally developed in an iterative manner. After the first version of the model is developed, the model is executed, results from the execution run are studied, the model is revised, and more iterations are carried out until an adequate level of understanding is developed. The process of developing a model involves the following general steps:

1. Definition of the *problem statement* for the computational model. This statement must provide the description of the purpose for building the model, the questions it must help to answer, and the type of expected results relevant to these questions.

2. Definition of the *model specification* to help define the conceptual model of the problem to be solved. This is a description of what is to be accomplished with the computational model to be constructed, and the assumptions (constraints) and domain laws to be followed. Ideally, the model specification should be clear, precise, complete, concise, and understandable. This description includes the list of relevant components, the interactions among the components, the relationships among the components, and the dynamic behavior of the model.

3. Definition of the *mathematical model*. This stage involves deriving a representation of the problem solution using mathematical entities and expressions and the details of the algorithms for the relationships and dynamic behavior of the model.

4. *Model implementation*. The implementation of the model can be carried out with a software environment such as MATLAB and Octave, in a simulation language, or in a general-purpose high-level programming language, such as Ada, C, C++, or Java. The simulation software to use is also an important practical decision. The main tasks in this phase are coding, debugging, and testing the software model.

5. *Verification* of the model. From different runs of the implementation of the model (or the model program), this stage compares the output results with those that would have been produced by correct implementation of the conceptual and mathematical models. This stage concentrates on attempting to document and prove the correctness of the model implementation.

6. *Validation* of the model. This stage compares the outputs of the verified model with the outputs of a real system (or a similar already developed model). This stage compares the model data and properties with the available knowledge and data about the real system. Model validation attempts to evaluate the extent to which the model promotes understanding.

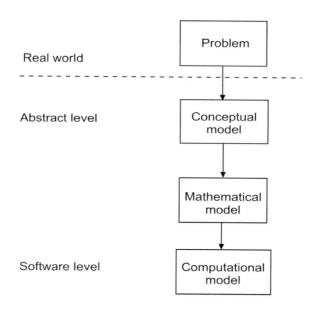

Figure 1.3 Model development and abstract levels.

A conceptual model can be considered a high-level specification of the problem and it is a descriptive model. It is usually described with some formal or semi-formal notation. For example, discrete-event simulation models are described with UML (the Unified Modeling Language) and/or extended simulation activity diagrams.

The conceptual model is formulated from the initial problem statement, informal user requirements, and data and knowledge gathered from analysis of previously developed models. The stages mentioned in the model development process are carried out at different levels of abstraction. Figure 1.3 illustrates the relationship between the various stages of model development and their abstraction level.

1.5 TEMPERATURE CONVERSION

The process of developing a computational model is illustrated in this section with an extremely simple problem: the temperature conversion problem. A basic sequence of steps is discussed for solving this problem and for developing a computational model.

1.5.1 Initial Problem Statement

American tourists visiting Europe do not usually understand the units of temperature used in weather reports. The problem is to devise some mechanism for indicating the temperature in Fahrenheit from a known temperature in Celsius.

1.5.2 Analysis and Conceptual Model

A brief analysis of the problem involves:

1. Understanding the problem. The main goal of the problem is to develop a temperature conversion facility from Celsius to Fahrenheit.

2. Finding the mathematical representation or formulas for the conversion of temperature from Celsius to Fahrenheit. Without this knowledge, we cannot derive a solution to this problem. The conversion formula is the mathematical model of the problem.

3. Knowledge of how to implement the mathematical model in a computer. We need to express the model in a particular computer tool or a programming language. The computer implementation must closely represent the model in order for it to be correct and useful.

4. Knowledge of how to test the program for correctness.

1.5.3 Mathematical Model

The mathematical representation of the solution to the problem is the formula expressing a temperature measurement F in Fahrenheit in terms of the temperature measurement C in Celsius, which is:

$$F = \frac{9}{5} C + 32. \qquad (1.1)$$

Here, C is a variable that represents the given temperature in degrees Celsius, and F is a derived variable, whose value depends on C.

A formal definition of a function is beyond the scope of this chapter. Informally, a *function* is a computation on elements in a set called the *domain* of the function, producing results that constitute a set called the *range* of the function. The elements in the domain are sometimes known as the input parameters. The elements in the range are called the output results.

Basically, a function defines a relationship between two (or more variables), x and y. This relation is expressed as $y = f(x)$, so y is a function of x. Normally, for every value of x, there is a corresponding value of y. Variable x is the independent variable and y is the dependent variable.

The mathematical model is the mathematical expression for the conversion of a temperature measurement in Celsius to the corresponding value in Fahrenheit. The mathematical formula expressing the conversion assigns a value to the desired temperature in the variable F, the dependent variable. The values of the variable C can change arbitrarily because it is the independent variable. The model uses real numbers to represent the temperature readings in various temperature units.

1.6 AREA AND PERIMETER OF A CIRCLE

In this section, another simple problem is formulated: the mathematical model(s) and the algorithm. This problem requires computing the area and circumference of a circle, given its radius. The mathematical model is:

$$cir = 2\pi r$$
$$area = \pi r^2$$

The high-level algorithm description in informal pseudo-code notation is:

1. Read the value of the radius of a circle, from the input device.

2. Compute the area of the circle.

3. Compute the circumference of the circle.

4. Print or display the value of the area of the circle to the output device.

5. Print or display the value of the circumference of the circle to the output device.

A more detailed algorithm description follows:

1. Read the value of the radius r of a circle, from the input device.

2. Establish the constant π with value 3.14159.

3. Compute the area of the circle.

4. Compute the circumference of the circle.

5. Print or display the value of *area* of the circle to the output device.

6. Print or display the value of *cir* of the circle to the output device.

The previous algorithm now can be implemented by a program that calculates the circumference and the area of a circle.

1.7 CATEGORIES OF COMPUTATIONAL MODELS

From the perspective of how the model changes state in time, computational models can be divided into two general categories:

1. Continuous models

2. Discrete models

A *continuous model* is one in which the changes of state in the model occur continuously with time. Often the *state variables* in the model are represented as continuous functions of time. These types of models are usually modeled as sets of difference or differential equations.

For example, a model that represents the temperature in a boiler as part of a power plant can be considered a continuous model because the state variable that represents the temperature of the boiler is implemented as a continuous function of time.

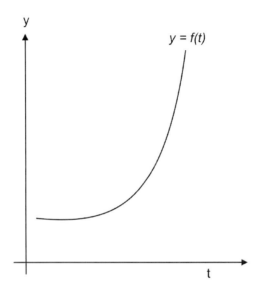

Figure 1.4   Continuous model.

In scientific and engineering practice, a computational model of a real physical system is often formulated as a continuous model and solved numerically by applying numerical methods implemented in a programming language. These models can also be simulated with software tools, such as Simulink and Scilab, which are computer programs designed for numeric computations and visualization. Figure 1.4 illustrates how the a variable changes with time.

A *discrete model* represents a system that changes its states at discrete points in time, i.e., at specific instants. The model of a simple car-wash system is a discrete-event model because an arrival event occurs, and causes a change in the state variable that represents the number of cars in the queue that are waiting to receive service from the machine (the server). This state variable and any other only change its values when an event occurs, i.e., at discrete instants. Figure 1.5 illustrates the changes in the number of cars in the queue of the model for the simple car-wash system.

Depending on the variability of some parameters, computational models can be separated into two categories:

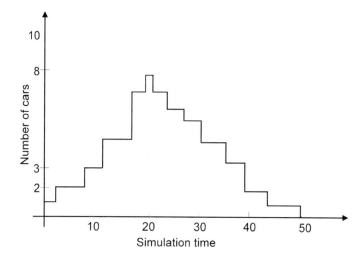

Figure 1.5 Discrete changes of number of cars in the queue.

1. Deterministic models

2. Stochastic models

A deterministic model exhibits a completely predictable behavior. A stochastic model includes some uncertainty implemented with random variables, whose values follow a probabilistic distribution. In practice, a significant number of models are stochastic because the real systems modeled usually include properties that are inherently random.

An example of a deterministic simulation model is a model of a simple car-wash system. In this model, cars arrive at exact specified instants (but at the same instants), and all have exact specified service periods (wash periods); the behavior of the model can be completely and exactly determined.

The simple car-wash system with varying car arrivals and varying service demand from each car, is a stochastic system. In a model of this system, only the averages of these parameters are specified together with a probability distribution for the variability of these parameters. Uncertainty is included in this model because these parameter values cannot be exactly determined.

1.8 GENERAL PROCESS OF SOFTWARE DEVELOPMENT

For large software systems, a general software development process involves carrying out a sequence of well-defined phases or activities. The process is also known as the *software life cycle*.

The simplest approach for using the software life cycle is the *waterfall model*. This model represents the sequence of phases or activities needed to develop the software system through installation and maintenance of the software. In this model, the activity in a given phase cannot be started until the activity of the previous phase has been completed.

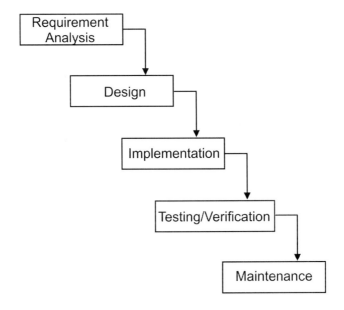

Figure 1.6 The waterfall model.

Figure 1.6 illustrates the sequence of phases that are performed in the waterfall software life cycle. The various phases of the software life cycle are the following:

1. *Analysis*, which results in documenting the problem description and what the problem solution is supposed to accomplish.

2. *Design*, which involves describing and documenting the detailed structure and behavior of the system model.

3. *Implementation* of the software using a programming language.

4. *Testing* and verification of the programs.

5. *Installation*, which results in delivery and installation of the programs.

6. *Maintenance.*

There are some variations of the waterfall model of the life cycle. These include returning to the previous phase when necessary. More recent trends in system development have emphasized an iterative approach, in which previous stages can be revised and enhanced.

A more complete model of the software life cycle is the *spiral model* that incorporates the construction of *prototypes* in the early stages. A prototype is an early version of the application that does not have all the final characteristics. Other development approaches involve prototyping and rapid application development (RAD).

1.9 MODULAR DESIGN

To design and implement large and complex problems, two principles are essential: decomposition and abstraction. A problem may be too large and complex to solve as a single unit. An important strategy in problem solving is *divide and conquer*. It consists of partitioning a large problem into *subproblems* that are smaller, easier to solve, and easier to manage. Each subproblem can be solved individually. These subproblems or modules need to be well organized in order to achieve the overall solution to the problem.

The technique used in modular design is *top-down design*. A more technical term for this technique is *decomposition*. The top module is the main module and includes the high-level solution, or the *big picture* design. The modules at the next lower level include more detailed design, and the modules at the bottom level include the maximum detail of design.

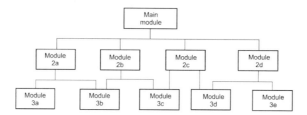

Figure 1.7 Modular design.

The *abstraction* principle is applied in this technique. The top-level module includes the design at the highest level of abstraction, and is the easiest to understand and describe because it includes no details of the design. The lowest-level modules contain the all the necessary detail, at the lowest level of abstraction. Applying top-down design results in a hierarchical solution to a problem and includes multiple levels of abstraction. Figure 1.7 shows the module hierarchy used in top-down design.

1.10 PROGRAMMING LANGUAGES

A programming language is used by programmers to write programs. This language contains a defined set of syntax and semantic rules. The syntax rules describe how to write well-defined statements. The semantic rules describe the meaning of the statements.

1.10.1 High-Level Programming Languages

The solution to a problem is implemented by a program written in an appropriate programming language. This program is known as the *source program* and is written in a high-level programming language.

A high-level programming language is a formal notation in which to write instructions to the computer in the form of a program. A programming language helps programmers in the writing of programs for a large family of problems.

High-level programming languages are hardware-independent and are problem-oriented (for a given family of problems). These languages allow more readable programs, and are easy to write and maintain. Examples of these languages are C, Fortran, Ada, C++, Eiffel, and Java.

Programming languages like C++ and Java can require considerable effort to learn and master. The Python programming language is much easier to learn and use.

There are several integrated development environments (IDE) that facilitate the development of programs. Examples of these are: Eclipse, Netbeans, CodeBlocks, and Codelite. Other tools include IDEs that are designed for numerical and scientific problem solving that have their own programming language. Some of these computational tools are MATLAB, Octave, Mathematica, Scilab, Stella, and Maple.

There are several IDEs available for developing scientific applications with Python. Some of these are Spyder, IEP, Eric, PyDev, WingIDE, Canopy, Komodo IDE, and Pycharm. The last four are commercial products but the companies provide an academic version with limited capabilities.

For some programming languages, the source program is compiled (translated or converted) into an equivalent program in *machine code*, which is the only programming language that the computer can understand. The computer can only execute instructions that are in machine code.

For other programming languages (such as Python), the source program is interpreted. This means every command in the source program is analyzed for correct syntax and then it is executed immediately by the interpreter.

The program executing in the computer usually reads input data from the input device and after carrying out some computations, it writes results to the output device(s).

1.10.2 Interpreters and Python

An interpreter is a special program that performs syntax checking of a command in a user program written in a high-level programming language and immediately executes the command. It repeats this processing for every command in the program. Examples of interpreters are the ones used for the following languages: Python, MATLAB, Octave, PHP, and Perl.

As mentioned previously, the Python interpreter reads a command written in the Python programming language and immediately executes the command. A Python program is a file of Python commands, so the program is also known as a *script*. Figure 1.8 illustrates the interpretation of a Python command and the response to the command. In a terminal or command window, all the interaction with a user takes place by typing Python commands.

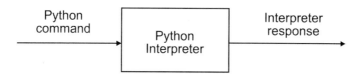

Figure 1.8 Python interpreter.

The **command prompt** is the symbol that the Python interpreter displays on the window to alert the user that it is waiting for a command. The Python prompt is >>> and is used for interactive mode of computing with Python. In a terminal window, the user starts the Python interpreter simply by typing python at the Linux prompt (Bash shell). After the interpreter starts, the Python prompt >>> is displayed.

The following commands were typed on a PC with Linux. The first command typed is a simple assignment of 15 to variable *num*. By typing the name of the variable, the interpreter displays the value associated with it (15). The next command typed assigns the value 20.6 to variable *y* and the command after that assigns the value 2.56 to variable *x*. After that, another command assigns the value 200 to variable *j*. Note how the Python interpreter immediately responds to a command; this is the interactive mode of using Python.

```
$ python
Python 2.7.8 (default, Oct 20 2014, 19:24:18)
[GCC 4.9.1] on linux2
Type "help", "copyright", "credits" or "license" for more
    information.
>>> num = 15
>>> num
15
>>> y = 20.6
>>> y
20.6
>>> x = 2.56
>>> j = 200
>>> j
200
>>>
```

The Python interpreter assumes that y is a non-integer numeric variable since the constant value 20.6 is assigned to y. The same applies to variable x. Variable j is an integer variable because the integer constant 200 is assigned to it. The following command performs a multiplication of the values of variables y and x, the intermediate result is added with constant 125.25, and the final resulting value of this computation is assigned to variable z. When there is no assignment to a variable, the Python interpreter displays the value of the variable. The next few lines show slightly more complex calculations.

```
>>> z = y * x + 125.25
>>> z
177.986
>>> z / 12.5
14.23888
>>>
```

1.10.3 Compilers

A compiler is a special program that translates another program written in a programming language into an equivalent program in binary or machine code, which is the only language that the computer accepts for processing.

In addition to *compilation*, an additional step known as *linking* is required before a program can be executed. Examples of programming languages that require compilation and linking are C, C++, Eiffel, Ada, and Fortran. Other programming languages, such as Java, require compilation and interpretation.

1.10.4 Compiling and Execution of Java Programs

Figure 1.9 Compiling a Java source program.

To compile and execute programs written in the Java programming language, two special programs are required, the compiler and the interpreter. The Java compiler checks for syntax errors in the source program and translates it into *bytecode*, which is the program in an intermediate form. The Java bytecode is not dependent on any particular platform or computer system. To execute this bytecode, the Java Virtual Machine (JVM), carries out the interpretation of the bytecode.

Figure 1.9 shows what is involved in compilation of a source program in Java. The Java compiler checks for syntax errors in the source program and then translates it into a program in bytecode, which is the program in an intermediate form.

Figure 1.10 Executing a Java program.

The Java bytecode is not dependent on any particular platform or computer system. This makes the bytecode very portable from one machine to another.

Figure 1.10 shows how to execute a program in bytecode. The Java virtual machine (JVM) carries out the interpretation of the program in bytecode.

1.10.5 Compiling and Executing C Programs

Figure 1.11 Compiling a C program.

Programs written in C must be compiled, linked, and loaded into memory before executing. An executable program file is produced as a result of linking. The libraries are a collection of additional code modules needed by the program. Figure 1.11 illustrates the compilation of a C program. Figure 1.12 illustrates the linkage of the program. The executable program is the final form of the program that is produced. Before a program starts to execute in the computer, it must be loaded into the memory of the computer.

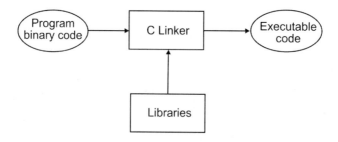

Figure 1.12 Linking a C program.

1.11 PRECISION, ACCURACY, AND ERRORS

Performing numerical computations involves considering several important concepts:

- Number representation

- The number of significant digits

- Precision and accuracy

- Errors

1.11.1 Number Representation

Numeric values use numbers that are represented in number systems. Numbers based on decimal representation use base 10. This is the common number representation used by humans. There are three other relevant number systems: binary (base 2), octal (base 8), and hexadecimal (base 16). Digital computers use the base 2, or binary, system. In a digital computer, a binary number consists of a number of binary digits or *bits*.

Any number z that is not zero ($z \neq 0$) can be written in scientific notation using decimal representation in the following manner:

$$z = \pm \, . \, d_1 d_2 d_3 \, \cdots \, d_s \times 10^p.$$

Each d_i is a decimal digit (has a value from $0, 1 \ldots 9$). Assuming that $d_1 > 0$, the part of the number $d_1 d_2 d_3 \cdots d_s$ is known as the *fraction*, or *mantissa* or *significand* of z. The quantity p is known as the *exponent* and its value is a signed integer.

The number z may be written using a binary representation, which uses binary digits or *bits* and base 2.

$$z = \pm \, . \, b_1 b_2 b_3 \, \cdots \, b_r \times 2^q$$

To represent the significand using a binary representation, the number of bits is different than the number of decimal digits in the decimal representation used previously. The exponent q also has a different value than the exponent p that is used in the decimal representation.

In a computer, most computations with numbers are performed in floating-point arithmetic in which the value of a real number is approximated with a finite number of digits in its mantissa, and a finite number of digits in its exponent. In this number system, a real number that is too big cannot be represented and causes an *overflow*. In a similar manner, a real number that is too small cannot be represented and causes an *underflow*.

1.11.2 Number of Significant Digits

The number of *significant digits* is the number of digits in a numeric value that defines it to be correct. In engineering and scientific computing, it is convenient and necessary to be able to estimate how many significant digits are needed in the computed result.

The number of bits in a binary number determines the precision with which the binary number represents a decimal number. A 32-bit number can represent approximately seven digits of a decimal number. A 64-bit binary number can represent 13 to 14 decimal digits.

1.11.3 Precision and Accuracy

Precision refers to how closely a numeric value used represents the value it is representing. *Accuracy* refers to how closely a number agrees with the true value of the number it is representing. Precision is governed by the number of digits being carried in the numerical calculations. Accuracy is governed by the errors in the numerical *approximation*; precision and accuracy are quantified by the errors in a numerical calculation.

1.11.4 Errors

An *error* is the difference between the true value *tv* of a number and its approximate value *av*. The *relative error* is the proportion of the error with respect to the true value.

$$error \quad = \quad tv - av$$

$$rel\ error \quad = \quad \frac{error}{tv}$$

The accuracy of a numerical calculation is quantified by the error of the calculation. Several types of errors can occur in numerical calculations.

- Errors in the parameters of the problem

- Algebraic errors in the calculations

- Iteration errors

- Approximation errors

- Roundoff errors

An *iteration error* is the error in an iterative method that approaches the exact solution of an exact problem asymptotically. Iteration errors must decrease toward zero as the iterative process progresses. The iteration error itself may be used to determine the successive approximations to the exact solution. Iteration errors can be reduced to the limit of the computing device.

An *approximation error* is the difference between the exact solution of an exact problem and the exact solution of an approximation of the exact problem. Approximation error can be reduced only by choosing a more accurate approximation of the exact problem.

A *roundoff error* is the error caused by the finite word length employed in the calculations. Roundoff error is more significant when small differences between large numbers are calculated. Most computers have either 32-bit or 64-bit word length, corresponding to approximately 7 or 13 significant decimal digits, respectively. Some computers have extended precision capability, which increases the number of bits to 128. Care must be exercised to ensure that enough significant digits are maintained in numerical calculations so that roundoff is not significant.

In many engineering and scientific calculations, 32-bit arithmetic is sufficient. However, in many other applications, 64-bit arithmetic is required. In a few special situations, 128-bit arithmetic maybe required. Such calculations are called *double precision* or *quad precision*, respectively. Many computations are evaluated using 64-bit arithmetic to minimize roundoff errors.

A floating-point number has three components: *sign*, the *exponent*, and the *significand*. The exponent is a signed integer represented in biased format (a fixed bias is added to it to make it into an unsigned number). The significand is a fixed-point number in the range $[1, 2)$. Because the binary representation of the significand always starts with one (1) and dot, this is fixed and hidden and only the fractional part of the significand is explicitly represented.

Except for integers and some fractions, all binary representations of decimal numbers are *approximations*, because of the finite number of bits used. Thus, some loss of precision in the binary representation of a decimal number is unavoidable. When binary numbers are combined in arithmetic operations such as addition, multiplication, etc., the true result is typically a longer binary number which cannot be represented exactly with the number of available bits in the binary number capability of the digital computer. Thus, the results are rounded off in the last available binary bit. This rounding off gives rise to roundoff errors, which can accumulate as the number of calculations increases.

The most common representation of numbers for computations dealing with values in a wide range is the floating-point format. The IEEE floating-point standard format (ANSI/IEEE Standard 754-1985) is used by the computer industry. Other formats will differ in their parameters and representation details, but the basic tradeoffs and algorithms remain the same.

Two of the most common floating-point formats are short (32-bit) and long (64-bit) floating-point formats. The short format has adequate range and precision for most common applications (magnitudes ranging from 1.2×10^{38} to 3.4×10^{38}). The long format is used for highly precise computations or those involving extreme variations in magnitude (from about 2.2×10^{308} to 1.8×10^{308}).

The value zero has no proper representation. For zero and other special values, the smallest and largest exponent codes (all 0s and all 1s in the biased exponent field) are not used for ordinary numbers. An all-0s word (0s in sign, exponent, and significand fields) represents +0; similarly, 0 and $\pm\infty$ have special representations, as does any nonsensical or indeterminate value, known as "not a number" (NaN).

When an arithmetic operation produces a result that is not exactly representable in the format being used, the result must be rounded to some representable value. The ANSI/IEEE standard prescribes several rounding options.

1.12 SUMMARY

Application programs are ones with which the user interacts to solve particular problems. Computational models are used to solve large and complex problems in the various scientific and engineering disciplines. These models are implemented by computer programs coded in a particular programming language. There are several standard programming languages, such as C, C++, Eiffel, Ada, Java. Compilation is the task of translating a program from its source language to an equivalent program in machine code. Other languages used in scientific computing such as Python, MATLAB, and Octave are interpreted. Computations are carried out on input data by executing individual commands or complete programs.

Key Terms

computational model	mathematical model	abstraction
algorithm	conceptual model	model development
compilers	linkers	interpreters
programs	commands	instructions
programming language	Java	C
C++	Eiffel	Ada
Python	interpreter	program execution
data definition	Source code	high-level language
simulation model	keywords	identifiers

1.13 EXERCISES

1.1 Explain the differences between a computational model and a mathematical model.

1.2 Explain the reason why the concept of abstraction is important in developing computational models.

1.3 Investigate and write a short report on the programming languages used to implement computational models.

1.4 What is a programming language? Why are they needed?

1.5 Explain why there are many programming languages.

1.6 What are the differences between compilation and interpretation in high-level programming languages?

1.7 Explain the purpose of compilation. How many compilers are necessary for a given application? What is the difference between program compilation and program execution? Explain.

1.8 What is the real purpose of developing a program? Can we just use a spreadsheet program to solve numerical problems? Explain.

1.9 Explain the differences between data definitions and instructions in a program written in a high-level programming language.

1.10 For developing small programs, is it still necessary to use a software development process? Explain. What are the main advantages in using a process for program development? What are the disadvantages?

## CHAPTER 2

Simple Python Programs

2.1 INTRODUCTION

This chapter presents an overview of the structure of a computer program, which include data definitions and basic instructions using the Python programming language. Because functions are one of the building blocks and the fundamental components of Python programs, the concepts of function definitions and function invocations are gradually explained, and complete Python programs are introduced that illustrate further the role of functions. This chapter also presents concepts and principles that are used in developing computational models by implementing two mathematical models with Python programs.

2.2 COMPUTING WITH PYTHON

There two modes of computing with Python:

- *Interactive mode*, which provides the Python prompt indicated by >>>. Each Python command is directly entered and the Python interpreter responds immediately to the command.

- *Script mode*, which is a sequence of Python commands in a file with a .py extension. This file is passed to the Python interpreter to process and is known as a *script*. A typical Python program is edited and stored as a script and an appropriate text editor is used to build a program as a Python script.

2.2.1 Using Interactive Mode with Simple Operations

The following Python command listing on Linux shows several Python instructions in interactive mode. These instructions set the value 34.5 to variable y, the value 12.48 to variable x, and then adds the values of variables x and y; the result is assigned to variable z. By typing the name of the variable, the Python interpreter responds displaying its value.

```
$ linux
Python 2.7.6 (default, Mar 22 2014, 22:59:38)
[GCC 4.8.2] on linux2
Type "help", "copyright", "credits" or "license" for more
        information.
>>> y = 34.5
>>> y
34.5
>>> x = 12.48
>>> x
12.48
>>> z = x + y
>>> z
46.980000000000004
>>>
```

2.2.2 Mathematical Operations

To increment the value of a variable that has an integer value is straightforward; the instruction adds the previous value of the variable with the increment value and the result becomes the new value of the variable. In the following example, variable ix is given a value of 6. Then the next command increments the value of the ix variable by 1, and therefore the new value of variable ix is 7. Note that a variable must be defined (assigned a value to it) before it can be used in an operation.

```
>>> ix = 6
>>> ix = ix + 1
>>> ix
7
```

Integer division only produces an integer quotient; for example, variable ix has a value of 7 and when is divided by 3 results in 2. To get the remainder, the modulus operator (%) is used.

```
>>> id = ix / 3
>>> id
2
>>> iy = ix % 3
>>> iy
1
```

Non-integer numbers are known as *floating-point* numbers. In the following example, the first statement assigns the constant value 21.5 to variable y. The second statement assigns the value 4.5 to variable x. In the third assignment statement, the expression $y + 1.5 \times x^3$ is evaluated and the resulting value is assigned to variable z.

```
>>> y = 21.5
>>> y
21.5
>>> x = 4.5
>>> x
4.5
>>> z = y + 1.5 * x**3
>>> z
158.1875
```

Complex numbers are represented by a pair of values known as the *real* part and the *imaginary* part, which is written with a suffix of j . In Python, these are always floating-point numbers. The following lines of Python commands illustrate the use of complex numbers. Variable z is defined as a complex variable, and its value is a complex number. Function *complex()* is used to create the complex number. The value of z is $(2.5 - 5.32j)$. The real part of a complex number can be retrieved by using the attribute *real* of the complex number and the imaginary part by using the attribute *imag* after the dot sign.

```
>>> x = 2.5
>>> y = -5.32
>>> z = complex(x, y)
>>> z
(2.5-5.32j)
>>> rp = z.real
>>> rp
2.5
>>> ip = z.imag
>>> ip
-5.32
```

2.2.3 More Advanced Mathematical Expressions

Simple arithmetic expressions are used in assignment statements. These are addition, subtraction, multiplication, division, and exponentiation. More complex calculations use various numerical functions, such as square root and trigonometric functions. These expressions apply the mathematical functions *cos* and *sqrt* that are defined in the *math* module. The *import* command is used to gain access to a module.

```
>>> from math import *
>>> pi
3.141592653589793
>>> PI = pi
>>> ss6 = cos(0.5*PI)
>>> ss6
6.123233995736766e-17
>>> ss7 = cos(0.25*PI)
>>> ss7
0.7071067811865476
>>> yy = exp(x)
>>> yy
961965785544776.4
>>> zz = exp(1.0e-5) - 1
>>> zz
1.0000050000069649e-05
>>> ff = acos(0.45)
>>> ff
1.1040309877476002
>>> mfact = factorial(5)
>>> mfact
120
```

In the following example, the value of the expression $\cos p + q$ is assigned to variable y and the value of $\sqrt{x - y}$ is assigned to variable q.

```
>>> p = 0.2 * PI
>>> q = 2.34
>>> y = cos(p) + q
>>> y
3.1490169943749473
>>> q = sqrt(x - y)
>>> q
1.1623179451531551
```

In the following example, the value of the mathematical expression $x^n \times y \times sin^{2m} x$ is assigned to variable z:

```
>>> x = 2.5
>>> y = -5.32
>>> n = 3.75
>>> m = 4
>>> z = (x**n) * y * sin(x)**(2*m)
>>> z
-2.719807659910173
```

2.2.4 Scientific Notation

Scientific notation is used to display very large and very small floating-point values. It is written with a letter e after the floating-point value followed by an integer exponent. In the following example, the mathematical equivalent for the first value of variable y is 5.77262×10^{12}. Scientific notation can also be used in mathematical expressions with assignments.

```
>>> y = 5.77262e+12
>>> y
5772620000000.0
>>> x = 5.4e8 + y
>>> x
5773160000000.0
>>> y = x * 126.5e10
>>> y
7.3030474e+24
```

2.3 PROGRAMS

A *program* consists of data definitions and instructions that manipulate the data. These are:

- *Data definitions*, which indicate the data to be manipulated by the instructions.

- A *sequence of instructions*, which perform the computations on the data in order to produce the desired results.

2.4 DATA DEFINITIONS

The data in a program consists of one or more data items. These are manipulated or transformed by the computations (computer operations). In Python, each data definition is specified by assigning a value to a variable and has:

- a reference, which is a *variable* with a unique *name* to refer to the data item, and

- a *value* associated with it.

The name of the reference (variable) to a data item is an *identifier* and is defined by the programmer; it must be different from any *keyword* in the programming language.

2.4.1 Data Objects

In Python, the data items are known as *data objects* and every variable references a data object. If the value associated with a data object does not change, then the data object is said to be *immutable*, otherwise it is *mutable*.

The three most important attributes of a data object are:

- the identity, which is the location (address) of the data object in the computer memory;

- the type, which defines the operations are allowed for the data object; and

- the value, which can be changed (mutable) or not (immutable).

2.4.2 Variables

As mentioned previously, a variable is a reference to a data object and the name of the variable is used in a program for uniquely identifying the variable and is known as an *identifier*. The special text words or symbols that indicate essential parts of a programming language are known as *keywords*. These are reserved words and cannot be used for any other purpose.

A problem that calculates the area of a triangle uses four variables. Examples of the names for these variables are *a, b, c*, and *area*.

2.4.3 Using Data Objects and Variables

In the following listing of Python commands, the first three commands include three assignments to variables x, y, and z. The fourth Python command uses the Python function *id()* to get the address of the data object referenced by variable x and this address is 19290088. Note that the address of the referenced object with variables y and z is the same, because these two variables refer to the same data object. After changing the value of variable y, the reference is different because now variable y refers to a different data object. Note that the # symbol is used to include a comment on a source line and has no effect on the instruction.

```
>>> x = 5.33
>>> y = 6
>>> z = y      # these now refer to the same data object
>>> id(x)      # get identity of data object
19290088
>>> id(y)
19257084
>>> id(z)
19257084
```

```
>>> y = y + 1
>>> id(y)
19257072
>>> id(z)
19257084
>>> z = z + 1
>>> id(z)
19257072
>>> type(x)
<type 'float'>
>>> type(y)
<type 'int'>
>>>
```

2.4.4 Basic Data Types

The fundamental data types are classified into the three categories:

- Numeric

- Text

- Boolean

The numeric types are further divided into two basic types, *integer*, and *float*. Values of *integer* type are those that are countable to a finite value, for example, age, number of parts, number of students enrolled in a course, and so on. Values of type *float* have a decimal point; for example, cost of a part, the height of a tower, current temperature in a boiler, a time interval. These values cannot be expressed as integers.

In the Python commands of the previous example, the Python function *type()* is used to get the type variable x and of variable y. Note that the type of variable x is *float* and the type of variable y is *int*.

Text data items are of type *string* and consist of a sequence of characters. The values for this types of data items are text values. An example of a string is the text value: 'Welcome!'.

The third data type is used for data objects whose values can take any of two truth values (*True* or *False*); these data objects are of type *bool*.

2.5 SIMPLE PYTHON PROGRAMS

A very simple program consists of data definitions and a sequence of instructions. The *script mode* is normally used for writing Python programs. Instructions are written into a text file using an appropriate text editor such as *gedit* on Linux and *Notepad++* on Windows. The text file with the source code is known as a *script* and has a .py extension.

An instruction performs a specific manipulation or computation on the data, it is written as a language statement in the program.

2.5.1 The Assignment Statement

As discussed in previous examples, the *assignment statement* is the most fundamental statement (high-level instruction); its general form is:

⟨ *variable_name* ⟩ = ⟨ expression ⟩

The assignment *operator* is denoted by the = symbol and on the left side of this operator a variable name must always be written. On the right side of the assignment operator, an expression is written. The Python interpreter evaluates the expression on the right-hand side of the assignment and the result is assigned to the variable on the left-hand side of the assignment operator.

In the following example, the first Python statement is a simple assignment that assigns the value 34.5 to variable x. The second assignment statement is a slightly more complex assignment that performs an addition of the value of variable x and the constant 11.38. The result of the addition is assigned to variable y.

```
x = 34.5
y = x + 11.38
```

2.5.2 Basic Input and Output Instructions

Input and output statements are used to read (input) data values from the input device (e.g., the keyboard) and write (output) data values to an output device (the computer screen).

2.5.2.1 Output Statement

In Python, the *print* statement is used for the output of variables and text strings. This output statement writes the value of one or more variables to the output device. The variables do not change their values. The general form of the output statement in Python is:

print ⟨ *data_list* ⟩

For example, in the following Python statements, the line will simply display the value of variable y on the screen. The second output displays the string literal "value of x= ", followed by the value of variable x.

```
print y
print "value of x= ", x
```

Note that the *print* instruction is a statement in Python 2; it is a function in Python 3 and is written as:

```
print (y)
print ("value of x= ", x)
```

2.5.2.2 Input Statements

The *input* statement reads a value of a variable from the input device (e.g., the keyboard). This statement is written with the function **input**, for of a single data value and assign to a variable. The following two lines of pseudo-code include the general form of the input statement and an example that uses the *read* statement to read a value of variable y.

$$\langle \; var\_name \; \rangle \; = \; \textbf{input} \; (\; \langle \; string\_lit \rangle \;)$$

The following example displays the string literal "Enter value of y: " and reads the value of variable y.

```
y = input ("Enter value of y: ")
```

2.5.3 Example Scripts with Input/Output

The following script computes 75% of the value of variable y. The name of the script is **prog01.py**.

```
y = 34.5
print (y)
y = y * 0.75
print(y)
```

At the Linux prompt, the Python interpreter is run with script **prog01.py**. The interpreter will execute every Python command in sequence (one after the other) and the results are displayed on the screen.

```
$ python prog01.py
34.5
25.875
```

The next script has more output but it carries out the same computations. The first line of the script is only a comment; it starts with the pound (#) symbol. The name of this second script is `prog02.py`.

```
# This script computes 75% of the value of y
y = 34.5
print "Initial value of y: ", y
y = y * 0.75
print "Final value of y: ", y
```

This script is started by invoking the Python interpreter with the name of the script, `prog02.py`.

```
$ python prog02.py
Initial value of y:   34.5
Final value of y:   25.875
```

The third script performs the same computations as the first two scripts. The main difference is that it inputs the value of variable y; in other words, the user of the program will enter the value of y.

```
# This script computes 75% of the value of y
y = input ("Enter initial value of y: ")
print "Initial value of y: ", y
y = y * 0.75
print "Final value of y: ", y
```

This script is started by invoking the Python interpreter with the name of the script, `prog03.py`.

```
$ python prog03.py
Enter initial value of y: 34.5
Initial value of y:   34.5
Final value of y:   25.875
```

2.6 A SIMPLE PROBLEM: TEMPERATURE CONVERSION

This section revisits and implements in Python the temperature conversion problem, which was discussed in the previous chapter. The solution and implementation is derived by following a basic sequence of steps.

The problem: given the value of the temperature in degrees Celsius, compute the corresponding value in degrees Fahrenheit and show this result.

2.6.1 Mathematical Model

The mathematical representation of the solution to the problem, the formula expressing a temperature measurement F in Fahrenheit in terms of the temperature measurement C in Celsius is:

$$F = \frac{9}{5} C + 32.$$

The solution to the problem is the mathematical expression for the conversion of a temperature measurement in Celsius to the corresponding value in Fahrenheit. The mathematical formula expressing the conversion assigns a value to the desired temperature in the variable F, the dependent variable. The values of the variable C can change arbitrarily because it is the independent variable. The mathematical model uses real numbers to represent the temperature readings in various temperature units.

2.6.2 Computational Model

The computational model is derived by implementing the mathematical model in a program using the Python programming language. This model is developed using a Terminal window Linux. In a similar manner to the previous examples, the computational model is developed by writing a Python program as a script using the *gedit* text editor, then executing the Python interpreter with the script.

The Python program is very simple and Listing 2.1 shows the complete source code.

Listing 2.1: Temperature conversion program.

```
1  """
2     Program    : tconvctof.py
3     Author     : Jose M Garrido
4     Date       : 5-12-2014
5     Description : Read value of temperature Celsius from
6     console, convert to degrees Fahrenheit, and display
7     value of this new temperature value on the output
8     console */
9  """
10
11 C = input("Enter value of temp in Celsius: ")
12 F = C * (9.0/5.0) + 32.0      # temperature in Fahrenheit
13 print "Value of temperature in Celsius: ", C
14 print "Temperature in Fahrenheit: ", F
```

Lines 1–9 are part of a multi-line comment. The computation of the temperature in Fahrenheit is performed in line 12 using an assignment statement.

The following listing shows the interpretation of the commands in the script *tconvctof.py* by executing the Python interpreter.

```
$ python tconvctof.py
Enter value of temp in Celsius: 25.0
Value of temp in Celsius:  25.0
Temperature in Fahrenheit:  77.0
```

This procedure can be repeated several times to compute the Fahrenheit temperature starting with a given value of 10.0 for the temperature in Celsius and then repeating in increments of 5.0 degrees Celsius. The last computation is for a given value of 45.0 degrees Celsius.

Table 2.1 shows the values of temperature in Celsius from 5.0 to 45.0 used to compute the corresponding temperature in Fahrenheit. This is a short set of results of the original problem. Figure 2.1 shows a plot of the values of temperature computed.

Table 2.1 Celsius and Fahrenheit temperatures.

Celsius	5	10	15	20	25	30	35	40	45
Fahrenheit	41	50	59	68	77	86	95	104	113

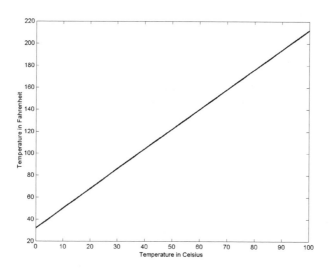

Figure 2.1 Plot of the values of temperature Celsius and Fahrenheit.

2.7 DISTANCE BETWEEN TWO POINTS

2.7.1 Problem Statement

The following problem requires computing the distance between two points in a Cartesian plane. A program is to be developed that computes this distance, given the values of the coordinates of the two points.

2.7.2 Analysis of the Problem

A Cartesian plane consists of two directed lines that perpendicularly intersect their respective zero points. The horizontal directed line is called the *x-axis* and the vertical directed line is called the *y-axis*. The point of intersection of the x-axis and the y-axis is known as the *origin* and is denoted by the letter O.

Figure 2.2 shows a Cartesian plane with two points, P_1 and P_2. Point P_1 is defined by two coordinate values (x_1, y_1) and point P_2 is defined by the coordinate values (x_2, y_2).

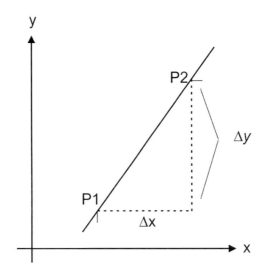

Figure 2.2 Horizontal and vertical distances between two points.

2.7.3 Design of the Solution

The horizontal distance between the two points, Δx, is computed by the difference $x_2 - x_1$. Similarly, the vertical distance between the two points is denoted by Δy and is computed by the difference $y_2 - y_1$. The distance, d, between two points P_1 and P_2 in a Cartesian plane, is calculated with the following mathematical expression:

$$d = \sqrt{\Delta x^2 + \Delta y^2}.$$

A detailed design in an algorithm follows:

1. Read the values of the coordinates for point P_1 from input device (keyboard).

2. Read the values of the coordinates for point P_2 from the input device.

3. Compute the horizontal distance, Δx, between the two points:

$$\Delta x = x_2 - x_1.$$

4. Compute the vertical distance, Δy, between the two points:

$$\Delta y = y_2 - y_1.$$

5. Compute the distance, d, between the two points:

$$d = \sqrt{\Delta x^2 + \Delta y^2}.$$

6. Display the value of the distance between the two points, on the output device (video screen).

2.7.4 Implementation

This phase implements the design by coding a program in Python, running the Python interpreter with the corresponding script, and testing the program. Listing 2.1a shows the source program, which is stored in file `distpoints.py`.

Listing 2.1a: A program to compute the distance between two points.

```
1  """
2     Program     : distpts.py
3     Author      : Jose M Garrido, January, 20, 2014.
4     Description : This program computes the distance
5        between two points in a Cartesian plane.
6  """
7
8  import math
9  x1 = input ("Enter value of x-coordinate of P1: ")
10 y1 = input ("Enter value of y-coordinate of P1: ")
11 print "Coordinates of P1: ", x1, y1
12 x2 = input ("Enter value of x-coordinate of P2: ")
13 y2 = input ("Enter value of y-coordinate of P2: ")
14 print "Coordinates of P2: ", x2, y2
```

```
15
16 # compute horizontal distance between points
17 dx = x2 - x1
18
19 # compute vertical distance between points
20 dy = y2 - y1
21 print "Horizontal and vertical distances: ", dx, dy
22
23 # compute the distance between the points
24 d = math.sqrt( dx ** 2 + dy ** 2 )
25
26 # display result
27 print "Distance between P1 and P2: ", d
```

The following listing shows the Python interpretation of the script *distpts.py* with the input values shown.

```
$ python distpts.py
Enter value of x-coordinate of P1: 2.25
Enter value of y-coordinate of P1: 1.5
Coordinates of P1:  2.25 1.5
Enter value of x-coordinate of P2: 1.3
Enter value of y-coordinate of P2: 0.45
Coordinates of P2:  1.3 0.45
Horizontal and vertical distances:  -0.95 -1.05
Distance between P1 and P2:  1.41598022585
```

2.8 GENERAL STRUCTURE OF A PYTHON PROGRAM

A typical program in the Python language has the general structure as shown in Figure 2.3. It consists of several parts:

1. The import commands are optional but they are present in almost all Python programs. Each of these uses the *import* statement and allows the program access to the definitions and code in the specified Python module.

2. Global data, which may consist of assignments of values to variables, in a similar manner as described previously. These are global data because they can be used by all functions in the program.

3. Definition of functions. This is an optional component of a Python program but it is almost always present. When present, one or more functions are defined in this part of the program. In Figure 2.3, a function is defined with name myfuncta.

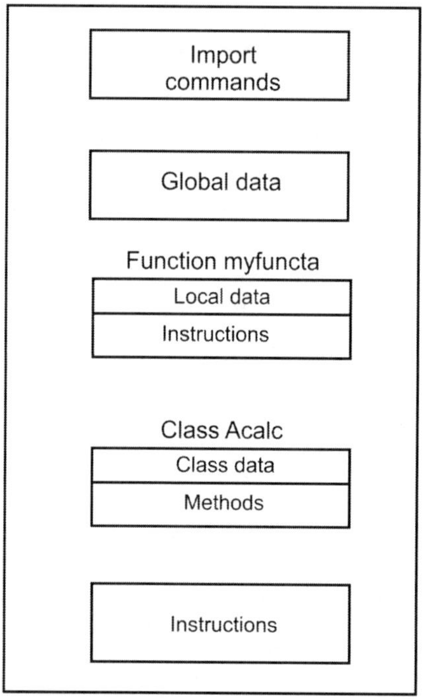

```
┌─────────────────────────────────────┐
│        ┌───────────────────┐        │
│        │      Import        │        │
│        │     commands       │        │
│        └───────────────────┘        │
│                                     │
│        ┌───────────────────┐        │
│        │    Global data     │        │
│        └───────────────────┘        │
│                                     │
│          Function myfuncta          │
│        ┌───────────────────┐        │
│        │     Local data     │        │
│        ├───────────────────┤        │
│        │    Instructions    │        │
│        └───────────────────┘        │
│                                     │
│             Class Acalc             │
│        ┌───────────────────┐        │
│        │     Class data     │        │
│        ├───────────────────┤        │
│        │      Methods       │        │
│        └───────────────────┘        │
│                                     │
│        ┌───────────────────┐        │
│        │    Instructions    │        │
│        └───────────────────┘        │
└─────────────────────────────────────┘
```

Figure 2.3 General structure of a Python program.

4. Definition of classes. This is another optional component in a Python program. A class definition allows the program to create objects of that class. When present, one or more classes are defined in this part of the program.

5. The instructions are Python statements that invoke or call the functions in the program and/or in the imported modules. These instructions can also create and manipulate objects using the class definitions in the program and/or in the imported modules.

Function definitions can be programmer-defined functions that are invoked (or called) in the program. The other functions that can be called are the built-in functions provided by standard Python interpreter *libraries* or by other Python modules. A library is a collection of related function definitions and/or class definitions that may also include data.

A function starts executing when it is called by another function or by an instruction in the program. Before a function can be called in a Python program, a function definition is required. Once a function is defined, it can be called or invoked by any other function or by an instruction in the program.

2.9 SIMPLE FUNCTIONS

A Python program is normally *decomposed* into modules, and these are divided into classes and functions. A function carries out a specific task in a program.

As mentioned previously, data in a function is known only to that function—the scope of the data is *local* to the function. The local data in a function has a limited lifetime; it only exists during execution of the function.

2.9.1 Function Definitions

A simple Python program consists of functions and instructions that call or invoke the various functions. Figure 2.4 illustrates the general structure of a function in the Python language.

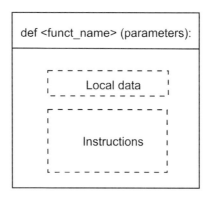

Figure 2.4 Structure of a python function.

In the source code, the general syntactical form of a function in the Python programming language is written as follows:

```
def function_name ( [parameters] ) :
        [ local declarations ]
        [ executable language statements ]
```

The relevant internal documentation of the function definition is described in one or more lines of comments, which begins with the characters (""") and ends with (""").

The local data definitions in the function are optional. The instructions implement the body of the function. The following Python source code shows a simple function for displaying a text message on the screen.

```
def show_message () :
        """
        This function displays a message
```

```
    on the screen.
    """

    print("Computing data")
```

This is a very simple function and its only purpose is to display a text message on the screen. This function does not declare parameters and the type of this function is *void* to indicate that this function does not return a value.

2.9.2   Function Calls

The name of the function is used when calling or invoking the function by some other function. The function that calls another function is known as the calling function; the second function is known as the called function. When a function calls or invokes another function, the flow of control is altered and the second function starts execution immediately.

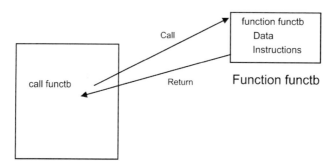

Figure 2.5   A function calling another function.

When the called function completes execution, the flow of control is transferred back (returned) to the calling function and it continues execution from the point after it called the second function.

Figure 2.5 illustrates an instruction calling function *functb*. After completing its execution, function *functb* returns the flow of control to the instruction that performed the call.

An example of this kind of function call is the call to function *show_message*, discussed previously. In Python, the statement that calls a simple function uses the function name and an empty parentheses pair. For example, the call to function *show_message* is written as:

```
show_message()
```

Listing 2.2 shows a Python program that defines function *show_message* then calls the function. This program is stored in file **shmessp.py**.

Listing 2.2: Python program that defines and calls a function.

```
 2 # Program     : shmessp.py
 3 # Author      : Jose M Garrido, May 28 2014.
 4 # Description : Define and call a simple function.
 5
 6 def show_message():
 7     """
 8         This function displays a message
 9         on the screen
10     """
11     print "Computing results ..... "
12
13 y = input("Enter a number: ")
14 sqy = y * y
15 show_message()
16 print "square of the number is: ", sqy
```

The function is defined in lines 6–11 and the function is called in line 15. The following listing shows the Python interpretation of the script *shmessp.py*.

```
$ python shmessp.py
Enter a number: 12
Computing results .....
square of the number is:   144
```

2.10 SUMMARY

The structure of a Python computer program includes data definitions, function definitions, class definitions, and basic instructions that manipulate the data. Functions are one of the building blocks and fundamental components of Python programs. Functions are first defined and then called in Python programs. These basic programming constructs are used in developing computational models by implementing the corresponding mathematical models with Python programs.

Key Terms

programs	functions	function invocation
function call	assignment statement	assignment operator
local declaration	variables	constants
function definition	class definitions	scope

2.11   EXERCISES

2.1   Why are functions defined and used in programs? Explain.

2.2   Develop a computational model (with a Python program) that computes the area of a right triangle given values of the altitude and the base.

2.3   Develop a computational model (with a Python program) that computes the distance between two points in a plane: P_1 with coordinates (x_1, y_1), and P_2 with coordinates (x_2, y_2). Use the coordinate values: $(2, 3)$ and $(4, 7)$.

2.4   Develop a computational model that computes the temperature in Celsius, given the values of the temperature in Fahrenheit.

2.5   Develop a computational model that computes the circumference and area of a square, given the values of its sides.

2.6   Develop a computational model (with a Python program) that computes the slope of a line between two points in a plane: P_1 with coordinates (x_1, y_1), and P_2 with coordinates (x_2, y_2). Use the coordinate values: $(0, -3/2)$ and $(2, 0)$.

II

Basic Programming Principles
with Python

## CHAPTER 3

Modules and Functions

3.1   INTRODUCTION

A program is usually partitioned into modular units, and in Python, these are modules, classes, and functions. A function is the most fundamental module or decomposition unit in Python programs. When called (invoked), a function carries out a specific task in a program and can receive input data from another function; these input data are known as arguments. The function can also return output data when it completes execution.

This chapter provides details on function definitions, invocation, and decomposition. It also discusses the basic mechanisms for data transfer between two functions; several examples are included that call mathematical built-in functions.

3.2   MODULAR DECOMPOSITION

A problem is often too large and complex to deal with as a single unit. In problem solving and algorithmic design, the problem is partitioned into smaller problems that are easier to solve. The final solution consists of an assembly of these smaller solutions. The partitioning of a problem into smaller parts is known as *decomposition*. These small parts are sometimes known as *modular units*, which are much easier to develop and manage.

System design usually emphasizes modular structuring, also called modular decomposition. With this approach, the solution to a problem consists of several smaller solutions corresponding to each of the subproblems. A problem is divided into smaller problems (or subproblems), and a solution is designed for each subproblem. These modular units are considered building blocks for constructing larger and more complex algorithms.

In addition to calling programmer-defined functions, the instructions in the program can call built-in functions provided by standard Python interpreter *libraries* or by other Python modules. A library is a collection of related function definitions and/or class definitions that may also include data.

As mentioned in the previous chapter, a function starts executing when it is called by an instruction in the program. Before a function can be called in a Python program, a function definition is required. Once a functon is defined, it can be called or invoked one or more times.

3.3 FUNCTIONS

A Python program is often *decomposed* into modules, and these are divided into classes and functions. A function carries out a specific task in a program.

The data in a function is known only to that function—the scope of the data is *local* to the function. The local data in a function has a limited lifetime; it only exists during execution of the function.

A Python program typically consists of functions and instructions that call or invoke the various functions. In the source code, the general syntactical form of a function definition in the Python programming language is written as follows:

```
def function_name ( [parameters] ) :
        [ local declarations ]
        [ executable language statements ]
```

The relevant internal documentation of the function definition is described in one or more lines of comments, which begin with the characters (""") and ends with (""").

The local data definitions in the function are optional. The instructions implement the body of the function.

3.3.1 Function Calls

After a function is defined, it can be called (invoked) and the name of the function is used by the instruction in the program that calls the function. When a function is called, the normal sequential flow of control is altered and the (called) function starts execution immediately.

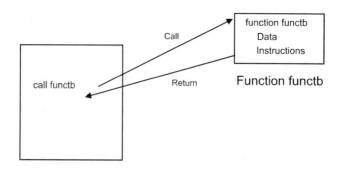

Figure 3.1 A function call.

When the called function completes execution, the flow of control is transferred back (returned) to the instruction that called the function and it continues execution from this point. Figure 3.1 illustrates a call to function *functb*. After completing its execution, function *functb* returns the flow of control to the instruction that called it. In Python, the statement that calls a simple function uses the function name and an empty parentheses pair.

3.4 CATEGORIES OF FUNCTIONS

Data transfer occurs between the instruction that calls a function and the called function. This data transfer may occur from the calling instruction to the called function, from the called function to the calling instruction, or in both directions. With respect to data transfer, there are four categories of functions:

1. Simple functions, which do not allow data transfer when they are called. The previous example, function *show_message*, is a simple function because there is no data transfer involved.

2. Functions that return a single value after completion.

3. Functions that specify one or more parameters, which are data items transferred to the function.

4. Functions that allow data transfers in both directions. These functions specify one or more parameters and return a value to the calling function.

3.4.1 Simple Function Calls

Simple functions do not return a value to the calling function. There is no data transfer to or from the function. Figure 3.1 shows an instruction that calls function *functb*. After completing its execution, the called function *functb* returns the flow of control to the instruction that called it. An example of this kind of function is *show_message*, discussed previously.

3.4.2 Calling Functions that Return Data

Value-returning functions transfer data back to the calling instruction. Typically, a single value is computed and is returned to the calling instruction.
These functions can be called in one of two ways:

• Call the function in a simple assignment statement.

• Call the function in an assignment statement with an expression.

These function definitions must include at least one return statement, which is written with the keyword **return** followed by an expression. The value in the return statement can be any valid expression, following the **return** keyword. The expression can include constants, variables, or a combination of these.

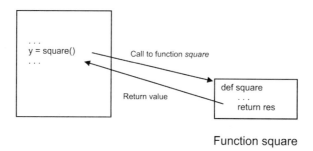

Figure 3.2 Calling function *square*.

Figure 3.2 shows an instruction that calls function *square*. The return value from function *square* is used in an assignment statement that assigns the value to variable *y*.

Listing 3.1 shows the source code of a Python program that includes in lines 2–10, the definition of function *square*. This function returns the value of the square of variable *x* that is assigned a value of 3.15. Lines 12–13 include the instructions that call the function and display the value of variable *y*. These are included after the function definition of *square*.

Lines 15–18 show instructions that involve a function call in a more complex expression of an assignment statement. The value of the expression is assigned to variable *fres* and the value of this variable is displayed. The Python instructions of this example are stored in file `prog04.py`.

Listing 3.1: Python program that computes the square of a value.

```
1 # This script computes the square of the value 3.15
    by calling function 'square'
2 def square():
3     """
4   description
5       This function returns the square of variable x
6       with value 3.15
7   """
8       x = 3.15
9       res = x**2
10      return res
11
12 y = square()
13 print "Value of y is: ", y
```

```
14
15 w = 2.35
16 q = 12.75
17 fres = q + w * square() + 23.45
18 print "Value of fres is: ", fres
```

The following listing shows the interpretation of the commands in the script *prog04.py* by executing the Python interpreter.

```
$ python prog04.py
Value of y is:  9.9225
Value of fres is:  59.517875
```

3.4.2.1 Including the Function Definition in Another Module

Sometimes it may be more convenient to include one or more function definitions in another module. Before a function can be called, the corresponding module that contains its definition has to be imported.

In this particular example, the function definition of function *square* is edited in the module *myf.py*. This example is performed in interactive mode with the Python interpreter and consists of a statement that imports module *myf* and an assignment statement that calls function *square*. The returned value is assigned to variable y. Note that the name of the function is preceded by the name of the module that contains its definition and a dot.

```
>>> import myf
>>> y = myf.square()
>>> y
9.9225
```

The function call can occur in a more complex expression of an assignment statement. In the following example, the value of the expression is assigned to variable *fres*.

```
>>> w = 2.35
>>> y = 12.75
>>> fres = y + w * myf.square() + 23.45
>>> fres
59.517875000000004
```

3.4.3 Calling Functions with Arguments

A function defined with one or more parameter specifications allows data transfer to the function when called. The parameters specified in the function definition are treated as local data and have local scope. The data values used when calling these functions are known as *arguments*.

Every argument in a function call must correspond to a parameter specification in the function definition. A function that is called with two arguments must be defined with two parameters.

Figure 3.3 illustrates the example in which one parameter is specified in the function definition of *squared*. The function call requires an argument and the function computes the square of the value in the argument; the called function returns the value computed.

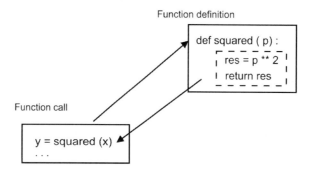

Figure 3.3 Calling function *squared* with an argument.

Listing 3.2 shows the Python program with the definition of function *squared* that includes one parameter *p*, and returns the value of the local variable *result*. The source code is stored in file *prog05.py*.

Listing 3.2: Python program that calls *squared* with an argument.

```
1 # This script computes the square value of a variable
2 #  by calling function 'squared'
3 def squared (p) :
4       """
5       description
6       This function returns the square of p
7       """
8       result = p ** 2
9       return result
10
11 x = 3.0
12 y1 = squared(x)
13 print "Value of the argument x: ", x
14 print "Value of y1: ", y1
15
16 y2 = x * 2.55 + squared(x)
17 print "Value of y2: ", y2
```

The following listing shows the interpretation of the commands in the script *prog05.py* by executing the Python interpreter.

```
$ python prog05.py
Value of the argument x:   3.0
Value of y1:   9.0
Value of y2:   16.65
```

3.4.3.1   Including Function squared in Another Module

If function *squared* is stored in module *myf.py*, then calling the function may be carried out in the script *prog05.py* or in interactive mode with Python as follows:

```
>>> y = myf.squared(3.0)
>>> y
9.0
>>> x = 3.5
>>> y = myf.squared(x)
>>> y
12.25
>>>
```

3.5   BUILT-IN MATHEMATICAL FUNCTIONS

Python provides a wide variety of functions organized and stored in various libraries of standard modules with many pre-defined functions. One such module is the mathematical module *math*. In Python programs, the access to this is achieved by importing the module with command `import math` at the top of a Python script (program).

The following language statements in interactive mode, include calls to function *sin* applied to variable x, which is the *argument* written in parentheses and its value is assumed in radians. The value returned by the function is the sine of x.

```
>>> from math import *

>>> x = 0.175 * pi
>>> y = 2.16 + sin(x)
>>> y
2.682498564715949
>>> j = 0.335
>>> z = x * sin(j * 0.32)
>>> z
0.05882346197763754
```

To compute the square root of the value of a variable, z expressed mathematically as: $\sqrt{z}$, the mathematical library provides the *sqrt()* function. For example, given the following mathematical expression:

$$var = \sqrt{\sin^2 x + \cos^2 y}.$$

The assignment statement to compute the value of variable *var* uses three functions: *sqrt()*, *sin()*, and *cos()* and is coded as:

```
>>> y = 0.335
>>> var = sqrt (sin(x) ** 2 + cos(y) ** 2)
>>> var
1.079312551150338
```

The exponential function *exp()* computes e raised to the given power. The following statement computes $q = y + x\,e^k$.

```
>>> k = 0.335
>>> q = y + x * exp(k)
>>> q
1.1035578677649929
```

To compute the logarithm base e of x, denoted mathematically as $\ln x$ or $\log_e x$, function *log()* is called. For example:

```
>>> t = log((q-y)/x)
>>> t
0.33499999999999996
```

The following table lists the basic mathematical functions available in module *math*.

Function	Description
fabs(x)	Returns the absolute value of x
sqrt(x)	Returns the square root of x, $x \geq 0$
pow(x, y)	Returns x to the power of y
ceil(x)	Returns the nearest integer larger than x
floor(x)	Returns the nearest integer less than x
exp(x)	Returns the value e^x, e is the base for natural logarithms
log(x)	Returns the natural logarithm of x (base e), $x > 0$
log10(x)	Returns the logarithm base 10 of x, $x > 0$

The following table lists the trigonometric and hyperbolic functions available.

Function	Description
`sin(x)`	Returns the sine of x, where x is in radians
`cos(x)`	Returns the cosine of x, where x is in radians
`tan(x)`	Returns the tangent of x, where x is in radians
`asin(x)`	Returns the arcsine of x, where $-1 < x < 1$
`acos(x)`	Retirns the arccosine of x, where $-1 < x < 1$
`atan(x)`	Returns the arctangent of x
`atan2(y, x)`	Returns the arctangent of the value y/x
`sinh(x)`	Returns the hyperbolic sine of x
`cosh(x)`	Returns the hyperbolic cosine of x
`tanh(x)`	Returns the hyperbolic tangent of x

3.6 SUMMARY

Functions are fundamental modular units in a program. A function has to be defined first in order to be called and calling functions involves several mechanisms for data transfer. Calling simple functions does not involve data transfer between the calling function and the called function. Value-returning functions return a value to the calling function. Calling functions that define one or more parameters involve values sent by the calling function and used as input in the called function.

Key Terms

modules	functions	function definition
function call	local declaration	arguments
return value	assignment	parameters
pre-defined functions	built-in functions	libraries

3.7 EXERCISES

3.1 Why is a function a decomposition unit? Explain.

3.2 Explain variations of data transfer among functions.

3.3 Write the Python code of a function defined with more than two parameters.

3.4 Write the Python code that calls the function that was defined with more than two parameters.

3.5 Develop a Python program that defines two functions, one to compute the area of a triangle, the other function to compute the circumference of a triangle. The program must call these functions from instructions that input the corresponding values.

3.6 Develop a Python program that defines two functions, one to compute the area of a circle, the other function to compute the circumference of a circle. The program must call these functions from the instructions that input the corresponding values.

3.7 Develop a Python program that defines a function that computes the volume of a cylinder. The program must call these functions from instructions that input the corresponding values.

3.8 Develop a Python program that defines a function that computes the volume of a sphere. The program must call these functions from instructions that input the corresponding values.

3.9 Develop a Python program that defines two functions, one to compute the area of an ellipse, the other function to compute the circumference of an ellipse. The program must call these functions from instructions that input the corresponding values.

3.10 Develop a Python program that computes the slope of a line between two points in a plane: P_1 with coordinates (x_1, y_1), and P_2 with coordinates (x_2, y_2). The program should include a function *slopef* and the parameters of this function (*slopef*) are the coordinates of the points in a plane. The program must input the coordinate values.

## CHAPTER 4

Program Structures

4.1 INTRODUCTION

Program structures, also known as design structures, are discussed along with how they are used in designing program logic to implement the solution to a problem. This chapter presents general concepts of algorithms, flowcharts, and pseudo-code.

As mentioned in a previous chapter, the purpose of computer problem solving is to design a solution to a problem; an algorithm describes precisely this design and is implemented in a computer program. Analyzing the problem includes understanding the problem, identifying the given (input) data and the required results. Developing a program involves implementing a computer solution to solve some real-world problem. Design of a solution to the problem requires finding some method to solve the problem.

Designing a solution to the problem consists of defining the necessary computations to be carried out in an appropriate sequence on the given data to produce the final required results.

The design of the solution to a problem is described by an *algorithm*, which is a complete and precise sequence of steps that need to be carried out to achieve the solution to a problem.

After the algorithm has been formulated and written, its implementation and the corresponding data definitions are carried out with a programming language such as Python.

4.2 ALGORITHMS

An algorithm is a clear, detailed, precise, and complete description of the sequence of steps to be performed in order to produce the desired results. An algorithm can be considered the transformation on the given data and involves a sequence of commands or operations that are to be carried out on the data in order to produce the desired results.

The algorithm is usually broken down into smaller tasks; the overall algorithm for a problem solution is decomposed into smaller algorithms, each defined to solve a subtask.

A computer implementation of an algorithm consists of a group of data definitions and one or more sequences of instructions to the computer for producing correct results when given appropriate input data. The implementation of an algorithm is in the form of a program, which is written in a programming language and it indicates to the computer how to transform the given data into correct results.

An algorithm is often described in a semiformal notation such as pseudo-code and flowcharts.

4.3 IMPLEMENTING ALGORITHMS

Programming languages have well-defined syntax and semantic rules. The syntax is defined by a set of grammar rules and a vocabulary (a set of words). The legal sentences are constructed using sentences in the form of *statements*. There are two groups of words that are used to write the statements: *reserved words* also known as *keywords* and *identifiers*.

Reserved words are the keywords of the language and have a predefined purpose. These are used in most statements. Examples are: **for**, **def**, **while**, and **if**. Identifiers are names for variables, constants, and functions that the programmer chooses, for example, *height*, *temperature*, *pressure*, *number_units*, and so on.

4.4 ALGORITHM DESCRIPTION

Designing a solution to a problem consists of designing and defining an algorithm, which will be as general as possible in order to solve a family or group of similar problems. An algorithm can be described at several levels of abstraction. Starting from a very high and general level of description of a preliminary design, to a much lower level that has a more detailed description of the design.

Several notations are used to describe an algorithm. An algorithmic notation is a set of general and informal rules used to describe an algorithm. Two widely used notations are:

- Flowcharts

- Pseudo-code

4.4.1 Flowcharts

The flow of control in a program is the order in which the operations will be executed. A flowchart is a visual representation of the flow of the data and the operations on this data. A flowchart consists of a set of symbolic blocks connected by arrows. The arrows that connect the blocks show the order for

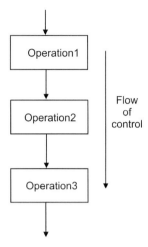

Figure 4.1 Flow of control.

describing a sequence of design or action steps. The arrows also show the flow of data.

The most basic flow of control is sequential—the operations are executed in sequence, as seen in Figure 4.1. Several basic flowchart blocks are shown in Figure 4.2. Every flowchart block has a specific symbol. A flowchart always begins with a *start* symbol, which has an arrow pointing from it. A flowchart ends with a *stop* symbol, which has one arrow pointing to it.

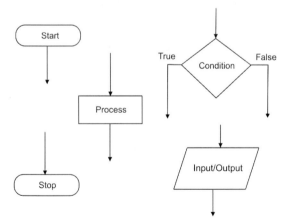

Figure 4.2 Basic flowchart symbols.

The *process* block or *transformation* block symbol is the most common and general symbol, shown as a rectangular box. A process block represents

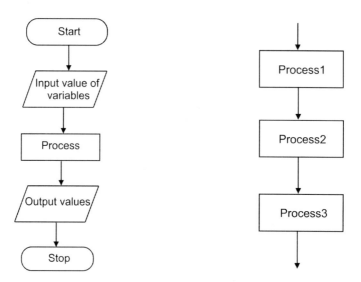

Figure 4.3 A simple flowchart example.

Figure 4.4 A flowchart with a sequence.

one or more operations of computation. This symbol is used to represent any computation or sequence of computations carried out on some data. There is one arrow pointing to it and one arrow pointing out from it.

The *selection* flowchart symbol has the shape of a vertical diamond and represents a *selection* of alternate paths in the sequence of design steps. It is shown in Figure 4.2 with a condition that is evaluated to *True* or *False*. This symbol is also known as a *decision block* because the flow of control of instructions can take one of two (or more) directions in the flowchart, based on the evaluation of a condition.

The *input-output* flowchart symbol is used for a data input or output operation. There is one arrow pointing into the block and one arrow pointing out from the block.

An example of a simple flowchart with several basic symbols in shown in Figure 4.3. For larger or more complex algorithms, flowcharts are used mainly for the high-level description of the algorithms and pseudo-code is used for describing the details.

4.4.2 Pseudo-Code

Pseudo-code is an informal notation that uses a few simple rules and English for describing the algorithm that defines a problem solution. It can be used to describe relatively large and complex algorithms. It is relatively easy to convert the pseudo-code description of an algorithm to a computer implementation in a high-level programming language.

4.5 DESIGN STRUCTURES

The flow of control of an algorithm can be completely defined with only four fundamental design structures. These structures can be specified using flowcharts and/or pseudo-code notations. The design structures are:

1. *Sequence*: Describes a sequence of operations.

2. *Selection*: This part of the algorithm takes a decision and selects one of several alternate paths of flow of actions. This structure is also known as alternation or conditional branch.

3. *Repetition*: This part of the algorithm has a sequence of steps that are to be executed zero, one, or more times.

4. *Input-output*: The values of variables are read from an input device (such as the keyboard) or the values of the variables (results) are written to an output device (such as the screen).

4.5.1 Sequence

A sequence structure consists of a group of operations that are to be executed one after the other, in the specified order, as shown in Figure 4.1. The symbol for a sequence can be directly represented by two or more *process* blocks connected by arrows in a flowchart. Figure 4.4 illustrates the sequence structure with several blocks. The sequence structure is the most common and basic structure used in algorithmic design.

4.5.2 Selection

With the selection structure, one of several alternate paths of the algorithm will be chosen based on the evaluation of a condition. Figure 4.5 illustrates the selection structure in flowchart form. In the figure, the actions or instructions in *Process1* are executed when the condition is True. The instructions in *Process2* are executed when the condition is False.

A concrete and simple flowchart example of the selection structure is shown in Figure 4.6. The condition of the selection structure is $len > 0$ and when this condition evaluates to True, the block with the action `add 3 to k` will execute. Otherwise, the block with the action `decrement k` will execute.

4.5.3 Repetition

The repetition structure indicates that a set of action steps are to be repeated several times. Figure 4.7 shows this structure. The execution of the actions in the *Process* block are repeated while the condition is True. This structure is also known as the *while-loop*.

A variation of the repetition structure is shown in Figure 4.8. The actions

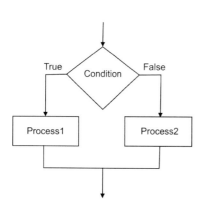

Figure 4.5 Selection structure.

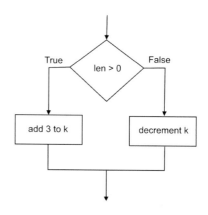

Figure 4.6 An example of the selection structure.

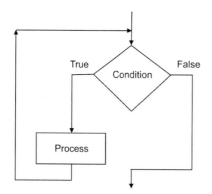

Figure 4.7 While-loop of the repetition structure.

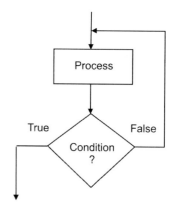

Figure 4.8 Repeat-until loop of the repetition structure.

in the *Process* block are repeated until the condition becomes True. This structure is also known as the *repeat-until* loop.

4.5.4 Simple Input/Output

Input and output statements are used to read (input) data values from the input device (e.g., the keyboard) and write (output) data values to an output device (mainly to the computer screen). The flowchart symbol is shown in Figure 4.9.

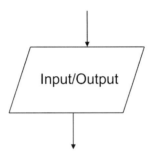

Figure 4.9 Flowchart data input/output symbol.

4.5.4.1 Output

In Python, the output statement is used for the output of a list of variables and literals; it is written with the keyword **print**. The output statement writes the value of one or more variables to the output device. The variables do not change their values. The general form of the output statement is:

print ⟨ *data_list* ⟩

In the following line of Python code, the *print* statement is used to output four data items that include the value of variables x and y:

```
print "value of x= ", x, "value of y = ", y
```

4.5.4.2 Input

The *input* function in Python reads a value of a variable from the input device (e.g., the keyboard). This input implies an assignment statement for the variable because the variable changes its value to the new value that is read from the input device. A text string is typically included to prompt the user for input of a data value. For example, in interactive mode, the following lines of code read the value of variable q:

```
>>> q = input ("Enter the value of q: ")
Enter the value of q: 45.32
>>> q
45.32
```

To read several values and assign them to corresponding variables, the general form of the input uses the *raw_input* and *split* functions (Python 2.7). Function *split* is used to separate the values inputted.

$$\langle \ var\_list \ \rangle \ = \ raw\_input().split()$$

The following example in interactive mode reads two values separated by a space, assigns these to variables x and y, then converts each to the appropriate type.

```
>>> x, y = raw_input().split()
12 36.8
>>> x = int(x)
>>> x
12
>>> y = float(y)
>>> y
36.8
```

4.6   COMPUTING AREA AND CIRCUMFERENCE

For this example, a computational model is developed that computes the area and circumference of a circle. The input value of the radius is read from the keyboard and the results written to the screen.

4.6.1   Specification

The specification of the problem can be described as a high-level algorithm in informal pseudo-code notation:

1. Read the value of the radius of a circle, from the input device.

2. Compute the area of the circle.

3. Compute the circumference of the circle.

4. Output or display the value of the area of the circle to the output device.

5. Output or display the value of the circumference of the circle to the output device.

4.6.2 Algorithm with the Mathematical Model

A detailed description of the algorithm and the corresponding mathematical model follows:

1. Read the value of the radius r of a circle, from the input device.

2. Establish the constant π with value 3.14159.

3. Compute the area of the circle, $area = \pi r^2$.

4. Compute the circumference of the circle, $cir = 2 \pi r$.

5. Print or display the value of $area$ of the circle to the output device.

6. Print or display the value of cir of the circle to the output device.

The following lines of pseudo-code completely define the algorithm.

$$read\ r$$
$$\pi = 3.1416$$
$$area = \pi r^2$$
$$cir = 2 \pi r$$
$$display\ \text{"Area = ", area, " Circumference = ", cir}$$

Listing 4.1 shows the Python program that implements the computational model; this program stored in file *areacir.py*.

Listing 4.1: Python program for computing the area and circumference.

```
 1 # Program     : areacir.py
 2 # Description : Read value of the radius of a circle,
 3 #     compute the area and circumference, display value of
 4 #     of these on the output console.
 5 # Author      : Jose M Garrido, May 27 2014.
 6
 7 from math import *
 8
 9 print "Compute area and circumference of a circle"
10 r = input("Enter value of radius: ")
11 print "Value of radius: ", r
12 area = pi * r ** 2
13 cir = 2.0 * pi * r
14 print "Value of area: ", area
15 print "Value of circumference: ", cir
```

The following listing shows the Linux shell commands that start the Python interpreter with the file `areacir.py`.

```
$ python areacir.py
Compute area and circumference of a circle
Enter value of radius: 3.15
Value of radius:  3.15
Value of area:  31.1724531052
Value of circumference:  19.7920337176
```

4.7 SUMMARY

An algorithm is a precise, detailed, and complete description of a solution to a problem. The notations to describe algorithms are flowcharts and pseudo-code. Flowcharts are a visual representation of the execution flow of the various instructions in the algorithm. Pseudo-code is an English-like notation to describe algorithms.

The design structures are sequence, selection, repetition, and input-output. These algorithmic structures are used to specify and describe any algorithm.

Key Terms

algorithm	flowcharts	pseudo-code	variables
constants	action step	structure	sequence
statements	selection	repetition	input/output
identifier	design		

4.8 EXERCISES

4.1 Write the algorithm for computing the area of a triangle. Use flowcharts.

4.2 Write the algorithm for computing the area of a triangle. Use pseudo-code.

4.3 Develop a computational model for computing the area of a triangle.

4.4 Write the algorithm for computing the perimeter of a triangle. Use pseudo-code.

4.5 Write the algorithm for computing the perimeter of a triangle. Use flowcharts.

4.6 Develop a computational model for computing the perimeter of a triangle.

4.7 Write the algorithmic description in flowchart and in pseudo-code to compute the conversion from a temperature reading in degrees Fahrenheit to Centigrade. The algorithm should also compute the conversion from Centigrade to Fahrenheit.

4.8 Develop a computational model to compute the conversion from a temperature reading in degrees Fahrenheit to Centigrade. The program should also compute the conversion from Centigrade to Fahrenheit.

4.9 Write an algorithm and data descriptions in flowchart and pseudo-code to compute the conversion from inches to centimeters and from centimeters to inches.

4.10 Develop a computational model to compute the conversion from inches to centimeters and from centimeters to inches.

## CHAPTER 5

The Selection Program Structure

5.1 INTRODUCTION

In the previous chapter, it was discussed that to completely describe an algorithm, four design structures are used: sequence, selection, repetition, and input/output. This chapter explains the selection program structure using pseudo-code, flowcharts, and the corresponding statements in the Python programming language for implementing computational models.

Conditions are expressions that evaluate to a truth value (*True* or *False*). Conditions are used in the selection statements. Simple conditions are formed with relational operators for comparing two data items. Compound conditions are formed by joining two or more simple conditions with *logical* operators.

The solution to a quadratic equation is discussed as an example of applying the selection statements.

5.2 CONDITIONAL EXPRESSIONS

A conditional expression, also known as a Boolean expression, consists of an expression that evaluates to a truth value, *True* or *False*.

5.2.1 Relational Operators

A conditional expression can be constructed by comparing the values of two data items and using a relational operator. The following list of relational operators in arithmetic notation can be used in a condition:

Arithmetic Operator	Description
>	Greater than
<	Less than
=	Equal to
≤	Less or equal to
≥	Greater or equal to
≠	Not equal to

These relational operators are used to construct conditions as in the following examples in algebraic notation:

$$y \leq 20.15$$
$$p \geq q$$
$$a = b$$

In Python, the relational operators are used with the following notation:

Relational Operator	Description	Arithmetic Notation
>	Greater than	>
<	Less than	<
==	Equal to	=
>=	greater than or equal to	≥
<=	Less than or equal to	≤
!=	Not equal to	≠

The previous examples of conditional expressions can be written as follows:

```
y <= 20.15
p >= q
a == b
```

Arithmetic expressions can be used as part of conditional expression. For example:

```
y >= (x + 45.6)
```

When this conditional expression is evaluated, the arithmetic expression $(x+45.6)$ is evaluated first, then the relational operators are applied. The order in which these operators are evaluated is specified in the following table:

Operator	Order of Evaluation
( )	1
*, /, %	2
+, -	3
=, <, >, <=, >=, !=	4

A conditional expression can be assigned to a variable and its value is of type **bool**. For example, the following statement assigns the value of the previous conditional expression to variable *y_flag*:

```
y_flag = y >= (x + 45.6)
```

5.2.2 Logical Operators

A compound conditional expression consists of one or more simple conditional expressions. The logical operators **and**, **or**, and **not** are used to construct compound conditional expressions from simpler conditions.

The **not** logical operator is applied with one simple condition, *cond*. If a compound condition is defined as **not** *cond*, the truth value of the compound condition is simply the negation of the truth value of the simple condition. For example, when the simple condition *cond* has truth value **False**, the resulting compound condition has a truth value of **True**. The opposite also applies, when the truth value of *cond* is **True**, the truth value of the compound condition is **False**. These rules are summarized in the following table.

cond	**not** cond
True	False
False	True

The **and** and **or** logical operators are used with two simple conditions *cond1* and *cond2*. The rules that apply for the truth value of the resulting compound condition are summarized in the following table.

cond1	cond2	cond1 **and** cond2	cond1 **or** cond2
True	True	True	True
True	False	False	True
False	True	False	True
False	False	False	False

The table shows that when the **and** logical operator is applied to two simple conditions, the truth value of the resulting compound condition is **True** when the truth values of both simple conditions are **True**. When the **or** logical operator is applied, the truth value of the compound condition is **True** when either or both simple conditions have value **True**.

The general forms of compound conditions from the simple conditions, *cond1* and *cond2* in Python are:

```
cond1 and cond2
cond1 or cond2
not cond1
```

The following examples include the **or** and the **and** logical operators:

```
(y > 12) or (x < 23)
(p==q) and (z <= r)
```

The following example includes the **not** operator:

```
not (y > 12)
```

5.3   THE SELECTION STRUCTURE

The selection program structure is also known as *alternation*, because alternate paths are considered, based on the evaluation of a condition. This section describes the selection structure with flowcharts, the concepts associated with conditional expressions, and the implementation in the Python programming language.

5.3.1   Selection Structure with Flowcharts and Pseudo-code

The selection structure is used for decision making in the logic of a program. Figure 5.1 shows the selection design structure using a flowchart. Two possible paths for the execution flow are shown. The condition is evaluated, and one of the paths is selected. If the condition is *True*, then the left path is selected and *Process1* is performed. If the condition is *False*, the other path is selected and *Process2* is performed. In pseudo-code, the selection structure is written as:

> **if** ⟨ *condition* ⟩
> > **then**
> > > ⟨ *statements in Process1* ⟩
> > **else**
> > > ⟨ *statements in Process2* ⟩
> **endif**

5.3.2   Selection with Python

With the Python language, the selection structure is written with an **if** statement and includes three sections: the *condition*, the *then-section*, and the *else-section*. The else-section is optional. The keywords that are used in this statement are: **if** and **else**. The general form of the **if** statement is:

> **if** ⟨ *condition* ⟩ :
> > ⟨ *statements in Process1* ⟩
> **else** :
> > ⟨ *statements in Process2* ⟩

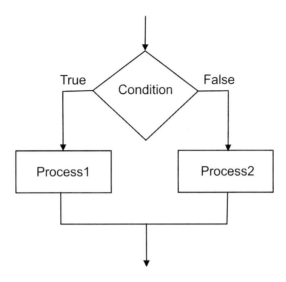

Figure 5.1 Flowchart of the selection structure.

When the condition is evaluated, only one of the two alternatives will be carried out: the one with the statements in *Process1* if the condition is *True*, or the one with the statements in *Process2* if the condition is *False*.

5.3.3   Example with Selection

The following example evaluates the condition $len > 0$, to select which operation is to be performed on variable k. Figure 5.2 shows the flowchart for part of the algorithm that includes this selection structure. In Python, this example is written as:

```
if (len > 0) :
        k= k + 3
else :
        k=k - 1
```

A second example in interactive mode is the following:

```
>>> y = 15
>>> if y >= 11 :
. . .      print "value of y is: ", y
. . else:
. . .      print "Value of y too small"
. . .
value of y is:   15
```

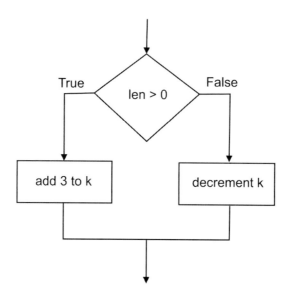

Figure 5.2 Example of selection structure.

The following example is a selection statement in Python that includes a compound condition.

```
if a < b or x >= y :
    a = x + 23.45
else :
    a = y
```

5.4 A COMPUTATIONAL MODEL WITH SELECTION

The following problem involves developing a computational model that includes a quadratic equation, which is a simple mathematical model of a second-degree equation. The solution to the quadratic equation involves complex numbers.

5.4.1 Analysis and Mathematical Model

The goal of the solution to the problem is to compute the two roots of the equation. The mathematical model is defined in the general form of the quadratic equation (second-degree equation):

$$ax^2 + bx + c = 0.$$

The given data for this problem are the values of the coefficients of the quadratic equation: a, b, and c. Because this mathematical model is a second-degree equation, the solution consists of the value of two roots: x_1 and x_2.

5.4.2 Algorithm for General Solution

The general solution gives the value of the two roots of the quadratic equation, when the value of the coefficient a is not zero $(a \neq 0)$. The values of the two roots are:

$$x_1 = \frac{-b + \sqrt{b^2 - 4ac}}{2a} \qquad x_2 = \frac{-b - \sqrt{b^2 - 4ac}}{2a}.$$

The expression inside the square root, $b^2 - 4ac$, is known as the *discriminant*. If the discriminant is negative, the solution will involve complex roots. Figure 5.3 shows the flowchart for the general solution and the following listing is a high-level pseudo-code version of the algorithm.

```
Input the values of coefficients  a,  b, and  c
Calculate value of the discriminant
if the value of the discriminant is less than zero
      then calculate the two complex roots
      else calculate the two real roots
endif
display the value of the roots
```

5.4.3 Detailed Algorithm

The algorithm in pseudo-code notation for the solution of the quadratic equation is:

```
read the value of a from the input device
read the value of b from the input device
read the value of c from the input device
compute the discriminant, disc = b² − 4ac
if discriminant less than zero
then
        // roots are complex
        compute x1 = (−b + √disc)/2a
        compute x2 = (−b − √disc)/2a
else
        // roots are real
        compute x1 = (−b + √disc)/2a
        compute x2 = (−b − √disc)/2a
endif
display values of the roots: x1 and x2
```

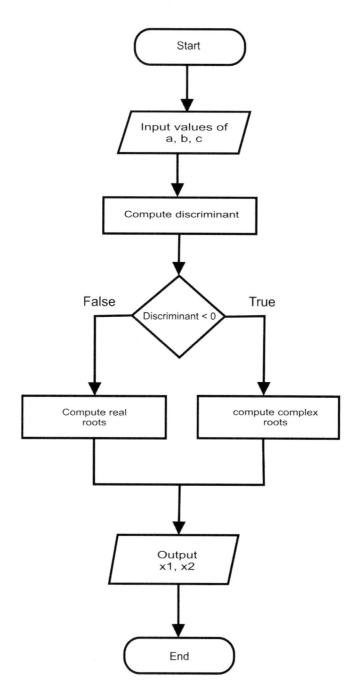

Figure 5.3 High-level flowchart for solving a quadratic equation.

Listing 5.1 shows the Python program that implements the algorithm for the solution of the quadratic equation, which is stored in the file `solquad.py`.

Listing 5.1 Program to compute the roots of a quadratic equation.

```python
 1 # Program     : solquad.py
 3 # Author      : Jose M Garrido, May 21 2014.
 4 # Description : Compute the roots of a quadratic equation.
 5 #     Read the value of the coefficients: a, b, and c from
 6 #     the input console, display value of roots.
 7
 8 from math import *
 9
10 a = input ("Enter value of coefficient a: ")
11 print "Value of a: ", a
12 b = input ("Enter value of coefficient b: ")
13 print "Value of a: ", b
14 c = input ("Enter value of coefficient c: ")
15 print "Value of a: ", c
16
17 disc = b ** 2 - 4.0 * a * c
18 print "discriminant: ", disc
19 if (disc < 0.0) :
20     # complex roots
21     disc = -disc
22     x1r = -b/(2.0 * a)
23     x1i = sqrt(disc)/(2.0 * a)
24     x2r = x1r
25     x2i = -x1i
26     print "Complex roots "
27     # print "x1r: ", x1r, " x1i: ",  x1i
28     x1 = complex( x1r, x1i)
29     #print "x2r: ", x2r, " x2i: ", x2i
30     x2 = complex (x2r, x2i)
31     print "x1: ", x1
32     print "x2: ", x2
33 else :
34     # real roots
35     x1r = (-b + sqrt(disc))/(2.0 * a)
36     x2r = (-b - sqrt(disc))/(2.0 * a)
37     print "Real roots:"
38     print "x1: ", x1r, " x2: ", x2r
```

The following shell commands start the Python interpreter and it processes the program *solquadra.py*. The program prompts the user for the three values of the coefficients, calculates the roots, then displays the value of the roots. Note that the roots are complex.

```
$ python solquad.py
Enter value of coefficient a: 1.25
Value of a:   1.25
Enter value of coefficient b: 2.5
Value of a:   2.5
Enter value of coefficient c: 2.85
Value of a:   2.85
discriminant:  -8.0
Complex roots
x1:  (-1+1.1313708499j)
x2:  (-1-1.1313708499j)
```

The following commands show the Python interpreter processing the program with a different set of input values. Note that in this case, the roots computed are real.

```
$ python solquad.py
Enter value of cofficient a: 2
Value of a:   2
Enter value of cofficient b: -20
Value of a:   -20
Enter value of cofficient c: 5
Value of a:   5
discriminant:  360.0
Real roots:
x1:  9.74341649025  x2:  0.256583509747
```

5.5 MULTI-LEVEL SELECTION

The multi-path selection involves more than two alternatives. The general **if** statement with multiple paths is used to implement this structure. In Python, the **elif** clause is used to expand the number of alternatives. The **if** statement with n alternative paths has the general form:

```
if ⟨ condition ⟩ :
        ⟨ block1 ⟩
elif ⟨ condition2 ⟩ :
        ⟨ block2 ⟩
elif ⟨ condition3 ⟩ :
        ⟨ block3 ⟩
...
else :
        ⟨ blockn ⟩
```

Each block of statements is executed when that particular path of logic is selected. This selection depends on the conditions in the multiple-path **if** statement that are evaluated from top to bottom until one of the conditions evaluates to True. The **elif** parts and the **else** part are optional. The following example shows the **if** statement with several paths.

```
print "Testing multi-path selection in a Python script"
y = 4.25
x = 2.55
if y > 15.50 :
      x = x + 1
      print "x: ", x
elif y > 4.5 :
      x = x + 7.85
      print "x: ", x
elif y > 3.85 :
      x = y * 3.25
      print "x: ", x
elif y > 2.98 :
      x = y + z*454.7
      print "x: ", x
else :
      x = y
      print "x: ", x
```

This portion of Python code is stored in file *test1.py*. The following lines of shell commands start the Python interpreter and process the script.

```
$ python test1.py
Testing multi-path selection in a Python script
x:   13.8125
```

5.6 SUMMARY

The selection structure is also known as alternation. It evaluates a condition and then follows one of two (or more) paths. The two general selection statements with **if** are explained in Python. The first one is applied when there are two or more possible paths in the algorithm, depending on how the condition evaluates. The multi-path selection statement is applied when the value of a single variable or expression is evaluated, and there are multiple possible values.

The condition in the **if** statement consists of a conditional expression, which evaluates to a truth value (True or False). Relational operators and logical operators are used to form more complex conditional expressions.

Key Terms

selection	alternation	condition	if statement
case statement	relational operator	logical operator	truth-value
then	else	endif	elif
elseif	end	multi-path selection	

5.7 EXERCISES

5.1 Develop a Python program that computes the conversion from gallons to liters and from liters to gallons. Include a flowchart, pseudo-code design, and a complete implementation in Python. The user inputs the string: "gallons" or "liters"; the model then computes the corresponding conversion.

5.2 Develop a Python program to calculate the total amount to pay for movie rental. Include a flowchart, pseudo-code design, and a complete implementation in Python. The movie rental store charges $3.50 per day for every DVD movie. For every additional period of 24 hours, the customer must pay $0.75.

5.3 Develop a Python program that finds and displays the largest of several numbers, which are read from the input device. Include a flowchart, pseudo-code design, and a complete implementation in Python.

5.4 Develop a Python program that finds and displays the smallest of several numbers, which are read from the input device. Include a flowchart, pseudo-code design, and a complete implementation in Python.

5.5 Develop a Python program that computes the gross and net pay of several employees. The input quantities are employee name, hourly rate, number of hours, percentage of tax (use 14.5%). The tax bracket is $115.00. When the number of hours is greater than 40, the (overtime) hourly rate is 40% higher. Include a flowchart, pseudo-code design, and a complete implementation in Python.

5.6 Develop a Python program that computes the fare in a ferry transport for passengers with motor vehicles. Include a flowchart, pseudo-code design, and a complete implementation in Python. Passengers pay an extra fare based on the vehicle's weight. Use the following data: vehicles with weight up to 780 lb pay $80.00, up to 1100 lb pay $127.50, and up to 2200 lb pay $210.50.

5.7 Develop a Python program that computes the average of student grades. The input data are the four letter grades for various work submitted by the students. Include a flowchart, pseudo-code design, and a complete implementation in the Python programming language.

## CHAPTER 6

The Repetition Program Structure

6.1 INTRODUCTION

This chapter presents the repetition program structure and using it for specifying, describing, and implementing algorithms in developing computational models. This structure and the corresponding statements are discussed with flowcharts, pseudo-code, and implementation in the Python programming language. The repetition structure specifies that a block of statements be executed repeatedly based on a given condition. Basically, the statements in the process block of code are executed several times, so this structure is often called a *loop* structure. A program segment that includes the repetition structure has three major parts in its form:

1. the initial conditions,

2. the steps that are to be repeated, and

3. the final results.

There are three general forms of the repetition structure: the *while*-loop, the *repeat-until* loop, and the *for*-loop. The first form of the repetition structure, the *while* construct, is the most flexible. The other two forms of the repetition structure can be expressed with the *while* construct.

6.2 REPETITION WITH THE WHILE-LOOP

The *while*-loop consists of a conditional expression and block of statements. This construct evaluates the condition before the process block of statements is executed. If the condition is true, the statements in the block are executed. This repeats while the condition evaluates to true; when the condition evaluates to false, the loop terminates.

6.2.1 While-Loop Flowchart

A flowchart with the *while*-loop structure is shown in Figure 6.1. The process block consists of a sequence of actions.

The actions in the *process* block are performed while the condition is true. After the actions in the process block are performed, the condition is again evaluated, and the actions are again performed if the condition is still true; otherwise, the loop terminates.

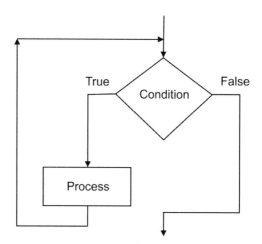

Figure 6.1 A flowchart with a while-loop.

The condition is tested first, and then the process block is performed. If this condition is initially false, the actions in the block are not performed.

The number of times that the loop is performed is normally a finite number. A well-defined loop will eventually terminate, unless it has been specified as a non-terminating loop. The condition is also known as the *loop condition*, and it determines when the loop terminates. A non-terminating loop is defined in special cases and will repeat the actions forever.

6.2.2 While Structure in Pseudo-Code

The form of the *while* statement includes the condition, the actions in the process block written as statements, and the keywords *while, do,* and *endwhile*. The block of statements is placed after the *do* keyword and before the *endwhile* keyword. The following lines of pseudo-code show the general form of the while-loop statement that is shown in the flowchart of Figure 6.1.

```
while ( condition )  do
     ( block of statements )
endwhile
```

The following example has a *while* statement and the block of statements is performed repeatedly while the condition j <= MAX_NUM is true.

```
while j <= MAX_NUM do
      set sum = sum + 12.5
      set y = x * 2.5
      add 3 to j
endwhile
display "Value of sum: ", sum
display "Value of y: ", y
```

6.2.3 While-Loop in the Python Language

The following lines of code show the general form of the while-loop statement in Python; it is similar to the pseudo-code statement and follows the loop definition shown in the flowchart of Figure 6.1.

> **while** ⟨ *condition* ⟩ :
> ⟨ *block of statements* ⟩

The previous example has a *while* statement with a condition that checks the value of variable *j*. The block of statements that are repeated are always indented (four columns to the right); these statements are repeated while the condition j <= MAX_NUM is true. The following lines of code show the Python implementation, which is stored in file *test2.py*. The *while* statement appears in line 6, the block of statements that are repeated are in lines 7–9. The print statement in line 10 is at the end and is outside the loop.

```
 1 # Script for testing while-loop
 2 x = 12.35
 3 MAX_NUM = 15
 4 j = 0
 5 sum = 0.0
 6 while ( j <= MAX_NUM)   :
 7         sum = sum + 12.5
 8         y = x * 2.5
 9         j = j + 3
10 print 'Value of sum: ', sum
```

The following Linux shell command starts the Python interpreter with the script *test2.py*.

```
$ python test2.py
Value of sum:   75.0
```

6.2.4 Loop Counter

As mentioned previously, in the while-loop construct, the condition is tested first and then the statements in the loop block are performed. If this condition is initially false, the statements are not performed.

The number of times that the loop is performed is normally a finite integer value. For this, the condition will eventually be evaluated to false, that is, the loop will terminate. This condition is often known as the *loop condition*, and it determines when the loop terminates. Only in some very special cases, the programmer can decide to write an infinite loop; this will repeat the statements in the repeat loop forever.

A *counter* variable stores the number of times (also known as iterations) that the loop executes. The counter variable is incremented every time the statements in the loop are performed. The variable must be initialized to a given value, typically to 0 or 1.

In the following pseudo-code listing, there is a counter variable with name *loop_counter* in the *while* statement. This counter variable is used to control the number of times the block statement is performed. The counter variable is initially set to 1, and is incremented every time through the loop.

```
Max_Num = 25      // maximum number of times to execute
set loop_counter = 1   // initial value of counter
while-loop_counter < Max_Num do
        display "Value of counter: ", loop_counter
        increment loop_counter
endwhile
```

The first time the statements in the block are performed, the loop counter variable *loop_counter* starts with a value equal to 1. The second time through the loop, variable *loop_counter* has a value equal to 2. The third time through the loop, it has a value of 3, and so on. Eventually, the counter variable will have a value equal to the value of *Max_Num* and the loop terminates. The following listing is the Python code, which is stored in file *test3.py*.

```
# Script for testing a loop counter in a while-loop

Max_Num = 15      # maximum number of times to execute
loop_counter = 1  # initial value of counter
while-loop_counter < Max_Num :
      print "Value of counter: ", loop_counter
      loop_counter = loop_counter + 1
```

The following Linux shell command starts the Python interpreter with the script *test3.py*.

```
$ python test3.py
Value of counter:  1
Value of counter:  2
Value of counter:  3
Value of counter:  4
Value of counter:  5
Value of counter:  6
Value of counter:  7
Value of counter:  8
Value of counter:  9
Value of counter:  10
Value of counter:  11
Value of counter:  12
Value of counter:  13
Value of counter:  14
Value of counter:  15
```

6.2.5 Accumulator Variables

An *accumulator* variable stores partial results of repeated calculations. The initial value of an accumulator variable is normally set to zero.

For example, the following algorithm in pseudo-code calculates the summation of numbers from input, and includes an accumulator variable. The statement accumulates the values of *cval* in variable *total* and it is included in the while-loop:

```
total = 0.0
while j < MAX_NUM
    set cval = j * 1.25
    add cval to total
    increment j
endwhile
display "Total accumulated: ", total
```

After the *endwhile* statement, the value of the accumulator variable *total* is displayed. The following code is the corresponding Python code, which is stored in file *test4.py*.

```
# Script for testing an accumulator variable in a while-loop

total = 0.0
j = 1
MAX_NUM = 15
while j < Max_num :
```

```
        cval = j * 1.25
        total = cval + total
        j = j + 1
print "Total accumulated: ", total
```

The following Linux shell command starts the Python interpreter with the script *test4.py*.

```
$ python test4.py
Total accumulated:  131.25
```

In programming, each counter and accumulator variable serves a specific purpose and these variables should be well documented.

6.2.6 Summation of Input Numbers

The following simple problem applies the concepts and implementation of while-loop and accumulator variable. The problem computes the summation of numeric values inputed from the main input device. Computing the summation should proceed while the input values are greater than 1.

The pseudo-code that describes the algorithm uses an input variable, an accumulator variable, a loop counter variable, and a conditional expression that evaluates whether the input value is greater than zero.

```
set innumber = 1.5  // number with dummy initial value
set loop_counter = 0
set sum = 0.0         // initialize accumulator variable
display "Enter a number: "
read innumer          // read first value
while innumber > 1.0  do
      add innumber to sum
      increment loop_counter
      display "Value of counter: ", loop_counter
      display "Enter a number: "
      read innumer
endwhile
display "Value of sum: ", sum
```

Listing 6.1 shows the Python program that implements the summation problem. The program is stored in file `summa.py`.

Listing 6.1 Python program for computing the summation.

```
 1 # Script for summation of input values a while-loop
 2 # Script: summa.py
 3
 4 loop_counter = 0
 5 sum = 0.0          # initial value of accumulator variable
 6 innumber = input( "Type number: ")  # read first value
 7 while innumber > 1.0 :
 8         sum = sum + innumber
 9         loop_counter = loop_counter + 1
10         print "Value of counter: ", loop_counter
11         innumber = input( "Enter a number: ")
```

The following output listing shows the shell commands that start the Python interpreter with file summa.py.

```
$ python summa1.py
Type number: 1.5
Value of counter:  1
Type number: 2.55
Value of counter:  2
Type number: 1.055
Value of counter:  3
Type number: 4.12
Value of counter:  4
Type number: 1.25
Value of counter:  5
Type number: 0.0
Value of sum:  10.475
```

6.3   REPEAT-UNTIL LOOP

The *repeat-until* loop is a control flow structure that allows actions to be executed repeatedly based on a given condition. The actions within the process block are executed first, and then the condition is evaluated. If the condition is not true the actions within the process block are executed again. This repeats until the condition becomes true.

Repeat-until structures check the condition after the block is executed; this is an important difference from the while-loop, which tests the condition before the actions within the block are executed. Figure 6.2 shows the flowchart for the repeat-until structure.

The pseudo-code statement of the repeat-until structure corresponds directly with the flowchart in Figure 6.2 and uses the keywords *repeat, until,* and *endrepeat.* The following lines of code shows the general form of the repeat-until statement.

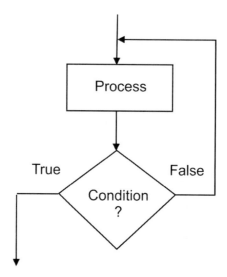

Figure 6.2 A flowchart with a repeat-until structure.

```
repeat
        ⟨ statements in block ⟩
until ⟨ condition ⟩
endrepeat
```

The following listing shows the pseudo-code of a repeat-until statement for the problem discussed in the previous section.

```
set innumber = 1.0  // dummy initial value
set l_counter = 0
set sum = 0.0        // accumulator variable
repeat
        add innumber to sum
        increment l_counter
        display "Value of counter: ", l_counter
        display "Type number: "
        read innumer
until innumber <= 0.0
endrepeat
display "Value of sum: ", sum
```

In Python, the repeat-until loop is not directly supported by a syntactic construct. However, it can be implemented with a while statement. Listing 6.2 shows the Python program that implements a problem with a loop counter. Note that a Boolean variable (*loop_cond*) is used to reference the value of the loop condition. The while statement in line 8 always checks for the reversed condition using the *not* operator that precedes the Boolean variable. The body of the loop includes the statements in lines 9−11 and the condition is evaluated at the end. The program is stored in file `test5.py`.

Listing 6.2 Python program with a loop counter.

```
1 # Script: test5.py
2 # This script tests a loop counter in a repeat-until loop
3 #    implemented with a while statement
4
5 Max_Num = 15        # maximum number of times to execute
6 loop_counter = 1    # initial value of counter
7 loop_cond = False
8 while not loop_cond :
9       print "Value of counter: ", loop_counter
10      loop_counter = loop_counter + 1
11      loop_cond = loop_counter >= Max_Num # until true
```

The following output listing shows the shell commands that start the Python interpreter with file `test5.py`.

```
$ python test5.py
Value of counter:  0
Value of counter:  1
Value of counter:  2
Value of counter:  3
Value of counter:  4
Value of counter:  5
Value of counter:  6
Value of counter:  7
Value of counter:  8
Value of counter:  9
Value of counter:  10
Value of counter:  11
Value of counter:  12
Value of counter:  13
Value of counter:  14
```

Listing 6.3 shows the Python program that implements the summation problem. The program is stored in file `summrep.py`.

Listing 6.3 Python program that computes a summation.

```
1 # Script: summrep.py
2 # This script computes a summation using a repeat-until
```

```
 3 #      loop implemented with a while statement
 4
 5 sum = 0.0
 6 loop_counter = 0
 7 innumber = input( "Enter a number: ") # first number
 8 lcond = innumber <= 0.0
 9 while not lcond:
10        sum = sum + innumber
11        loop_counter = loop_counter + 1
12        print "Value of counter: ", loop_counter
13        innumber = input( "Enter a number: ")
14        lcond = innumber <= 0.0
15
16 print "Value of sum: ", sum
```

The following output listing shows the shell commands that start the Python interpreter with file summrep.py.

```
$ python summrep.py
Enter a number: 4.7
Value of counter:   1
Enter a number: 7.88
Value of counter:   2
Enter a number: 0.8
Value of counter:   3
Enter a number: 2.145
Value of counter:   4
Enter a number: 0.0
Value of sum:   15.525
```

6.4 FOR-LOOP STRUCTURE

The *for*-loop structure explicitly uses a loop counter; the initial value and the final value of the loop counter are specified. The *for*-loop is most useful when the number of times that the loop is carried out is known in advance. In pseudo-code, the *for* statement has the following general form:

> *for* ⟨ *counter* ⟩ = ⟨ *initial_val* ⟩ *to* ⟨ *final_val* ⟩
> *do*
>   Block of statements
> *endfor*

On every iteration, the loop counter is automatically incremented. The last time through the loop, the loop counter reaches its final value and the loop terminates. The *for*-loop is similar to the *while*-loop in that the condition is evaluated before carrying out the operations in the repeat loop.

The following listing in pseudo-code uses a *for*-loop for the repetition part of the summation problem. Variable *j* is the counter variable, which is automatically incremented and is used to control the number of times the statements in a block is to be performed.

```
for j = 1 to num do
        set sum = sum + 12.5
        set y = x * 2.5
endfor
display "Value of sum: ", sum
display "Value of y: ", y
```

In Python, the simple use of the *for* statement uses function *range*. The first, last, and the increment values of the loop counter are specified. The last value specified is not really included as one of the values of the loop counter. The increment is optional; if not included, its value is 1.

Listing 6.4 shows the Python program that includes a simple for-loop. The program is stored in file `test6.py`. The loop statement is in line 6. Note that values of *j* are: 1, ..., 9 and the value of this loop counter is displayed in line 9 in every iteration of the loop.

Listing 6.4 Python program that includes a for-loop.

```
 1 # Script: test6.py
 2 # This script tests a for-loop
 3 x = 3.45
 4 num = 10
 5 sum = 0.0
 6 for j in range(1, num) :
 7         sum = sum + 12.5
 8         y = x * 2.5
 9         print "Loop counter: ", j
10
11 print "Value of sum: ", sum
12 print "Value of y: ", y
```

The following output listing shows the shell commands that start the Python interpreter with file `test6.py`.

```
$ python test6.py
Loop counter:  1
Loop counter:  2
Loop counter:  3
```

```
Loop counter:  4
Loop counter:  5
Loop counter:  6
Loop counter:  7
Loop counter:  8
Loop counter:  9
Value of sum:   112.5
Value of y:   8.625
```

6.4.1 Summation Problem with a For-Loop

Using the for-loop construct of the repetition structure, the algorithm for the summation of input data can be defined in a relatively straightforward manner with pseudo-code. The most significant difference from the previous design is that the number of data inputs from the input device is included at the beginning of the algorithm. As in the previous case, the input value is added to variable *sum* only if the value entered is greater than zero.

```
set innumber = 1.0   // number with dummy initial value
set sum = 0.0        // initialize accumulator variable
display "Number of input data to read: "
read MaxNum
for-loop_counter = 1 to MaxNum do
      display "Type number: "
      read innumber
      if innumber > 0.0
      then
          add innumber to sum
          display "Value of counter: ", loop_counter
      endif
endfor
display "Value of sum: ", sum
```

Listing 6.5 shows the Python source program that implements the summation problem with a for-loop. The program is stored in file summfor.c. Note that in line 5, the upper bound value specified for *loop_counter* is *MaxNum+1*.

Listing 6.5 Python program for computing the summation with for-loop.

```
1 # Script: summfor.py
2 # This script computes a summation using a for-loop
3 sum = 0.0        # initialize value of accumulator variable
4 MaxNum = input ("Enter number of input data to read: ")
5 for-loop_counter in range(1, MaxNum+1) :
6     innumber = input ("Enter number: ")
7     if innumber > 0.0 :
```

```
 8                  sum = sum + innumber
 9                  print "Value of counter: ", loop_counter
10
11 print "Value of sum: ", sum
```

The following output listing shows the shell commands that start the Python interpreter with file `summfor.py`.

```
$ python summfor.py
Enter number of input data to read: 5
Enter number: 12.66
Value of counter:  1
Enter number: 2.432
Value of counter:  2
Enter number: 5.78
Value of counter:  3
Enter number: 23.85
Value of counter:  4
Enter number: 22.12
Value of counter:  5
Value of sum:  66.842
```

6.4.2 Factorial Problem

The *factorial* operation, denoted by the symbol !, can be defined in a general and informal manner as follows:

$$y! = y\,(y-1)\,(y-2)\,(y-3)\,\ldots\,1.$$

For example, the factorial of 5 is:

$$5! = 5 \times 4 \times 3 \times 2 \times 1.$$

6.4.2.1 Mathematical Specification of Factorial

A mathematical specification of the factorial function is as follows, for $y \geq 0$:

$$y! = \begin{cases} 1 & \text{when } y = 0 \\ y\,(y-1)! & \text{when } y > 0. \end{cases}$$

The base case in this definition is the value of 1 for the function if the argument has value zero, which is $0! = 1$. The general (recursive) case is $y! = y\,(y-1)!$, if the value of the argument is greater than zero. This function is not defined for negative values of the argument.

6.4.2.2 Computing Factorial

In the following Python program, the factorial function *mfact* has one parameter: the value for which the factorial is to be computed. Listing 6.6 shows a Python program, *factp.py*, that includes function function *mfact*. This function is called in line 22 to compute the factorial of a number and the result value is displayed on the console.

Listing 6.6 Python source program for computing factorial.

```
 1 #
 2 # Program     : factp.py
 3 # Author      : Jose M Garrido, May 28 2014.
 4 # Description : Compute the factorial of a number.
 5
 6 def mfact(num):
 7     """
 8         This function computes the factorial of num >= 0
 9         it multiplies num * (num-1) * num-2 * ...1
10     """
11     res = 1
12     if num > 0:
13         for num in range(num, 1, -1):
14             res = res * num
15         return res
16     elif num == 0:
17         return 1
18     else :
19         return -1
20
21 y = input("Enter a number to compute factorial: ")
22 fy = mfact(y)
23 print "Factorial is: ", fy
```

Note that this implementation returns -1 for negative values of the argument. The following shell commands execute the Python interpreter with program *factp.py* and computes the factorial for several values of the input number.

```
$ python factp.py
Enter a number to compute factorial: 5
Factorial is:   120

$ python factp.py
Enter a number to compute factorial: 0
Factorial is:   1
```

```
$ python factp.py
Enter a number to compute factorial: 1
Factorial is:  1
```

6.5   SUMMARY

The repetition structure is used in algorithms in order to perform repeatedly a group of action steps (instructions) in the process block. There are three types of loop structures: *while*-loop, *repeat-until*, and *for*-loop. In the *while* construct, the loop condition is tested first, and then the block of statements is performed if the condition is true. The loop terminates when the condition is false.

In the *repeat-until* construct, the group of statements in the block is carried out first, and then the loop condition is tested. If the loop condition is true, the loop terminates; otherwise the statements in the block are performed again.

The number of times the statements in the block are carried out depends on the condition of the loop. In the *for*-loop, the number of times to repeat execution is explicitly indicated by using the initial and final values of the loop counter. Accumulator variables are also very useful with algorithms and programs that include loops. Several examples of programs in Python are shown.

Key Terms

repetition	loop	while	loop condition
do	endrepeat	block	loop termination
loop counter	endwhile	accumulator	repeat-until
for	to	downto	endfor
end	iterations	summation	factorial

6.6   EXERCISES

6.1   Develop a computational model that computes the maximum value from a set of input numbers. Use a while-loop in the algorithm and implement in Python.

6.2   Develop a computational model that computes the maximum value from a set of input numbers. Use a for-loop in the algorithm and implement in Python.

6.3   Develop a computational model that computes the maximum value from a set of input numbers. Use a repeat-until loop in the algorithm and implement in Python.

6.4 Develop a computational model that finds the minimum value from a set of input values. Use a while-loop in the algorithm and implement in Python.

6.5 Develop a computational model that finds the minimum value from a set of input values. Use a for-loop in the algorithm and implement in Python.

6.6 Develop a computational model that finds the minimum value from a set of input values. Use a repeat-until loop in the algorithm and implement in Python.

6.7 Develop a computational model that computes the average of a set of input values. Use a while-loop in the algorithm and implement in Python.

6.8 Develop a computational model that computes the average of a set of input values. Use a for-loop in the algorithm and implement in Python.

6.9 Develop a computational model that computes the average of a set of input values. Use a repeat-until loop in the algorithm and implement in Python.

6.10 Develop a computational model that computes the student group average, maximum, and minimum grade. The computational model uses the input grade for every student. Use a while-loop in the algorithm and implement in Python.

6.11 Develop a computational model that computes the student group average, maximum, and minimum grade. The computational model uses the input grade for every student. Use a for-loop in the algorithm and implement in Python.

6.12 Develop a computational model that reads rainfall data in inches for yearly quarters from the last five years. The computational model is to compute the average rainfall per quarter, the average rainfall per year, and the maximum rainfall per quarter and for each year. Implement in Python.

6.13 Develop a computational model that computes the total inventory value amount and total per item. The computational model is to read item code, cost, and description for each item. The number of items to process is not known. Implement in Python.

III

Data Structures, Object Orientation, and Recursion

## CHAPTER 7

Python Lists, Strings, and Other Data Sequences

7.1 INTRODUCTION

Most programming languages support *arrays*, which are one the most fundamental data structures. With an array, multiple values can be stored and each is referenced with the name of the array and specifying an *index* value. The individual values of an array are known as *elements*.

Python uses *lists*, which are more general data structures that can be used to represent arrays. This chapter discusses lists and other data sequences used in Python. In a subsequent chapter, the extensive array handling operations and objects in the *NumPy* library will be discussed.

7.2 LISTS

A list in Python is simply an ordered collection of items each of which can be of any type. A list is a dynamic mutable data structure and this means that items can be added to and deleted from it. The list data structure is the most common data *sequence* in Python. A sequence is a set of values identified by integer indices.

To define a list in Python, the items are separated by commas and in square brackets. A simple list with name *vv* and *n* items is defined as follows:

$$vv = [p_1, \ p_2, \ p_3, \ \ldots, \ p_n].$$

For example, the command that follows defines a list with name *vals* and six data items:

```
vals = [1,2,3,4,5,6]
```

7.2.1 Indexing Lists

An individual item in the list can be referenced by using an index, which is an integer number that indicates the relative position of the item in the list. The values of index numbers always start at zero. In the list *vals* defined previously, the index values are: 0, 1, 2, 3, 4, and 5.

In Python, the reference to an individual item of a list is written with the name of the list and the index value or an index variable within brackets. The following Python commands in interactive mode define the list *vals*, reference the first item on the list with index value 0, reference the fourth item with index value 3, then use an index variable *idx* with a value of 4 to reference an item of list *vals*.

```
>>> vals = [1, 2, 3, 4, 5, 6]
>>> vals
[1, 2, 3, 4, 5, 6]
>>> vals[0]
1
>>> vals[3]
4
>>> idx = 4
>>> vals[idx]
5
```

In non-interactive mode, the command **print vals[idx]** is used to display the value of the list item indexed with variable *idx*.

An index value of −1 is used to reference the last item of a list and an index value of −2 is used to reference the item that is previous to the last item of the list. The following commands also in interactive mode illustrate this.

```
>>> vals[-1]
6
>>> vals[-2]
5
```

Because a list is a mutable data structure, the items of the list can change value by performing assignment on them. The second of the following Python commands assigns the new value of 23.55 to the item that has index value 3.

```
>>> vals
[1, 2, 3, 4, 5, 6]
>>> vals [3] = 23.55
>>> vals
[1, 2, 3, 23.55, 5, 6]
```

7.2.2 Slicing Operations

The slicing operations are used to access a sublist of the list. The colon notation is used to specify the range of index values of the items. The first index
value is written before the colon and the last index value is written after the
colon. This indicates the range of index values from the start index value up
to but not including the last index value specified.

In the following example, which uses Python in interactive mode, the second command specifies a sublist of list *vals*, that includes the items starting
with index value 0 up to but not including the item with index value 4. The
third command assigns the sublist `vals[2:5]` to variable *y*; so this command
creates a new sublist and assigns it to *y*.

```
>>> vals
[1, 2, 3, 4, 5, 6]
>>> vals[0:4]
[1, 2, 3, 4]
>>> y = vals[2:5]
>>> y
[3, 4, 5]
```

A range of items can be updated using slicing and assignment. For example,
the following command changes the values of items with index 0 and up to
but not including the item with index value 2.

```
>>> vals[0:2] = vals[1:3]
>>> vals
[2, 3, 3, 23.55, 5, 6]
```

Using slicing, the second index value can be left out and implies that the
range of index values starts from the item with the index value specified to
the last item of the list. In a similar manner, the first index value can be left
out and implies that the range of items starts with the first item of the list.

```
>>> vals[1:]
[3, 3, 23.55, 5, 6]
>>> vals[:5]
[2, 3, 3, 23.55, 5]
```

The first useful operation on a list is to get the number of items in a list.
Function *len* is called to get the number of items from the list specified in
parenthesis. In the following commands, the first command gets the length of
list *vals* and assigns this value to variable *n*. The next command shows the
value of *n*, which is 6 because *vals* has six items. The next command calls

function *range* to generate another list starting at 0 and the last value is 5 (one before 6). Recall that function *range* was used in the for-loop discussed previously. The next command combines functions *range* and *len* to produce the same result as the previous command.

```
>>> n = len(vals)
>>> n
6
>>> range(n)
[0, 1, 2, 3, 4, 5]
>>> range(len(vals))
[0, 1, 2, 3, 4, 5]
```

7.2.3 Iterating over a List with a Loop

Indexing is very useful to access the items of a list iteratively in a loop. A for-loop accesses the items of a list one by one by iterating over the index values of the list. Listing 7.1 computes the summation of the items in list *vals2* and selects only the ones that have a value ≤ 3.15. The Python script is stored in file sumlist.py.

Listing 7.1 Python program for computing the summation on a list.

```
1 # Script: sumlist.py
2 # Compute the summation of the values in a list
3 # that are less or equal to 3.15 using a loop
4 #
5 vals2 = [2, 3.45, 1.22, 4.87, 0.78, 2.45, 8.76]
6 nl = range ( len(vals2))
7 sum = 0.0
8 for i in nl:
9     if vals2[i] <= 3.15 :
10         print "Index: ", i
11         sum = sum + vals2[i]
12 print "Summation: ", sum
```

The following output listing shows the shell commands that start the Python interpreter with file sumlist.py, and the results computed.

```
$ python sumlist.py
Index:  0
Index:  2
Index:  4
Index:  5
Summation:  6.45
```

Python supports iterating directly over the items of a list using the *for* statement. Listing 7.2 shows a Python program that computes the same summation of the list discussed in the previous problem. Instead of using indexing, this program iterate over the items of the list *vals2* and the final result is the same as in the previous program. Note that the variable *item* refers to the value of an individual item of the list. This Python script is stored in file sumlistb.py.

Listing 7.2 Python program for computing the summation on a list.

```
 1 # Script: sumlistb.py
 2 # Compute the summation of the values in a list
 3 # that are less or equal to 3.15 by
 4 # using a loop to iterate over the items of a list
 5 #
 6 vals2 = [2, 3.45, 1.22, 4.87, 0.78, 2.45, 8.76]
 7 sum = 0.0
 8 for item in vals2:
 9     if item <= 3.15 :
10         print "item ", item
11         sum = sum + item
12 print "Summation: ", sum
```

The following output listing shows the shell commands that start the Python interpreter with file sumlistb.py and the results.

```
$ python sumlistb.py
item   2
item   1.22
item   0.78
item   2.45
Summation:   6.45
```

7.2.4 Creating a List Using a Loop

A list can be created starting with an empty list; items can be appended using a for-loop. The *append* method is an operation of a list and is very useful for creating a list. The following command appends the value of a new item *v* to a list *mlist*:

```
mlist.append(v)
```

The following example builds a list of items with values that are multiples of 5. Listing 7.3 shows a Python program that builds the list starting with an item with value 5. Line 8 has a *for* statement that defines the loop that iterates with loop counter *j* starting with 1 up to *SIZE*. The value of the current item is computed in line 9 and it is placed at the end of the current list using the *append* list method in line 10. Note that the variable *item* refers to the value of an individual item of the list. This Python script is stored in file `blist5.py`.

Listing 7.3 Python program with a list of values that are multiples of 5.

```
 1 # Script: blist5.py
 2 # Python script to build a list with
 3 # items with values multiple of 5
 4
 5 SIZE = 15      # number of items in list
 6 listmf = []    # create empty list
 7 # the list starts with 5
 8 for j in range (1, SIZE+1):
 9       item = j * 5
10       listmf.append(item)
11
12 print "List is: ", listmf
```

The following output listing shows the shell commands that start the Python interpreter with file `blist5.py` and the results.

```
$ python blist5.py
List is:   [5, 10, 15, 20, 25, 30, 35, 40, 45, 50]
```

7.2.5   Passing Lists to a Function

One or more lists can be passed to a function and the lists are used as arguments in the function call. The lists are specified as parameters in the function definition.

Listing 7.4 shows a Python program that includes a function definition *buildlf* starting in line 8. The function has two parameters: a list *llist* and simple variable *lsize*. The function builds the list by appending items into it. The function call with two arguments appears in line 15 and the results are displayed by the instruction in line 16.

Listing 7.4 Program that builds a list with values that are multiples of 5.

```
 1 # Script: blist5f.py
 2 # Python script to build a list with
 3 # items with values multiple of 5
 4 # using function buildlf
```

```
 5
 6 #Function that builds a list
 7 # the list starts with 5
 8 def buildlf (llist, lsize):
 9     for j in range (1, lsize+1):
10         item = j * 5
11         llist.append(item)
12
13 SIZE = 10      # number of items in list
14 listmf = []    # create empty list
15 buildlf(listmf, SIZE)
16 print "List is: ", listmf
```

The following output listing shows the shell commands that start the Python interpreter with file `blist5f.py` and the results.

```
$ python blist5f.py
List is:  [5, 10, 15, 20, 25, 30, 35, 40, 45, 50]
```

A list can be returned by a function and the list must be defined in the function. Listing 7.5 shows a Python program that includes a function definition *buildlg* starting in line 8. This function builds a list and then returns it in line 13. Note that the function includes only one parameter: *lsize*. The script calls in the function in an assignment statement in line 16.

Listing 7.5 Python program that builds a list of values multiples of 5.

```
 1 # Script: blist5g.py
 2 # Python script to build a list with
 3 # items with values multiple of 5
 4 # using function buildlg
 5
 6 #Function that builds a list
 7 # the list starts with 5
 8 def buildlg (lsize):
 9     llist = [] # create empty list
10     for j in range (1, lsize+1):
11         item = j * 5
12         llist.append(item)
13     return llist
14
15 SIZE = 10      # number of items in list
16 listmf = buildlg(SIZE)
17 print "List is: ", listmf
```

The results are the same as the previous two Python scripts. The following output listing shows the shell commands that start the Python interpreter with file blist5g.py and the results.

```
$ python blist5g.py
List is:   [5, 10, 15, 20, 25, 30, 35, 40, 45, 50]
```

7.2.6   Additional Operations on Lists

In addition to list method *append,* there are several methods for lists provided in Python. The *extend* method appends another list to the current list. The following commands using Python in interactive mode define two lists *listm* and *lst2,* then append the second list to the first list.

```
>>> listm = [5, 10, 15, 20, 25, 30]
>>> lst2 = [35, 40]
>>> listm.extend(lst2)
>>> listm
[5, 10, 15, 20, 25, 30, 35, 40]
```

List method *insert* places an item at a given position of a list. Calling this method requires two arguments, the first is the index value before which the new item is to be inserted, the second argument is the value of the item to be inserted in the list. The following command inserts an item with value 33 into list *listm* at position with index value 5.

```
>>> listm.insert(5, 33)
>>> listm
[5, 10, 15, 20, 25, 33, 30, 35, 40]
```

List method *remove* searches for the first item with the specified value and removes it from the list. The following command removes the item with value 35 from the list *listm.*

```
>>> listm.remove(35)
>>> listm
[5, 10, 15, 20, 25, 33, 30, 40]
```

List method *pop* removes the last item from the specified list and returns the value of the item. The following command removes and displays the last item from list *listm.*

```
>>> listm.pop()
40
>>> listm
[5, 10, 15, 20, 25, 33, 30]
```

List method *index* finds the first item with the value specified and returns its index value. The following command gets and returns the index value of the item that has value 33.

```
>>> listm.index(33)
5
```

List method *sort* rearranges in ascending order the items of a list. The following commands sort the items (in ascending order) in list *listm* and displays the list again.

```
>>> listm.sort()
>>> listm
[5, 10, 15, 20, 25, 30, 33]
```

List method *reverse* rearranges the items of a list in reverse order. The following command reverses the items in list *listm*.

```
>>> listm.reverse()
>>> listm
[33, 30, 25, 20, 15, 10, 5]
```

7.3 TEMPERATURE CONVERSION PROBLEM

The temperature conversion problem was discussed in the previous chapter. The description of the revised problem is: given a list of values of temperature in degrees Celsius, compute the corresponding values in degrees Fahrenheit and show this result.

7.3.1 Mathematical Model

The mathematical representation of the solution to the problem, the formula expressing a temperature measurement F in Fahrenheit in terms of the temperature measurement C in Celsius is:

$$F = \frac{9}{5}\,C + 32.$$

The solution to the problem applies the mathematical expression for the conversion of a temperature measurement in Celsius to the corresponding value in Fahrenheit. The mathematical formula expressing the conversion assigns a value to the desired temperature in the variable *itemF*, the dependent variable. The values of the variable *itemC* can change arbitrarily because it is the independent variable. The mathematical model uses floating-point numbers to represent the temperature readings in various temperature units.

7.3.2 The Python Implementation

The solution to his problem is implemented in Python using lists. Variable *itemC* refers to a value of the temperature in Celsius and variable *itemF* refers to the corresponding value of the temperature in Fahrenheit. All values of the temperatures in Celsius are placed in list *listC* and all values computed of the temperature in Fahrenheit are placed in list *listF*.

Listing 7.6 shows a Python program that computes the temperature in Fahrenheit for every value in the list of temperature in Celsius. This program uses a loop in which the two lists are built by appending a new item to the lists. This Python script is stored in file `tconvctfl.py`.

Listing 7.6 Python program for temperature conversion on a list.

```
1 # Program      : tconvctfl.py
2 # Author       : Jose M Garrido
3 # Date         : 6-02-2014
4 # Description : Read values of temperature in Celsius
5 # from console, convert to degrees Fahrenheit, and
6 # display corresponding values of temperature
7 # in fahrenheit on screen
8
9 SIZE = 15      # number of items in list
10 listC = []     # create empty list for temp in Celsius
11 listF = []     # create empty list for temp in Fahrenheit
12 # listC starts with 5
13 for j in range (1, SIZE+1):
14      itemC = j * 5
15      listC.append(itemC)
16      itemF = itemC * (9.0/5.0) + 32.0 # temp in Fahrenheit
17      listF.append(itemF)
18
19 print "Values of temperature in Celsius: "
20 print listC
21 print "Values of temperature in Fahrenheit: "
22 print listF
```

The following listing shows the shell command that starts the Python interpreter with file `tconvctfl.py` and the results.

```
$ python tconvctfl.py
Values of temperature in Celsius:
  [5, 10, 15, 20, 25, 30, 35, 40, 45, 50, 55, 60, 65, 70, 75]
Values of temperature in Fahrenheit:
  [41.0, 50.0, 59.0, 68.0, 77.0, 86.0, 95.0, 104.0, 113.0,
   122.0, 131.0, 140.0, 149.0, 158.0, 167.0]
```

7.3.3   Implementation Using a Function

Listing 7.5 shows a Python script that solves the same problem as the previous script. It defines function *tconvf* in lines 9–16. This function takes a list of values of temperature in Celsius, computes the temperature in Fahrenheit, and returns these values in a new list. This Python script creates a list of values of temperature in Celsius in line 18, calls function *tconvf* in line 19 using the list as the argument, then displays the two lists in lines 20–23. This script is stored in file `tconvs.py`.

Listing 7.5   Python program calls a function for temperature conversion.

```
 1 #  Program      : tconvs.py
 2 #  Author       : Jose M Garrido
 3 #  Date         : 6-02-2014
 4 #  Description : Given a list of values of temperature in
 5 #  Celsius, convert to degrees Fahrenheit, and return a
 6 #  list of values in Fahrenheit. This script defines and
 7 #  calls function 'tconvf'
 8
 9 def tconvf (listC):
10     # listC list of temperature values in Celsius
11     listF = []     # empty list for temp in Fahrenheit
12     size = len(listC)
13     for j in range (0, size):
14         itemF = listC[j] * (9.0/5.0) + 32.0
15         listF.append(itemF)
16     return listF
17
18 c = [0, 5, 10, 15, 20, 25, 30, 35, 40, 45, 50, 55]
19 f = tconvf(c)
20 print "Values of temperature in Celsius: "
21 print c
22 print "Values of temperature in Fahrenheit: "
23 print f
```

The following listing shows the shell command that starts the Python interpreter with file `tconvs.py` and the results.

```
$ python tconvs.py
Values of temperature in Celsius:
[0, 5, 10, 15, 20, 25, 30, 35, 40, 45, 50, 55]
Values of temperature in Fahrenheit:
[32.0, 41.0, 50.0, 59.0, 68.0, 77.0, 86.0, 95.0, 104.0, 113.0,
 122.0, 131.0]
```

7.4 LIST COMPREHENSIONS

A list comprehension is a compact notation in Python for generating a list of a given size and with the elements initialized according to the specified expression. The following example generates a list *ll* with 12 elements all initialized with value 1.

```
>>> lsize = 12
>>> ll = [ 1 for j in range(lsize) ]
>>> ll
[1, 1, 1, 1, 1, 1, 1, 1, 1, 1, 1, 1]
```

The same list can be generated using a for-loop and function *append*, as the following example shows.

```
>>> ll = []
>>> for j in range(lsize):
...     ll.append(1)
...
>>> ll
[1, 1, 1, 1, 1, 1, 1, 1, 1, 1, 1, 1]
```

In the notation for list comprehension, the expression appears first followed by one or more *for* clauses and all within brackets. The following example generates a list *ll* of size *lsize* with the elements initialized to a value from the expression $j + 12.5$.

```
>>> ll = [ j+12.5 for j in range(lsize)]
>>> ll
[12.5, 13.5, 14.5, 15.5, 16.5, 17.5, 18.5, 19.5, 20.5,
    21.5, 22.5, 23.5]
```

7.5   LISTS OF LISTS

Lists of lists are also known as *nested lists*, which means that one or more items in a list are also lists. Multidimensional arrays can be defined with nested lists. In the following example, the first command creates list *lst1* with four items. The second command creates list *lst2* and its third item is list *lst1*. The third command displays list *lst2* and the last command shows that the length of list *lst2* is 5.

```
>>> lst1 = [12, 54, 2, 9]
>>> lst2 = [99, 5, lst1, 20, 7]
>>> lst2
[99, 5, [12, 54, 2, 9], 20, 7]
>>> len (lst2)
5
```

To reference an item of a list that is part of a larger list, two indices are required. The first index refers to an item in the outer list, the second index refers to an item in the inner list. In the following example, the first command uses index value 2 to reference the third item of list *lst2* and this item is the inner list *lst1*. The second command uses two index values; the first index value, 2, indicates the third item of list *list2*, and the second index value, 3, references the fourth item of the inner list *lst1*, which has value 9.

```
>>> lst2 [2]
[12, 54, 2, 9]
>>> lst2 [2] [3]
9
```

The following commands create a small matrix *smatrix* with two rows and three columns, references the element of the second row and third column, and assign the value to variable *eval*. In a similar manner, the element of the first row and second column is referenced and its value is assigned to variable *fval*.

```
>>> smatrix = [[9, 2, 5], [4, 8, 6]]
>>> smatrix
[[9, 2, 5], [4, 8, 6]]
>>> eval = smatrix[1][2]
>>> eval
6
>>> fval = smatrix[0][1]
>>> fval
2
```

The following example generates a 3 by 5 list, that is, a list with three rows and five columns. The outer *for*-loop is used to generate a row list, and the inner *for*-loop generates all the elements in a row initialized to value 1.

```
>>> nll = [[]]
>>> for i in range(3):
...         row = []
...         for j in range (5):
...             row.append(1)
...         nll.append(row)
...
>>> nll
[[], [1, 1, 1, 1, 1], [1, 1, 1, 1, 1], [1, 1, 1, 1, 1]]
```

The following command calls list function *pop* to remove the first element from the list, which is an empty list.

```
>>> nll.pop(0)
[]
>>> nll
[[1, 1, 1, 1, 1], [1, 1, 1, 1, 1], [1, 1, 1, 1, 1]]
```

Nested list comprehensions are used to generate multi-dimensional lists initialized to a value according to a specified expression. The following example generates a 3 by 5 list, that is, a list with three rows and five columns. The inner (first) *for* clause is used to generate the values in a row, the second *for* clause generates all the rows.

```
>>> lll = [[1 for i in range(5)] for j in range (3)]
>>> lll
[[1, 1, 1, 1, 1], [1, 1, 1, 1, 1], [1, 1, 1, 1, 1]]
```

7.6 TUPLES

A tuple is a Python sequence similar to a list. To create a tuple of items, write the values separated by commas. It is often convenient to enclose the items in parenthesis. For example:

```
>>> xt = (4.5, 6, 78, 19)
>>> xt
(4.5, 6, 78, 19)
```

A tuple is immutable, which means that after creating a tuple it cannot be changed. The value of the elements in a tuple cannot be altered, and elements cannot be added or removed from the tuple.

As with lists, the individual elements of a tuple can be referenced by using an index value. In the following example, the third element of tuple *xt* is accessed and its value is assigned to variable *yy*.

```
>>> yy = xt[2]
>>> yy
78
```

Tuples can be nested, which means tuples of tuples can be created. For example, the following creates a tuple *xt2* that includes tuple *xt* as its second element.

```
>>> xt2 = (25, xt, 16)
>>> xt2
(25, (4.5, 6, 78, 19), 16)
```

Method *len* can be used to get the number of elements in a tuple. The following assignment statement gets the length of tuple *xt* and assigns this value to variable *lxt*.

```
>>> lxt = len (xt)
>>> lxt
4
```

A tuple can be converted to a list by calling method *list*. For example, the following command converts tuple *xt* to a list *llxt*.

```
>>> llxt = list (xt)
>>> llxt
[4.5, 6, 78, 19]
```

A list can be converted to a tuple by calling method *tuple*. For example, the following commands create list *vals* then convert the list to a tuple *mytuple*.

```
>>> vals = [1, 2, 3, 4, 5, 6]
>>> vals
[1, 2, 3, 4, 5, 6]
>>> mytuple = tuple (vals)
>>> mytuple
(1, 2, 3, 4, 5, 6)
```

It is possible to build lists of tuples and tuples of lists. The following command defines a list *myltt* of tuples.

```
>>> myltt = [(12, 45.25), (45, 68.5), (25, 78.95)]
>>> myltt
[(12, 45.25), (45, 68.5), (25, 78.95)]
```

The following command defines a tuple *mytup* of lists.

```
>>> mytup = ([14, 45.25], [55, 68.5], [28, 78.95])
>>> mytup
([14, 45.25], [55, 68.5], [28, 78.95])
```

7.7 DICTIONARIES

Dictionaries are also Python data structures except that these are indexed by *keys*. A dictionary is used as an unordered set of *key* and *value* pairs that are enclosed in curly braces. The key can be any immutable type and must be unique. The corresponding value associated with a key is written after a semicolon and following the key.

The following example creates a dictionary of three key-value pairs separated by commas. Note that the keys are strings in this example. The last command extracts the value of the pair that has given key *'price'*.

```
>>> mydict = {'desc': 'valve 5in', 'price': 23.75, 'quantity': 54}
>>> mydict
{'price': 23.75, 'quantity': 54, 'desc': 'valve 5in'}
>>> mydict['price']
23.75
```

The value of a pair in a dictionary can be updated given the key. The following example changes the value of price to 25.30.

```
>>> mydict['price'] = 25.30
>>> mydict
{'price': 25.3, 'quantity': 54, 'desc': 'valve 5in'}
```

The *in* keyword is used to check whether a key appears in the given dictionary. The following command checks the dictionary *mydict* for the key *'desc'*.

```
>>> 'desc' in mydict
True
```

The dictionary method *keys* is used to get a list of all the keys in a given dictionary. The following command gets a list of the keys in *mydict*.

```
>>> mydict.keys()
['price', 'quantity', 'desc']
```

A given list of two-tuples can be converted to a dictionary by calling function *dict*. The following commands define a list of two-tuples (tuples with two values) *myltt* and convert this list to a dictionary.

```
>>> myltt = [(12, 45.25), (45, 68.5), (25, 78.95)]
>>> myltt
[(12, 45.25), (45, 68.5), (25, 78.95)]
>>> mdict3 = dict (myltt)
>>> mdict3
{25: 78.95, 12: 45.25, 45: 68.5}
```

7.8   STRINGS

A string is a sequence of text characters in a particular character encoding. Syntactically, a string literal is enclosed in single or double quotes and can be assigned to a variable. Strings are immutable, and once defined, strings cannot be modified. The following command defines a string literal and assigns the reference to variable *mystr*.

```
>>> mystr = 'State University'
>>> mystr
'State University'
```

One of the most commonly used operations is concatenation. It joins two or more strings and creates a longer string. The concatenation string operator is the plus sign (+). The following command concatenates two strings and the newly created string is assigned to variable *str2*.

```
>>> str2 = 'Kennesaw ' + mystr
>>> str2
'Kennesaw State University'
```

Function *len* gets the length of a string, that is, the number of characters in the string. The following command gets the number of characters in the string referenced by variable *str2*.

```
>>> s = len (str2)
>>> s
25
```

Indexing is used to access particular characters of a string. Index values start at zero and are written within brackets. The following command prints the character with index 5 of string *str2*, which is s.

```
>>> print str2[5]
s
```

The character accessed in a string can be assigned to another variable, as the next example shows.

```
>>> x = str2[5]
>>> x
's'
```

The slicing operation is used to access a subset of a string. This operation requires the slice operator (:) and indexes that specify the range of characters to include. In the following example, the first command prints the subset of string *str2* specified from the second character (with index 1) up to the fourth character (with index 4). The second command assigns to variable *strx* a subset of string *str2* specified from the first character to the fifth character.

```
>>> print str2[1:4]
enn
>>> strx = str2[0:5]
>>> strx
'Kenne'
```

The following example creates a new string that consists of "Kansas" concatenated with a subset of *str2* that starts with the ninth character to the end of *str2*.

```
>>> nstrx = "Kansas" + str2[8:]
>>> nstrx
'Kansas State University'
```

The membership operator *in* is used to check whether a character or a substring is contained in another string and returns `True` or `False`. In the following example, the first command checks if the character 'K' belongs in string *nstrx*. The second command checks if the substring *strx* is contained in the string *str2*.

```
>>> 'K' in nstrx
True
>>> strx in str2
True
```

Some characters in a string and escape characters are written preceded with a backslash (\). For example, when a newline character is part of a string, that indicates that a change of line occurs at that point and the rest of the string appears on a newline. The following example shows a string that is to be displayed on two lines.

```
>>> message = "Start the program now\n click on the icon"
>>> message
'Start the program now\n click on the icon'
>>> print message
Start the program now
 click on the icon
```

When a string is enclosed in double quotes (") but one or more double quotes are also used as part of the string, these have to be escaped with a backslash as in the following example. The same applies with single quotes.

```
>>> mess1 = "The program responds with \"welcome\" then waits"
>>> print mess1
The program responds with "welcome" then waits
```

Python provides several string methods and the general form to call these methods is:

```
string_name.method_name(arguments)
```

For example, method *find* is used to find the index value in string *nstrx* where the substring `"State"` starts, as in the following example.

```
>>> nstrx
'Kansas State University'
>>> nstrx.find("State")
7
```

Another very useful string method is *isdigit* and is used to check whether all characters in a string are decimal digits. The following example checks whether string *nn* contains only decimal digits.

```
>>> nn = "54321"
>>> nn.isdigit()
True
```

7.9 SIMPLE NUMERICAL APPLICATIONS USING LISTS

This section discusses several simple applications of arrays as lists; a few of these applications perform simple manipulation of arrays, and other applications perform slightly more complex operations with arrays such as searching and sorting.

The problems discussed in this section compute the average value and the maximum value in an array named *varr*. The algorithms that solve these problems examine all the elements of the array.

7.9.1 The Average Value in an Array

To compute the average value in an array, the algorithm is designed to first compute the summation of all the elements in the array; the accumulator variable *sum* is used to store this. Second, the algorithm computes the average value by diving the value of *sum* by the number of elements in the array. The following listing has the pseudo-code description of the algorithm.

1. Initialize the value of the accumulator variable, *sum*, to zero.

2. For every element of the array, add its value to the accumulator variable *sum*.

3. Divide the value of the accumulator variable by the number of elements in the array, *num*.

The accumulator variable *sum* stores the summation of the element values in the array named *varr* with *num* elements. The average value, *ave*, of array *varr* using index j starting with $j = 1$ to $j = n$ is expressed mathematically as:

$$ave = \frac{1}{num} \sum_{j=1}^{num} varr_j.$$

Listing 7.7 shows the Python script that implement the algorithm that computes the average value of the elements in the array. This code is stored in the script file aver.py.

Listing 7.7: Python script file for computing the average in a list.

```
1 # Python script file to compute average value in a list
2 # This script inputs the array size
3 # and the elements of the array from the console
4 # Computes the average value in the array
5 # File: aver.py
6 num = input('Enter array size: ');
7 varr = []   # empty list
8 for j in range(0, num):
9       item = input('Enter array element: ')
10      varr.append(item)
11
12 # Now compute the average value in list
13 sum = 0.0
14 for j in range(0, num):
15      print "index: ", j, " value: ", varr[j]
16      sum = sum + varr[j]
17
18 ave = sum/num
19 print "Average value: ", ave
```

The following output listing shows the result of executing the script *aver.py* at the Linux prompt.

```
$ python aver.py
Enter array size: 4
Enter array element: 9.75
Enter array element: 8.34
Enter array element: 7.25
Enter array element: 6.77
index:  0  value:  9.75
index:  1  value:  8.34
index:  2  value:  7.25
index:  3  value:  6.77
Average value:  8.0275
```

7.9.2 Maximum Value in a List

Consider a problem that deals with finding the maximum value in an array named *varr*. The algorithm with the solution to this problem also examines all the elements of the array.

The variable *max_arr* stores the maximum value found so far. The name of the index variable is *j*. The algorithm description is:

1. Read the value of the array size, *num*, and the value of the array elements.

2. Initialize the variable *max_arr* that stores the current largest value found (so far). This initial value is the value of the first element of the array.

3. Initialize the index variable (value zero).

4. For each of the other elements of the array, compare the value of the next array element; if the value of the current element is greater than the value of *max_arr* (the largest value so far), change the value of *max_arr* to this element value, and store the index value of the element in variable *k*.

5. The index value of variable *k* is the index of the element with the largest value in the array.

Listing 7.8 contains the Python script that implements the algorithm for finding the maximum value in an array; the script is stored in the file arrmax.py. As in the previous examples, the list is first created by reading the values of the elements, in lines 6–10. Finding the maximum value in the list is performed in lines 13–18

Listing 7.8: Python script file for finding the maximum value in a list.

```
1 # Python script file to find the maximum value in a list
2 # This script inputs the array size
3 # and the elements of the array from the console
4 # Computes the maximum value in the array
5 # File: arrmax.py
6 num = input ('Enter array size: ')
7 varr = []   # empty list
8 for j in range(0, num):
9       item = input('Enter array element: ')
10      varr.append(item)
11
12 # Now find the maximum value in list
13 max_arr = varr[0] # initial value of max_arr
14 for j in range(1, num):
15      print "index: ", j, " value: ", varr[j]
```

```
16      if varr[j] > max_arr:
17          k = j
18          max_arr = varr[j]
19
20 print "Index of max value: ", k
21 print "Max value: ", varr[k]
```

Executing the script file `arrmax.py` with the Python interpreter produces the following output.

```
$ python arrmax.py
Enter array size: 4
Enter array element: 5.56
Enter array element: 7.87
Enter array element: 3.78
Enter array element: 2.7
index:   1   value:   7.87
index:   2   value:   3.78
index:   3   value:   2.7
Index of max value:   1
Max value:   7.87
```

7.9.3   Searching

Looking for an array element with a particular value, known as the *key*, is called searching and involves examining some or all elements of an array. The search ends when and if an element of the array has a value equal to the requested value. Two general techniques for searching are linear search and binary search.

7.9.3.1   *Linear Search*

A linear search algorithm examines the elements of an array in a *sequential* manner starting with the first element. The algorithm examines the first element of the array, then the next element, and so on until the last element of the array. Every array element is compared with the key value, and if an array element is equal to the requested value, the algorithm has found the element and the search terminates. This may occur before the algorithm has examined all the elements of the array.

The result of this search is the index of the element in the array that is equal to the key value given. If the key value is not found, the algorithm indicates this with a negative result or in some other manner. The following is an algorithm description of a general linear search using a search condition of an element equal to the value of a *key*.

1. Repeat for every element of the array:

    (a) Compare the current element with the requested value or key. If the value of the array element satisfies the condition, store the value of the index of the element found and terminate the search.

    (b) If values are not equal, continue search.

2. If no element with value equal to the value requested is found, set the result to value -1.

The algorithm outputs the index value of the element that satisfies the search condition, whose value is equal to the requested value *kval*. If no element is found that satisfies the search condition, the algorithm outputs a negative value.

The Python script is stored in file `lsearch.py`. Listing 7.9 shows the Python commands that implement the algorithm that searches the list *llist* for the key value, *key*. In line 18, list *llist* is created. In line 20, there is a function call to *lsearchf* using two arguments: the list and the key value to search. The function definition appears in lines 8–16. The function returns the index value of the element found that is equal to the key value, or -1 if not found.

Listing 7.9: Script file for computing a linear search in a list.

```
1 # Python script for linear search
2 # it performs a linear search of the array varr
3 # looking for the value kvar
4 #    The algorithm sets the result, the index value of
5 #    the element found, or -1 if not found.
6 # File: lsearch.py
7
8 def lsearchf (varr, kval):
9        # find the element in varr equal to kval
10       found = False
11       num = len (varr)
12       for j in range(0, num):
13              if found == False and varr [j] == kval:
14                     found = True
15                     return j
16       return -1
17
18 llist = [23, 12, 19, 35, 22, 81, 14, 8, 33]
19 key = input ("Enter the key value: ")
20 result = lsearchf(llist, key)
21 if result >= 0:
22        print 'Result index is: ', result
23 else:
24        print 'Key not found'
```

Executing the script file `lsearch.py` with the Python interpreter produces the following output.

```
$ python lsearch.py
Enter the key value: 33
Result index is:  8
$ python lsearch.py
Enter the key value: 45
Key not found
```

7.9.3.2 Binary Search

Binary search is a more complex search method and is very efficient, compared to linear search, because the number of comparisons is smaller.

A prerequisite for the binary search technique is that the element values in the array to be searched are sorted in ascending order. The array elements to search are split into two halves or partitions of about the same size. The middle element is compared with the key (requested) value. If the element with this value is not found, the search is continued on only one partition. This partition is again split into two smaller partitions until the element is found or until no more splits are possible because the element is not found.

With a search algorithm, the efficiency of the algorithm is determined by the number of compare operations with respect to the size of the array. The average number of comparisons with linear search for an array with N elements is $N/2$, and if the element is not found, the number of comparisons is N. With binary search, the number of comparisons is $\log_2 N$. The informal description of the algorithm is:

1. Assign the lower and upper bounds of the array to *lower* and *upper*.

2. While the lower value is less than the upper value, continue the search.

   (a) Split the array into two partitions. Compare the middle element with the key value.

   (b) If the value of the middle element is equal to the key value, terminate search and the result is the index of this element.

   (c) If the key value is less than the middle element, change the upper bound to the index of the middle element minus 1. Continue the search on the lower partition.

   (d) If the key value is greater than or equal to the middle element, change the lower bound to the index of the middle element plus 1. Continue the search on the upper partition.

3. If the key value is not found in the array, the result is -1.

Listing 7.10 shows the Python script that implements the binary search algorithm. The script commands are stored in command file `bsearch.py`.

Listing 7.10: Python script for searching for a key using binary search.

```
 1 # Python script that implements a binary search
 2 # of list (array) llist using key value key.
 3 # The result is the index value of
 4 # the element found, or -1 if not found.
 5 # File: bsearch.py
 6
 7 def bsearchf (varr, kval):
 8        # find the element in varr equal to kval
 9        num = len (varr)
10        lower = 0
11        upper = num - 1
12        while lower <= upper :
13               middle = (lower + upper) / 2
14               if kval == varr[middle]:
15                    return middle;  # result
16               else:
17                     if kval < varr[middle]:
18                          upper = middle - 1
19                     else:
20                          lower = middle + 1
21        return -1  # not found
22
23 llist = [9, 10, 13, 61, 72, 82, 89, 95, 102]
24 key = input('Enter key value: ')
25 result = bsearchf(llist, key)
26 if result >= 0:
27      print 'Result index is: ', result
28 else:
29      print 'Key not found'
```

Executing the *bsearch* script with the Python interpreter produces the following output listing.

```
$ python bsearch.py
Enter key value: 81
Key not found
$ python bsearch.py
Enter key value: 72
Result index is:  4
```

Note that because of the limited precision of digital computers, it is not recommended that two floating-point values be tested for equality. Instead, a small fixed constant value is used to compare with the absolute difference of the two values.

The following example defines the symbol *EPSIL* as a constant with a relatively small value 0.000001 and is used to compare with the difference of the values of variables *aa* and *bb*. In this case, the two variables are not considered equal because their difference is not less than or equal to *EPSIL*.

```
>>> EPSIL = 0.000001
>>> aa = 46.005
>>> bb = 46.0055
>>> aa - bb
-0.000499999999995282
>>> abs(aa - bb) <= EPSIL
False
```

7.10   SUMMARY

An array in Python is a *list* data structure that stores several values. Each of these values is known as an element. To refer to an individual element an index is used to indicate the relative position of the element in the array.

Searching an array consists of looking for a particular element value or key. Two common search algorithms are linear search and binary search. Computing an approximation of the rate of change and the area under a curve is much more convenient using arrays, as shown in the case study discussed.

A list is the most general data structure in Python. Other data structures are strings, tuples, and dictionaries.

Key Terms

creating lists	accessing elements	size of a list
index	array element	element reference
searching	linear search	binary search
key value	algorithm efficiency	summation
accumulator	strings	tuples
dictionaries	passing lists to functions	returning a list

7.11 EXERCISES

7.1 The definition of the *median* of a group of numbers arranged in ascending (or descending) order is the value in the middle. The *mode* is the number that appears most often. Write a Python program that finds the median and mode of a list.

7.2 Develop a Python program that reads the values of numbers and builds a list. Copy the even numbers into a separate list, the odd numbers into another list, and the negative numbers into another list. The program must print all of the lists separately.

7.3 Develop a Python program that solves the same problem described in the previous exercise. Instead of using the three lists for even, odd, and negative numbers, use a single nested list.

7.4 Develop a program that computes the standard deviation of values in an array. Implement using the Python programming language. The standard deviation measures the spread, or dispersion, of the values in the array with respect to the average value. The standard deviation of array X with n elements is defined as:

$$std = \sqrt{\frac{sqd}{n-1}},$$

where

$$sqd = \sum_{j=0}^{n-1}(X_j - Av)^2.$$

7.5 Develop a program that finds the minimum value element in an array and returns the index value of the element found. Implement using the Python programming language.

7.6 Develop a computational model that computes the average, minimum, and maximum rainfall per year and per quarter (for the last five years) from the rainfall data provided for the last five years. Four quarters of rainfall are provided, measured in inches. Use a matrix to store these values. Implement using the Python programming language.

7.7 Develop a program that sorts an array using the Insertion sort technique. This sort algorithm divides the array into two parts. The first is initially empty; it is the part of the array with the elements in order. The second part of the array has the elements in the array that still need to be sorted. The algorithm takes the element from the second part and determines the position for it in the first part. To insert this element in a particular position of the first part, the elements to right of this position need to be shifted one position to the right. Implement using the Python programming language.

## CHAPTER 8

Object Orientation

8.1 INTRODUCTION

The basic principles of object orientation are explained in this chapter. The concepts of objects, collection of objects, encapsulation, models, information hiding, and classes are explained.

The structure of a class and how to create objects or instances of a class are explained in a simplified manner. Other preliminary concepts in object-oriented programming such as inheritance, reuse, abstraction, and modularization are also discussed.

A model is an abstract representation of groups of objects, each one representing a real-world entity. Real-world applications consist of collections of real-world objects interacting with one another and with their surroundings.

System design usually emphasizes modular structuring, also called modular decomposition. A problem is often partitioned into smaller problems (or subproblems), and a solution is designed for each subproblem. Object-oriented design enhances modular design by providing classes as an important decomposition (modular) unit. In Python, many modules contain one or class definitions.

8.2 OBJECTS IN THE PROBLEM DOMAIN

Real-world entities or real-world objects are the fundamental components of a real-world system. Identifying and modeling real-world entities in the problem domain are the central focus of the object-oriented approach. A real-world entity has the responsibility of carrying out a specific task and is modeled as an object.

Abstraction is used to model the objects in a problem domain and involves the elimination of unessential characteristics. A model includes only the relevant aspects of the real-world system. Therefore, only the essential characteristics of the objects are included in the model. Several levels of detail are needed to completely define objects and the collections of objects in a model. Object-oriented modeling consists of:

1. Identifying the relevant objects for the model

2. Describing these objects using abstraction

3. Defining collections of similar objects

Objects with similar characteristics are grouped into collections, and these are modeled as *classes*. The Unified Modeling Language (UML) is a standard notation used to describe objects and classes in a problem domain.

8.3   DEFINING CLASSES

In modeling a real-world problem, collections of similar objects are identified. Classes are then defined as abstract descriptions of these collections of objects, which are objects with the same structure and behavior. A class defines the attributes and behavior for all the objects of the class. An object belongs to a class, and any object of the class is an *instance* of the class. A software implementation of class consists of:

- Data definitions that represent the attributes of the class

- Behavior representation as one or more operations (also known as methods)

Figure 8.1 shows two collections of objects in the problem domain and the corresponding model with two classes.

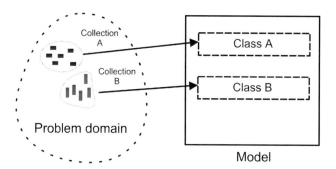

Figure 8.1   Collections of objects.

The following example shows a representation of class *Employee*, in a simplified UML diagram includes the structure and behavior for objects of this class. The diagram is basically a rectangle divided into three sections. The top section indicates the class of the object, the middle section includes the list of the attributes and their current values, and the bottom section includes the list of object operations.

Figure 8.2 shows the diagram that describes class *Employee*. The attributes defined for this class are *salary*, *emp_number*, *name*, and *emp_date*. The behavior is defined in methods: *get_name*, *increase_sal*, *get_age*, and *start_emp*.

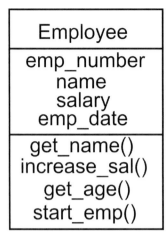

Figure 8.2 Diagram of class *Employee*.

8.4 DESCRIBING OBJECTS

Objects exhibit independent behavior and interact with one another. Objects communicate by sending messages to each other. Every object has:

- a state, represented by the set of properties (or attributes) and their associated values;

- behavior, represented by the operations, also known as methods, of the object;

- identity, which is an implicit or explicit property that can uniquely identify an object.

The state of an object is defined by the values of its attributes. Several objects of the same class would typically have different states because their attributes would have different values.

8.5 INTERACTION BETWEEN TWO OBJECTS

The interaction between two objects involves an object sending *messages* to another object. The object that sends the message is the requestor of a service that can be provided by the receiver object.

The first object sends a message to request a service, which is provided by the second object, which is the one receiving the message. The sender object is known as the *client* of a service, and the receiver object is known as the *supplier* of the service. In this simple scenario, objects perform operations in response to messages.

A message is always sent to a specific object and is also known as *method invocation*. A message normally contains the following components:

- the owner of the operation, which is the receiver of the message,

- the operation to be invoked or started,

- the input data required by the operation to perform, and

- the result of the operation.

The standard UML diagram is the *communication diagram* and is used to describe the general interaction between two (or more) objects (the sending of messages between objects).

The interaction between two diagrams is illustrated by Figure 8.3, which shows a simple communication diagram describing the interaction between an object of class *Person* with an object of class *Ball*. In this example, the object of class *Person* invokes the *move* operation of the object of class *Ball*. The first object sends a message to the second object. As a result of this message, the object of class *Ball* performs its *move* operation.

Figure 8.3 Interaction between two objects.

8.6   DESIGN WITH CLASSES

Object orientation provides enhanced modularity in developing software systems. An application is basically a set of well-structured and related *modules*. The class is one of the most basic modular units of a program. A Python program is decomposed into a set of modules, which consists of a related set of functions and classes.

This is the *static view* of a program (that implements an application). The *dynamic view* of the application is a set of objects performing their behavior and interacting among themselves.

8.6.1   Encapsulation

The encapsulation principle describes an object as the integration of its attributes and behavior in a single unit. There is an imaginary protecting wall surrounding the object. This is considered a protection mechanism. To protect the features of an object, an access mode is specified for every feature.

The access mode specifies which features of the object can be accessed from other objects. If access to a feature (attribute or operation) is not allowed, the access mode of the feature is specified to be *private*. If a feature of an object is *public*, it is accessible from any other objects.

8.6.2   Data Hiding

As mentioned previously, an object that provides a set of services to other objects is known as a *provider* object, and all other objects that request these services by sending messages are known as *client* objects. An object can be a service provider for some services, and it can also be a client for services that it requests from other (provider) objects.

The principle of data hiding (or information hiding) provides the description of a class and only shows the services the objects of the class provide to other objects and hides all implementation details. In this manner, a class description presents two views:

1. The *external view* of the objects of a class. This view consists of the list of services (or operations) that other objects can invoke. The list of services can be used as a service contract between the provider object and the client objects.

2. The *internal view* of the objects of the class. This view describes the implementation details of the data and the operations of the objects in a class. This information is hidden from other objects.

These two views of an object are described at two different levels of abstraction. The external view is at a higher level of abstraction. The external view is often known as the *class specification*, and the internal view as the *class implementation*.

With the external view, information about an object is limited to that only necessary for the object's features to be invoked by other objects. The rest of the knowledge about the object is not revealed.

In general, the external view of the objects should be kept separate from the internal view. The internal view of properties and operations of an object are hidden from other objects. The object presents its external view to other objects and shows what features (operations and attributes) are accessible.

8.7   SUMMARY

One of the important tasks in modeling object-oriented applications is to identify the objects and collections of similar objects in the problem domain. An object has properties and behaviors. The class is a definition of objects with the same characteristics.

A model is an abstract representation of a real system. Modeling involves selecting relevant objects and relevant characteristics of these objects.

Objects collaborate by sending messages to each other. A message is a request by an object to carry out a certain operation on another object, or on the same object. Information hiding emphasizes the separation of the list of operations that an object offers to other objects from the implementation details that are hidden to other objects.

Key Terms

models	abstraction	objects
collections	real-world entities	object state
object behavior	messages	attributes
operations	methods	functions
UML diagram	interactions	method invocation
classes	encapsulation	information hiding
private	public	class specification
class implementation	collaboration	

8.8 EXERCISES

8.1 What are the differences between classes and objects? Why is an object considered a dynamic concept? Why is the class considered a static concept? Explain.

8.2 Explain the differences and similarities of the UML class and object diagrams. What do these diagrams actually describe about an object and about a class? Explain.

8.3 Is object interaction the same as object behavior? Explain and give examples. What UML diagram describes this?

8.4 Briefly explain the differences between encapsulation and information hiding. How are these two concepts related? Explain.

8.5 Are the external view and internal view of a class at different levels of abstraction? Explain. Is this considered important in software development? Explain.

8.6 Identify the type and number of objects involved in an automobile rental office. For every type of object, list the properties and operations. Draw the class and object diagrams for this problem.

8.7 Describe the object interactions necessary for the objects in the automobile rental office. Draw the corresponding communication diagrams.

8.8 Identify the various objects in a supermarket. How many objects of each type are there? List the properties and the necessary operations of the objects. Draw the corresponding UML diagrams for this problem.

8.9 List the private and public characteristics (properties and operations) for every type of object in the two previous exercises. Why do you need to make this distinction? Explain.

## CHAPTER 9

Object-Oriented Programs

9.1 INTRODUCTION

The first part of this chapter presents and explains the basic structure of a class. An introduction to data descriptions and the general structure of methods is discussed. The construction of simple object-oriented programs is presented.

The second part of this chapter discusses inheritance as a class relationship among classes. The other basic class relationship is composition, which is a stronger form of association. These relationships are easily modeled in UML diagrams. Composition can be considered a horizontal relationship and inheritance a vertical relationship.

Inheritance is a facility provided by an object-oriented language for defining new classes from existing classes. The basic inheritance relationships and their applications are explained in some detail. Inheritance enhances class reuse, that is, the use of a class in more than one application.

9.2 PROGRAMS

An object-oriented program consists of the implementation of the classes and additional language statements in one or more modules. When the program is running, objects of these classes are created and made to interact among themselves. This is the dynamic view of a program, which describes the behavior of the program while it executes.

9.3 DEFINITION OF CLASSES

A class definition includes the data structures and the behavior of the objects in that class and consists of:

- The definitions of the attributes of the class

- Descriptions of the operations, known as methods of the class

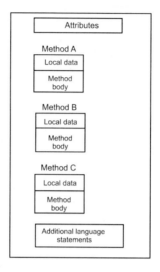

Figure 9.1 Structure of a typical class.

The general structure of a class is illustrated in Figure 9.1 and shows the attributes and the definitions of three methods: *Method A*, *Method B*, and *Method C*. Each of these operations consists of local data and its instructions.

The software implementation of a program is carried out by writing the code in a suitable programming language. Detailed design is often written in pseudo-code, which is a high-level notation at the level between the modeling diagrams and the programming language.

In object-oriented programming, there are two general categories of variables:

- Variables of elementary or primitive type

- Object reference variables

Reference variables are defined when creating objects of a class. An object reference variable refers to an object.

9.4   CLASS DEFINITIONS IN PYTHON

A Python program can include zero or more class definitions, function definitions, and instructions to create objects and to manipulate the objects created. A class definition is implemented with a class header and additional statements; these are:

1. The **class** statement is used for the class header in the class definition. This includes a name of the class and other information related to the class.

2. An optional comment string that serves as class documentation and which may be used to include a textual documentation or description of the class.

3. Data definitions.

4. One or more member functions of the class and these are known as *methods*.

In Python, the general syntax of a class definition is:

```
class ⟨ class_name ⟩ :
        [data definitions]
        [method definitions]
```

9.4.1 Data Definitions in a Class

In a class definition, two categories of variables can be defined:

- Class variables

- Instance variables

Class variables are defined usually at the top and their values are shared by all objects or instances of the class. When a class variable is used outside the class, its name has to be prefixed by the class name. For example, if a class variable xx is defined in class *Point*, the prefix `Point.` is used with the name of the variable. To display the value of xx the statement is:

```
print "Variable xx is: ", Point.xx
```

Instance variables are data definitions that appear inside a method and their values will be unique to each instance or object of the class. The name of all instance variables in a class have the prefix `self.`

9.4.2 Methods in a Class Definition

A method can have zero or more parameter declarations; the first parameter is always *self*. Four categories of methods can be defined in a class:

1. Constructor methods

2. Accessor methods

3. Mutator methods

4. Auxiliary methods

A class may have one or more *constructor methods* and these are used to assign initial values to instance variables and to perform any computation. A constructor method is used to create a new instance or object of the class. The name of this method is __init__ and is always the same for any class.

An *accessor method* returns the value of an instance variable or a class variable.

A *mutator method* changes the value of an instance variable and, in effect, changes the state of the object referred.

An *auxiliary method* is one that is only called internally in the class and it is called with the self. prefix.

9.4.3   Example of a Class Definition

In the following example, class *Circle* is defined and the statements that appear after the class are the instructions to perform computing with the objects created of the class.

Line 2 has a documentation string for the class. In line 3, class variable *circount* is defined with an initial value of zero. Lines 5–7 define a *constructor* method, which is used to initialize the instance variables and create a new instance of the class.

In line 6, the value of parameter *ir* is assigned to instance variable *radius*. The name of all instance variables in a class have the prefix self. and can be referenced in one or more methods of the class. In line 7, the class variable *circount* is incremented. The value of this variable is the number of objects of this class that have been created. The name of this variable is prefixed with the name of class *Circle*.

In line 9, a mutator method is defined that sets a new value to instance variable *radius*. In line 12, a mutator method is defined that computes the value of instance variable *cir*. In line 16, another mutator method is defined that computes the value of instance variable *area*. In line 20, an accessor method is defined that returns the value of instance variable *radius*.

```
1 class Circle:
2     'Circle for computing circumference and area'
3     circount = 0
4
5     def __init__(self, ir):
6         self.radius = ir
7         Circle.circount += 1
8
9     def setRadius(self, ir):
10         self.radius = ir
```

```
11
12      def compCircum(self):
13          self.cir = 2.0 * math.pi * self.radius
14          return self.cir
15
16      def compArea (self):
17          self.area = math.pi * self.radius * self.radius
18          return self.area
19
20      def getRadius(self):
21            return self.radius
```

9.5   CREATING AND MANIPULATING OBJECTS

After a class has been defined, objects of the class can be created by invoking one of the constructor methods of the class. The general form of the assignment statement used to create an object is:

⟨ ref_name ⟩ = ⟨ class_name ⟩ (arguments)

ref_name is a reference variable that is used as a reference to the newly created object. For example, to create an object of class *Circle* with a radius of 2.35 and a reference variable *cirobj*, the statement is:

```
cirobj = Circle(2.35)
```

The reference variable is used to manipulate the object by invoking one or more of its methods. For example, the following statements are used to compute the area of the object referenced by variable *cirobj* and to display this value.

```
area = cirobj.compArea()
print "Area of the circle: ", area
```

To change the value of an instance variable of an object, one of the mutator methods of the object is invoked. For example, to change the value of the radius to 4.55 of the object referenced by variable *cirobj*:

```
cirobj.setRadius(4.55)
```

9.6 COMPLETE PROGRAM WITH A CLASS

Listing 9.1 shows a Python program stored in file `circlep.py` that includes the class definition *Circle*. The statements following the class definition create and manipulate objects of class *Circle*.

Listing 9.1 Python program with definition of class *Circle*.

```
 2 # Program     : circlep.py
 3 # Author      : Jose M Garrido, May 21 2014.
 4 # Description : This program defines a class for circles
 5 #     Computes the area and circumference of circles
 7 #     Reads the value of the radius for several circles
 8 #     from the input console, display results.
 9
10 import math
11
12 class Circle:
13     'Circle for computing circumference and area'
14     circount = 0
15
16     def __init__(self, ir):
17         self.radius = ir
18         Circle.circount += 1
19
20     def setRadius(self, ir):
21         self.radius = ir
22
23     def compCircum(self):
24         self.cir = 2.0 * math.pi * self.radius
25         return self.cir
26
27     def compArea (self):
28         self.area = math.pi * self.radius * self.radius
29         return self.area
30
31     def getRadius(self):
32         return self.radius
33
34 r1 = input ("Enter value of radius 1: ")
35 r2 = input ("Enter value of radius 2: ")
36 r3 = input ("Enter value of radius 3: ")
37
38 cobj1 = Circle(r1)
39 cobj2 = Circle(r2)
40 cobj3 = Circle(r3)
```

```
41
42 print "Value of radius1: ", cobj1.getRadius()
43 print "Value of radius2: ", cobj2.getRadius()
44 print "Value of radius3: ", cobj3.getRadius()
45
46 print "Number of objects created of class Circle: ",
      Circle.circount
47 cperim1 = cobj1.compCircum()
48 print "Perimeter of first circle object: ", cperim1
49 carea1 = cobj1.compArea()
50 print "Area of first circle: ", carea1
51
52 r1 = input ("Enter new value of radius 1: ")
53 cobj1.setRadius(r1)
54 print "Radius of first circle: ", cobj1.getRadius()
55 cperim1 = cobj1.compCircum()
56 print "Perimeter of first circle object: ", cperim1
57 carea1 = cobj1.compArea()
58 print "Area of first circle: ", carea1
```

9.7 SCOPE OF VARIABLES

The *scope* of a variable is that portion of a program in which statements can reference that data item. Variables and constants declared as attributes of the class can be accessed from anywhere in the class. Instructions in any functioned of the class can use these data items. Local definitions define variables that can only be used by instructions in the function in which they have been declared.

The *persistence* of a variable is the interval of time that the data item exists—the lifetime of the data item. The lifetime of a variable declared as an attribute of a class, exists during the complete life of an object of the class. Variables declared locally will normally have a lifetime only during which the function executes.

9.8 CLASS HIERARCHY WITH INHERITANCE

Classes in an application that are related in some manner are organized in the form a hierarchy of classes. Others are completely independent because they do not have any relationship with other classes.

In a class hierarchy, the most general class is placed at the top and is known as the *base* class, *parent* class, and is also known as the *super* class. A *subclass* inherits the characteristics (all attributes and operations) of its parent class. These characteristics of a class are also known as *features*. A subclass can be further inherited by lower-level classes.

Figure 9.2 illustrates the inheritance class relationship among several classes. Class *University_employee* is the base class; the other three classes inherit the features of this base class.

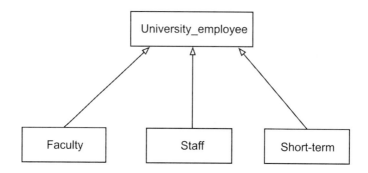

Figure 9.2 Class inheritance.

9.9 DEFINING CLASSES WITH INHERITANCE

The purpose of inheritance is to define a new class from an existing class and to shorten the time compared to the development of a class from scratch. Inheritance also enhances class reuse.

The base class is a more general class than its subclasses. A derived class can be defined by adding more features or modifying some of the inherited features and can be defined as:

- An *extension* of the base class, if in addition to the inherited features, it includes its own data and operations.

- A *specialized* version of the base class, if it overrides (redefines) one or more of the features inherited from its parent class.

- A combination of an extension and a specialization of the base class.

Multiple inheritance is the ability of a class to inherit the characteristics from more than one parent class. Most object-oriented programming languages support multiple inheritance.

In the simple class hierarchy with inheritance shown in Figure 9.2, the base class is *University_employee* and the subclasses are *Faculty*, *Staff*, and *Short-term*. All objects of class *Faculty* in Figure 9.2 are also objects of class *University_employee*, because this is the base class for the other classes. On the contrary, not all objects of class *University_employee* are objects of class *Faculty*.

9.9.1 Inheritance with Python

The definition of a subclass in Python include, one or more names of base classes. The general form of the Python statement for the header in the definition of a subclass is:

> **class** ⟨ *class_name* ⟩ (⟨ *base_class_list* ⟩) :
> . . .

The header of the subclasses Faculty and Staff in Figure 9.2 are written:

```
class Faculty (University_employee):
```

and

```
class Staff (University_employee):
```

9.9.2 Inheritance and Constructor Methods

The constructor methods of a base class are the only features that are not inherited by the subclasses. A constructor method of a subclass will normally invoke the constructor method of the base class.

The statement to call or invoke a constructor method of the base class from the subclass is:

> ⟨ *baseclass_name* ⟩.__init__(⟨ arguments ⟩)

The following portion of Python code from the file univemp.py defines the base class *University_employee*. The constructor method of this class sets initial values for the instance variables *name*, *date_start*, and *phone*.

```
 8 class University_employee:
 9     'Base class'
10     empcount = 0
11
12     def __init__(self, name, datas, phone):
13         self.name = name
14         self.date_start = datas
15         self.phone = phone
16         University_employee.empcount += 1
17
18     def setPhone(self, nphone):
19         self.phone = nphone
20
```

```
21    def getPhone(self):
22        return self.phone
23
24    def getName (self):
25        return self.name
26
27    def getDates(self):
28        return self.date_start
```

The following portion of code defines a subclass *Faculty* that inherits the features of an existing (base) class *University_employee*.

The subclass *Faculty* has one class variable *faccount* and two other instance variables *rank* and *tenure*. The constructor method of this class in lines 34–38 invokes the constructor method of the base class and sets initial values for its two instance variables. The following code shows this subclass.

```
30 class Faculty (University_employee):
31    'Subclass of University_employee'
32    faccount = 0
33
34    def __init__(self, name, datas, phone, rank, tenure):
35        University_employee.__init__(self, name, datas,
       phone)
36        self.rank = rank
37        self.tenure = tenure
38        Faculty.faccount += 1
39
40    def getRank(self):
41        return self.rank
42
43    def setRank(self, nrank):
44        self.rank = nrank
45
46    def getTenure(self):
47        return self.tenure
48
49    def setTenure(self, nten):
50        self.tenure = nten
```

In a similar manner to the previous subclass, *Staff* is a subclass that inherits the features of the (base) class *University_employee*.

The subclass *Staff* has one class variable *staffcount* and two other instance variables *position* and *train*. The constructor method of this class, defined in lines 56–60, invokes the constructor method of the base class and sets initial

values for its two instance variables. The following code shows the definition
of this subclass.

```
52 class Staff (University_employee):
53     'Subclass of University_employee'
54     staffcount = 0
55
56     def __init__(self, name, datas, phone, position,
    train_level):
57         University_employee.__init__(self, name, datas,
    phone)
58         self.position = position
59         self.train = train_level
60         Staff.staffcount += 1
61
62     def getPos(self):
63         return self.position
64
65     def setPos(self, npos):
66         self.position = npos
67
68     def getTrain(self):
69         return self.train
70
71     def setTrain(self, ntrainl):
72         self.train = ntrainl
```

9.9.3 Example Objects

In the following code, objects of the base class and the two subclasses are
created and methods of these objects are called. Notice that in line 78, the
object referenced by *femp* of subclass *Faculty* invokes method *setPhone*, which
is a method of the base class. This is perfectly legal because the features of
the base class are available to the subclasses. The complete program is stored
in file univemp.py.

```
74 gemp = University_employee("Jose Garrido", "10 Oct 2007",
    2138)
75 femp = Faculty("F. Hunt", "23 March 2010", 1121, 2, False)
76 semp = Staff("J Sommer", "12 April 1999", 6543, 12, 2)
77 print "Phone of Jose: ", gemp.getPhone()
78 femp.setPhone(4454)
79 print "Phone of Hunt: ", femp.getPhone()
```

```
80 print "Tenure status Hunt: ", femp.getTenure()
81 print "Training level Sommer: ", semp.getTrain()
```

9.10 OVERLOADING AND OVERRIDING METHODS

Overloading is an object-oriented facility that allows the definition of more than one method to be defined with the same name in a class definition. This facility is not directly supported in Python but there are some more advanced ways to implement this facility.

With inheritance, a class can be defined as a *specialized* subclass of the base class. To use this facility, one or more methods of the base class are *redefined* (or overridden) in the subclass. The subclass is said to re-implement one or more methods of the base class.

9.11 SUMMARY

A class is a collection of objects with similar characteristics. It is also a type for object reference variables. A class is a reusable unit; it can be reused in other applications. A program consists of an assembly of classes, function definitions, and language statements. This is the static view of a program.

The structure of a class consists of data definitions and method definitions. Data definitions in the class define the attributes of the class; data definitions also appear in the functions to define the function's local data. A function includes data definitions and instructions.

Data definitions consist of the declarations of constants, variables of simple types, and object reference variables.

The implementation of classes in Python and other object-oriented programming languages is accomplished by writing the program using the language statements and structuring the program according to the language syntax.

With inheritance, a subclass (derived class) inherits all the features of its base (parent) class. Inheritance is a vertical relationship among classes and enhances class reuse. The constructor method of the base classes is not inherited.

A subclass can be an extension and/or a specialization of the base class. If a subclass defines new features in addition to the ones that it inherits from the base class, then the subclass is said to be an extension to the base class. If a subclass redefines (overrides) one or more functions of the base class, then the subclass is said to be a specialization of the base class. In most cases, subclasses are both an extension and a specialization of the base class.

Key Terms

static view	dynamic view	decomposition
modules	units	class reuse
package	devices	data declaration
variables	constants	simple types
primitive types	class structure	object references
class description	local data	initial value
data types	scope	persistence
classification	parent class	super class
base class	subclass	derived class
horizontal relationship	vertical relationship	class hierarchy
inherit	extension	specialization
class reuse	inheritance	method overriding

9.12   EXERCISES

9.1   Is a class an appropriate decomposition unit? What other units are possible to consider?

9.2   The software implementation for a problem is decomposed into classes and functions. Explain this decomposition structure. What is the purpose of decomposing a problem into subproblems?

9.3   Is it convenient to include public attribute definitions in classes? What are the advantages and disadvantages? *Hint*: review the concepts of encapsulation and information hiding.

9.4   When a program executes, the objects of the program interact, collaborating to accomplish the overall solution to the problem. This is a dynamic view of a program. Where and when are these objects created and started? Explain.

9.5   Class reuse can be very useful. Explain and write an example.

9.6   The relationships among classes are identified early in software development. Explain horizontal and vertical relationships and how these are shown in UML diagrams.

9.7   Define one or two additional redefined methods in class *Faculty*.

9.8   Define one or two additional methods in extending class *Staff*.

9.9   Define a new class that inherits class *Person*. This class must include methods that define it as an extension and specialization of the base class *Person*.

9.10 Write a document that describes and explains how class reuse is used in the previous problem, Exercise 9.9.

9.11 Define two classes in addition to the ones shown in Figure 9.2.

9.12 Develop a program that manipulates complex numbers. A complex number has two attributes, the real part and the imaginary part of the number. The basic operations on complex numbers are complex addition, complex subtraction, complex multiplication, and complex division. *Hint*: in addition to the rectangular representation of complex numbers (x,y), it might be helpful to include attributes for the polar representation of complex numbers (module, angle).

9.13 Define the classes of a problem that deals with motor vehicles; two groups or collections of these vehicles are trucks and cars. Include class *Sport_car* in the class hierarchy as a subclass of *Car*. Design and implement a program with all these classes. Include attributes such as horse-power, maximum speed, passenger capacity, load capacity, and weight. Include the relevant functions.

## CHAPTER 10

Linked Lists

10.1 INTRODUCTION

A linked list is a data structure that consists of a sequence of data items of the same or similar types and each data item or *node* has one or more links to another node. This data structure is dynamic in the sense that the number of data items can change. A linked list can grow and shrink during the execution of the program that is manipulating it. Recall that an array is also a data structure that stores a collection of data items, but the array is static because once it is created, more elements cannot be added or removed.

This chapter discusses the basic forms of simple linked lists, double-ended linked lists, and multiple linked lists. Several classes are defined for nodes and the actual linked lists. The operations possible on linked lists and higher-level data structures, such as stacks and queues implemented with linked lists, are also discussed. Abstract data types (ADTs) are discussed and defined for queues and stacks.

10.2 NODES AND LINKED LISTS

A linked list is a data structure that consists of a chain or sequence of nodes connected in some manner. A *node* is a relatively smaller data structure that contains data and one or more links that are used to connect the node to one on more other nodes. In graphical form, a node may be depicted as a box, which is divided into two types of components:

- A data block that stores one or more *data components*

- One or more *link components* that are references to other nodes

A simple node has a simple data block and one reference to another node. Figure 10.1 shows a representation of a simple node. Figure 10.2 illustrates the general form of a simple linked list in which nodes contain a reference to the next node. Note H is a reference to the first node (the head) of the

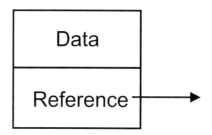

Figure 10.1 Structure of a node.

linked list. The last node (*Node 3* in Figure 10.2) has a link that refers to a black dot to indicate that the node has no connection to any other node and the reference of the node has a value *None*. When comparing linked lists with arrays, the main differences observed are:

- Linked lists are dynamic in size because they can grow and shrink; arrays are static in size.

- In linked lists, nodes are linked by references and based on many nodes; an array is a large block of memory with the elements located contiguously.

- The nodes in a linked list are referenced by relationship, not by position; to find a data item, always start from the first item (no direct access). Recall that access to the elements in an array is carried out using an index.

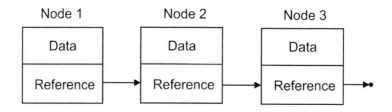

Figure 10.2 A simple linked list.

Linked lists and arrays are considered *low-level* data structures. These are used to implement *higher-level* data structures. Examples of simple higher-level data structures are *stacks* and *queues* and each one exhibits a different behavior implemented by an appropriate algorithm. More advanced and complex higher-level data structures are priority queues, trees, graphs, sets, and others.

10.2.1 Nodes

As mentioned previously, a simple node in a linked list has a data block and a reference that connects it to another node. These nodes can be located anywhere in memory and do not have to be stored contiguously in memory. The following listing shows the Python code with a class definition of a node. Class *Node* includes two attributes: the *data* and the reference *next* to another node. The class also defines two methods, and the constructor has one parameter with a default value of None.

```
class Node:
    def __init__(self, data = None):
        self.data = data
        self.next = None

    def strnode (self):
        print self.data
```

The following example includes several Python statements to create objects of class *None*, using the indicated data as an argument and the default. Note that *nd1* is the reference to a new node with the string "Hi there" as its data. Node object *nd2* is created with 24 as a its data.

```
nd1 = Node("Hi there")
nd2 = Node(24)
nd1.strnode()
nd2.strnode()
```

10.2.2 Definition of a Class for Linked Lists

A linked list is an object that creates, references, and manipulates node objects. A set of operations is defined for the linked list and some of these basic operations are:

- Create an empty linked list.
- Create and insert a new node at the front of the linked list.
- Insert a new node at the back of the linked list.
- Insert a new node at a specified position in the linked list.
- Get a copy of the data in the node at the front of the linked list.
- Get a copy of the data in the node at a specified position in the linked list.

- Remove the node at the front of the linked list.

- Remove the node at the back of the linked list.

- Remove the node at a specified position in the linked list.

- Traverse the list to display all the data in the nodes of the linked list.

- Check whether the linked list is empty.

- Check whether the linked list is full.

- Find a node of the linked list that contains a specified data item.

These operations are implemented as methods in class *LinkedList* and it is shown in Listing 10.1 and is stored file `linklistc.py`. In addition to these methods, two attributes are defined, *numnodes* and *head*. The value of the first attribute *numnodes* is the number of nodes in the linked list. The second attribute *head* is a reference to the first node of the linked list. This node is also known as the *head node* because it is the front of the linked list. In an empty list, the value of *numnodes* is zero and the value of *head* is `None`.

Listing 10.1: Python implementation for class *LinkedList*.

```
11 class LinkedList:
12     def __init__(self):
13         self.numnodes = 0
14         self.head = None
15
16     def insertFirst(self, data):
17         newnode = Node(data)
18         newnode.next = self.head
19         self.head = newnode
20         self.numnodes += 1
21
22     def insertLast(self, data):
23         newnode = Node(data)
24         newnode.next = None
25         if self.head == None:
26             self.head = newnode
27             return
28         lnode = self.head
29         while lnode.next != None :
30             lnode = lnode.next
31         lnode.next = newnode # new node is now the last node
32         self.numnodes += 1
33
34     def remFirst(self):
```

```
35          cnode = self.head
36          self.head = cnode.next # new head is second node
37          cnode.next = None
38          del cnode
39          self.numnodes -= 1
40
41      def remLast(self):
42          lnode = self.head
43          while lnode.next != None: #traversing list
44              pnode = lnode
45              lnode = lnode.next
46          pnode.next = None
47          del lnode
48          self.numnodes -= 1
49
50      def getFirst(self):
51          lnode = self.head  # first node
52          return lnode.data
53
54      def getLast(self):
55          lnode = self.head
56          while lnode.next != None: #traversing list
57              lnode = lnode.next
58              return lnode.data
59
60      def print_list(self):
61          lnode = self.head
62          while lnode:
63              lnode.strnode()  #print lnode.data
64              lnode = lnode.next
65
66      def getSize(self):
67          return self.numnodes
```

10.2.3 Creating and Manipulating a Linked List

To create an empty list, the constructor in class *LinkedList* is invoked as the following example shows. The assignment statement defines *listObj*, which now references an empty linked list object.

```
listObj = Linkedlist()
```

Method *empty* checks whether the list is empty by comparing the value of the head reference *head* with *None*. The following example checks the linked list referenced by *listObj* if empty.

```
if listObj.empty() == True:
    . . .
```

A node can be inserted in the linked list at the front, at the back, or in any other place specified. Method *insertFirst* creates and inserts a new node at the front of a linked list, given the data for the node. The new node becomes the head or front node of the linked list and the method increments the value of attribute *numnodes*. Figure 10.3 shows the insertion of a new node to the front of the list.

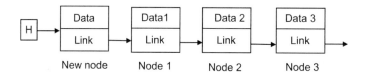

Figure 10.3 A new node inserted in the front of a linked list.

Assuming that *newData* refers to the data component for a new node, the following example invokes the method that creates and inserts the node:

```
llistObj.insertFirst (newData)
```

Method *getFirst* returns the data in the first node of the linked list. Method *remFirst* is called to remove and delete the node at the front of the linked list. The following example gets the data then removes the first node of the linked list.

```
data = listObj.getFirst()
listObj.remFirst()
```

Method *getLast* returns the data component of the last node in the linked list. Method *remLast* removes the last node of the linked list. The following example gets the data then removes the last node of the linked list.

```
data = listobj.getLast()
listObj.remLast()
```

Simple *traversal* of a linked list involves accessing every node in the linked list by following the links to the next node until the last node. Recall that the link of the last node is `None`. The following example calls method *print_llist*, which traverses a linked list to display the data of every node.

```
listObj.print_llist()
```

The following listing shows a Python script that imports class *Node* and class *LinkedList* to create and manipulate a linked list object. The script is stored in file `testlinklist.py`.

```
from linklistc import Node, LinkedList

print "New linked list"
listObj = LinkedList()
listObj.insertFirst("John")
listObj.insertFirst(99)
listObj.insertFirst(45)
listObj.insertLast(78)
listObj.insertLast(88)
listObj.insertLast("Mary")
print "Remove first node"
listObj.remFirst()
print "remove last node"
listObj.remLast()
listObj.print_list()
```

Using the Python interpreter to run the script produces the following output:

```
$ python testlinklist.py
New linked list
45
99
John
78
88
Mary
Remove first node
remove last node
99
John
78
88
```

More flexibility is obtained by including in the class an operation to insert a node at a specified position in the linked list. For example, insert a new node after current node 2. Figure 10.4 illustrates changing the links so that a new node is inserted after node 2. An enhanced implementation of class *Node* and class *LinkedList* is stored in file `linklist2c.py`.

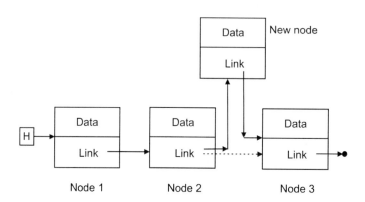

Figure 10.4 A new node inserted after node 2.

10.3 LINKED LISTS WITH TWO ENDS

The linked lists discussed previously have only one end, which includes a reference to the first node, and this reference is also known as the head of the linked list. In addition to the head node, providing a reference to the last node gives the linked list more flexibility for implementing some of the operations to manipulate linked list objects.

With two ends, a linked list has two references: one to the first node H, also known as the *head* or front of the list, and a reference to the last node L, also known as the *back* of the linked list. Figure 10.5 illustrates a linked list with a head reference H and a back reference L.

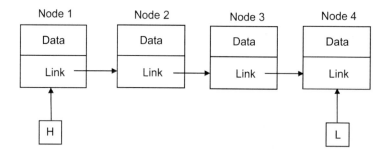

Figure 10.5 A linked list with two ends.

The class definition of a two-end linked list *TeLinkedList* includes an additional attribute, the reference to the last node (the *last*). An object of this class has the ability to directly add a new node to the back of the linked list without traversing it from the front. In a similar manner, the last node of a linked list can be removed without traversing it from the front. The implementation of this class is stored in file `telinklistc.py`.

Linked lists with two ends are very helpful and convenient for implementing higher-level data structures such as *stacks* and *queues*.

10.4 DOUBLE-LINKED LISTS

Linked lists that have nodes with only one link, a reference to the next node, can only traverse the linked list in one direction, starting at the front and toward the back of the list. A second link is included in the definition of the nodes that is a reference to the *previous* node. Figure 10.6 shows a linked list with nodes that have two links: a reference to the next node and a reference to the previous node. Such linked lists are known as *double linked lists*.

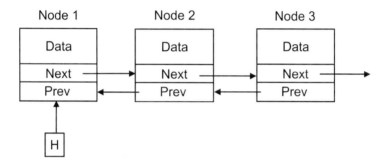

Figure 10.6 A linked list with two links per node.

The following listing of Python statements defines class *DNode*, which can be used for creating and manipulating nodes with two links, *next* that references the next node in the linked list, and *prev* that references the previous node in the linked list. Class *DNode* and class *DLinkedList* are implemented in module dlinklistc.py.

```
class DNode:
    def __init__(self, data = None):
        self.data = data
        self.next = None
        self.prev = None

    def strnode (self):
        print self.data
```

10.5 STACKS AND QUEUES DATA STRUCTURES

More practical data structures are used in problem solving and can be implemented with linked lists or with arrays. The structure and operations of two simple and widely known higher-level data structures, *queues* and *stacks*, are discussed here.

10.5.1 Stacks

A stack is a higher-level dynamical data structure that stores a collection of data items, each of which is stored in a node. Each node in a stack includes a data block and one or more links.

A stack has only one end: the *top* of the stack. The main characteristics of a stack are:

- Nodes can only be inserted at the top of the stack (TOS).

- Nodes can only be removed from the top of the stack.

- Nodes are removed in reverse order from that in which they are inserted into the stack. A stack is also known as a last in and first out (**LIFO**) data structure.

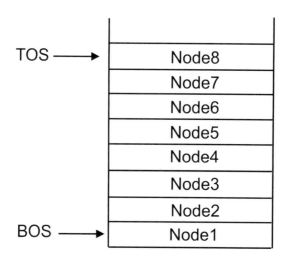

Figure 10.7 A stack as a dynamical data structure.

Figure 10.7 shows a stack and the top of the stack as the insertion point and the removal point. A class for stacks includes the following operations:

- *create_stack*, create an empty stack.

- *empty*, returns true if the stack is empty; otherwise returns false.

- *full*, returns true if the stack is full; otherwise returns false.

- *gettop*, returns a copy of the data block at the top of the stack without removing the node from the stack.

- *pop*, removes the node from the top of the stack.

- *push*, inserts a new node to the top of the stack.

- *getsize*, returns the number of nodes currently in the stack.

The most direct way to implement a stack is with a single-list linked list in which insertions and deletions are performed at the front of the linked list. The two-ended linked list class *TeLinkedList* is used to implement class *Stack*, which is stored in files `stack.py`. Listing 10.2 shows the Python source code of class *Stack*.

Listing 10.2: Python implementation of class *Stack* .

```
1  # A simple class for a stack using two-ended Linked List
2  from telinklistc import TeLinkedList
3
4  class Stack:
5      capacity = 100
6      def __init__(self):
7          self.list = TeLinkedList()
8
9      def empty (self):
10         if self.list.numnodes == 0:
11             return True
12         else:
13             return False
14
15     def full (self):
16         if self.list.numnodes == capacity:
17             return True
18         else:
19             return False
20
21     def push(self, data):
22         self.list.insertFirst(data)
23
24     def pop (self):
25         self.list.remFirst()
26
27     def get_top (self):
28         data = self.list.getFirst()
29         return data
30
```

```
31      def getSize(self):
32          lsize = self.list.numnodes
33          return lsize
34
35      def printStack(self):
36          self.list.print_list()
```

The Python commands that create a stack object and manipulate the stack are included in the following listing and stored in file `teststack.py`.

```
from stack import Stack

print "New stack"
listObj = Stack()
listObj.push("John")
listObj.push(99)
listObj.push(45)
print "TOS: ", listObj.get_top()
print "Stack empty? ", listObj.empty()
listObj.push(78)
listObj.push(88)
print "TOS: ", listObj.get_top()
listObj.pop()
print "TOS: ", listObj.get_top()
listObj.push(204)
print "TOS: ", listObj.get_top()
print "Size of stack: ", listObj.getSize()
listObj.printStack()
```

The following listing shows the Python interpreter running script `teststack.py` and the results produced.

```
$ python teststack.py
New stack
TOS:   45
Stack empty?  False
TOS:   88
TOS:   78
TOS:   204
Size of stack:   5
204
78
45
99
John
```

10.5.2 Queues

A queue is a dynamical data structure that stores a collection of data items or nodes and that has two ends: the *head* and the *tail*. The basic restrictions on manipulating a queue are:

- Nodes or data items are inserted at the tail of the queue.

- Nodes or data items are removed from the head of the queue.

- Nodes or data items are removed in the same order that they were inserted into the queue and is also known as a first in and first out (FIFO) data structure.

Figure 10.8 illustrates the form of a queue. It shows the insertion point at the tail and the removal point at the head of the queue. The relevant operations for manipulating a queue are:

- *empty*, returns true if the queue is empty; otherwise returns false.

- *full*, returns true if the queue is full; otherwise returns false.

- *getHead*, returns a copy of the data object at the head of the queue without removing the object from the queue.

- *removeHead*, removes the head item from the queue.

- *insertTail*, inserts a new data item into the tail of the queue.

- *getsize*, returns the number of data items currently in the queue.

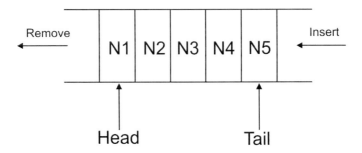

Figure 10.8 A queue as a dynamical data structure.

Queues can be implemented with single-linked lists, but a good way to implement a queue class is with a linked list with two ends. Class *Queue* is implemented with class *TeLinkedList*, which has already defined most of the needed operations. Listing 10.3 shows the Python source code of class *Queue*, which is stored in file `queue.py`.

Listing 10.3: Python implementation of class *Queue*.

```
1 # A simple class for a queue using two-ended Linked List
2 from telinklistc import TeLinkedList
3
4 class Queue:
5     capacity = 100
6     def __init__(self):
7         self.list = TeLinkedList()
8
9     def empty (self):
10         if self.list.numnodes == 0:
11             return True
12         else:
13             return False
14
15     def full (self):
16         if self.list.numnodes == capacity:
17             return True
18         else:
19             return False
20
21     def insertTail(self, data):
22         self.list.insertLast(data)
23
24     def getHead(self):
25         ldata = self.list.getFirst()
26         return ldata
27
28     def removeHead (self):
29         self.list.remFirst()
30
31     def getSize(self):
32         lsize = self.list.numnodes
33         return lsize
34
35     def printQueue(self):
36         self.list.print_list()
```

The following Python script is used to test class *Queue*. It creates an object of the class and inserts and removes several data items.

```
from queue import Queue

print "New queue"
listObj = Queue()
listObj.insertTail("John")
```

```
print "Head: ", listObj.getHead()
listObj.insertTail(99)
listObj.insertTail(45)
print "Queue empty? ", listObj.empty()
listObj.insertTail(78)
listObj.insertTail(88)
listObj.removeHead()
print "Head: ", listObj.getHead()
listObj.insertTail(204)
print "Size of queue: ", listObj.getSize()
listObj.printQueue()
```

The following listing shows the Linux shell commands that compile, link, and execute the program. The results produced by the program execution are also shown.

```
$ python testqueue.py
New queue
Head:  John
Queue empty?  False
Head:  99
Size of queue:  5
99
45
78
88
204
```

10.6   SUMMARY

Linked lists are dynamical data structures for storing a sequence of nodes. The data items are smaller data structures known as nodes. After the list object has been created, it can grow or shrink by adding or removing nodes. Lists can have one or two ends. Each node may have one or more links. A link is a reference to another node of the linked list. Classes of higher-level data structures are implemented with linked lists. Queues and stacks are examples of higher-level data structures and can be implemented with arrays or with linked lists.

Key Terms

linked lists	nodes	links
dynamic structure	low-level structures	high-level structures
next	previous	queues
stacks	data block	abstract data type

10.7   EXERCISES

10.1   Design and implement in Python a linked list in which each node represents a flight stop in a route to a destination. The data in each node is the airport code (an integer number). The airline can add or delete intermediate flight stops. It can also calculate and display the number of stops from the starting airport to the destination airport.

10.2   The class defined in this chapter for linked lists defines nodes with only one link and with only one end. Develop a class *Stack2* that uses double-linked lists with two ends. Develop a complete Python program that creates and manipulates three stacks.

10.3   Class *LinkedList* for linked lists defines nodes with only one link and with only one end. Given nodes that store person names and addresses, define an additional method that searches for a node with a given *name* and returns a copy of the data in the node. Develop a complete Python program that includes the necessary classes.

10.4   Develop a Python program that defines an array of N linked lists. This can be used if the linked list is used to implement a priority queue with N different priorities. A node will have an additional component, which is the priority defined by an integer variable.

10.5   Class *Queue* implements a queue data structure. Modify this class and implement it with a linked list that includes nodes of class *DNode*.

10.6   Class *Stack* implements a queue data structure. Modify this class and implement it with a linked list that includes nodes of class *DNode*.

## CHAPTER 11

Recursion

11.1 INTRODUCTION

Recursion is a design and programming technique used to implement a repetitive task or to implement a circular definition of a data structure. In methods and functions, recursion involves defining the method or function in terms of itself, which means that the method or function has a call to itself. Similarly, a structure is defined in terms of itself.

Recursion can be used to describe a solution in a simpler, clearer manner than with an iterative solution. Recursion can be used to describe complex algorithms by partitioning the problem into several smaller subproblems of the same kind and then combining the solutions to these subproblems. This chapter introduces the basic concepts and recursive problem-solving approach.

11.2 RECURSIVE APPROACH TO PROBLEM SOLVING

Most problems that can be solved recursively can also be solved iteratively. With iteration, a set of instructions is executed repeatedly until some *terminating condition* has been satisfied. Similarly with recursion, a set of instructions of a method or a function, is invoked repeatedly until a terminating condition becomes true.

A recursive definition of a method or a function consists of two parts:

1. One or more *base cases* that define the terminating conditions.

2. One or more *recursive cases*.

11.3 RECURSIVE DEFINITION OF FUNCTIONS

Three examples of recursive definition of methods are presented, and these are factorial, sum of squares, and reversing a list.

11.3.1 Factorial Problem

The *factorial* operation, denoted by the symbol !, can be defined in a general and informal manner as follows:

$$y! = y\,(y-1)\,(y-2)\,(y-3)\,\ldots 1.$$

For example, the factorial of 5 is: $5! = 5 \times 4 \times 3 \times 2 \times 1$.

Mathematical Specification of Factorial. A mathematical specification of the factorial function follows and is valid for $y \geq 0$:

$$y! = \begin{cases} 1 & \text{when } y = 0 \\ y\,(y-1)! & \text{when } y > 0. \end{cases}$$

The base case in this definition is the value of 1 for the function if the argument has value zero, that is, $0! = 1$. The general (recursive) case is $y! = y\,(y-1)!$ when the value of the argument is greater than zero. Note that this function is not defined for negative values of the argument.

Computing Factorial. In the following Python program, the factorial recursive function *rfact* has one parameter: the value for which the factorial is to be computed. Listing 11.1 shows a Python program that includes the recursive function, *rfact*, and the program is stored in file `rfactp.py`. The function is first called in line 21 to compute the factorial of a number. The recursive step in function *rfact* appears in lines 13–16.

Listing 11.1 Python program for computing factorial recursively.

```
2 # Program    : rfactp.py
3 # Author     : Jose M Garrido, May 29 2014.
4 # Description : Compute the recursive factorial of
        an integer number.
5
6 def rfact(n):
7       """
8       This function computes the factorial of num >= 0
9       it multiplies num * (num-1) * num-2 * ...1
10      """
11      if n < 0:           # negative values
12          return -1
13      if n > 0:           # recursive step
14          print "Factorial of ", n
15          res = n * rfact (n - 1)
16          return res
17      else:               # base case
```

```
18              return 1
19
20 y = input("Enter a number to compute factorial: ")
21 fy = rfact(y)
22 print "Factorial is: ", fy
```

Note that this implementation returns -1 for negative values of the argument. The following shell commands execute the Python interpreter with program `rfactp.py` and computes the factorial of an input number.

```
C:\python_progs>python rfactp.py
Enter a number to compute factorial: 5
Factorial of  5
Factorial of  4
Factorial of  3
Factorial of  2
Factorial of  1
Factorial is:   120
```

11.3.2 Sum of Squares

This is another example of a simple recursive function *sumsq*, which adds all the squares of numbers from m to k, with $m \le k$. The informal description of the function *sumsq* is:

$$sumsq(m, k) = m^2 + (m + 1)^2 + (m + 2)^2 + \ldots + k^2.$$

The solution approach is to decompose or break down the problem into smaller subproblems, such that the smaller problems can be solved with the same technique as that used to solve the overall problem. The final solution is computed by combining the solutions to the subproblems.

A mathematical specification of this recursive function is as follows, assuming that $m \ge 0$ and $k \ge 0$:

$$sumsq(m, k) = \begin{cases} m^2 + sumsq(m + 1, k), & \text{when } m < k \\ m^2, & \text{otherwise.} \end{cases}$$

Listing 11.2 shows a Python program that includes the recursive function, *rsumsq* and the program is stored in file `rsumsqp.py`.

Listing 11.2 Python program for computing the sum of squares.

```
1 # Program     : rsumsqp.py
2 # Author      : Jose M Garrido, June 3, 2014.
3 # Description : Compute the recursive sum of squares of
4 #     integer numbers from m to k.
5
```

```
 6 def rsumsq(m, k):
 7     """
 8        This function computes the sum of squares from m
 9        to k
10     """
11     if m < k:
12         res = m * m + rsumsq(m+1, k)   # recursive step
13         return res
14     else:               # base case
15         res = m * m
16         return res
17
18 x = input("Enter first number to compute sum of squares: ")
19 y = input("Enter second number: ")
20 r = rsumsq(x, y)
21 print "The sum of squares is: ", r
```

The following shell commands execute the Python interpreter with program rsumsqp.py and computes the sum of squares from m up to k.

```
$ python rsumsqp.py
Enter first number to compute sum of squares: 3
Enter second number: 12
The sum of squares is:   645
```

11.3.3 Reversing a Linked List

The recursive processing of a linked list is a very practical group of problems. Recursive reversal of a linked list is simpler and more elegant than performing the same operation iteratively.

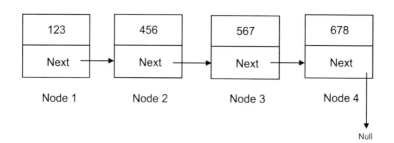

Figure 11.1 Linked list of characters.

Given the simple list of numbers, listobj = (123, 456, 567, 678), Figure 11.1

shows a linked list with the four nodes, each with a number. The problem is to change the list so that the nodes will appear in reverse order.

A recursive approach is to apply the divide-and-conquer approach. The first step is to partition the list into two sublists: the head and the tail. The head sublist contains only the first element of the list. The second sublist contains the rest of the list. The second step is to reverse (recursively) the tail sublist. The third step is to attach or concatenate the reversed tail sublist and the head.

The method of reversing a linked list recursively is shown in Figure 11.2. After partitioning the list, the head sublist contains only the first element of the list: $head = (123)$. The second sublist contains the rest of the list, $tail = (456, 567, 678)$.

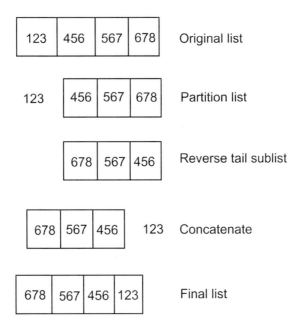

Figure 11.2 Reversing a linked list recursively.

The recursive operation continues to partition the tail sublist until a simple operation of reversing the sublist is found. Reversing a list containing a single element is easy; the reverse of the list is the same list. If the sublist is empty, the reverse of the list is an empty list. These will be included as the two base cases in the overall solution to the problem.

The overall solution is achieved by combining the concatenation or joining together the reversed tail sublist and the head sublist. Two auxiliary methods are used: the first one, *detach_head*, partitions the list into head and tail. The other method, *attach*, joins the reversed tail with the head.

Listing 11.3 shows two auxiliary methods in class *LinkedList* and these methods are *detach_head* and *attach*. These auxiliary methods are used in the recursive function *revList*. The class is stored in file `linklistc.py` and only the last part of the listing is shown.

Listing 11.3 Auxiliary methods in class *LinkedList*.

```
69      def detach_head(self):
70              #detach head node
71              lhead = self.head
72              lref = self.head.next
73              self.head = lref
74              lhead.next = None
75              self.numnodes -= 1
76              return lhead
77
78      def attach(self, subl2):
79              lnode = self.head
80              while lnode.next != None:
81                  lnode = lnode.next
82              lnode.next = subl2  # join last node
```

The following listing shows a Python program that includes the recursive function, *revList* and other instructions to test the recursive reversal of a linked list. The program is stored in file `testrevlist.py`.

```
from linklistc import Node, LinkedList

def revList(clist):
    if clist == None:
        return None
    if clist.getSize() == 1:
        return clist
    else:
        headl = clist.detach_head() # partition list
        taill = clist
        rlist = revList(taill)  # reverse tail
        rlist.attach(headl)     # attach head
        return rlist

print "New linked list"
listObj = LinkedList()
listObj.insertLast(123)
listObj.insertLast(456)
listObj.insertLast(567)
```

```
listObj.insertLast(678)
listObj.print_list()
print "Reverse list"
nlist = revList(listObj)
nlist.print_list()
```

The following shell commands execute the Python interpreter with program `testrevlist.py`; it creates and reverses a linked list.

```
$ python testrevlist.py
New linked list
123
456
567
678
Reverse list
678
567
456
123
```

11.4 ANALYZING RECURSION

Storage is provided for objects, constants, and functions that have local variables, constants, and parameter variables. Recursion uses a *runtime stack*. A block of memory serves as a medium for the runtime stack, which grows when a function is invoked and shrinks when the execution of a function is completed.

Whenever a function is invoked, the runtime system dynamically allocates storage for its local variables, constants, and parameter variables declared within the function from the allocated chunk of memory. This data is placed in a data structure known as a *frame* or *activation record* and inserted on the stack with a *push* operation. After the execution of the invoked function is completed, the frame created for that function is removed from the top of the stack with the *pop* operation.

A frame is created to provide storage for each function. The *runtime stack* refers to a stack-type data structure (LIFO) associated with each process that is provided and maintained by the system. The runtime stack holds data of all the functions that have been invoked but not yet completed processing. A runtime stack grows and shrinks depending on the number of functions involved in processing and the interaction between them.

11.5 SUMMARY

For several algorithms, the mathematical representations are recursive by default. One example is the factorial function, of which the recursive form is easier to understand. The same is true for the algorithm that reverses a list. A recursive function definition contains one or more calls to itself and in many cases, the recursive function is very powerful and compact in defining the solution to complex problems. The main disadvantages are that it demands a relatively large amount of memory and time to build its internal stack.

Defining a recursive function may require defining additional functions known as auxiliary functions or methods. Often, a software developer must decide between using iterative and recursive algorithms.

Key Terms

base cases	recursive case	recursive call
terminating condition	recursive solution	iterative solution
recursive functions	list functions	auxiliary functions
activation frame	activation record	runtime stack

11.6 EXERCISES

11.1 Develop a Python program with a recursive function that finds and displays n natural numbers that are even.

11.2 Develop a Python program with a recursive function that computes exponentiation of a number given and displays the result.

11.3 Develop a Python program with a recursive function that displays the letters in a string in reverse order.

11.4 Develop a Python program with a recursive function that reverses the letters in a string.

11.5 Develop a Python program with a recursive function that checks whether a given string is a palindrome. A palindrome is a string that does not change when it is reversed. For example: "madam," "radar," and so on.

11.6 Develop a Python program with a recursive function that performs a linear search in an array of integer values.

11.7 Develop a Python program with a recursive function that performs a binary search in an array of integer values.

IV

Fundamental Computational
Models with Python

## CHAPTER 12

Computational Models with Arithmetic Growth

12.1 INTRODUCTION

Given a real-world problem, a mathematical model is defined and formulated; then one or more techniques are used to implement this model in a computer to derive the corresponding computational model. This chapter presents an introduction to simple mathematical models that exhibit arithmetic growth and describes the overall behavior of simple mathematical and computational models. This includes definitions and explanations of several important concepts related to modeling.

Some of these concepts have been briefly introduced in preceding chapters. The derivation of difference and functional equations is explained; their use in mathematical modeling is illustrated with a few examples. A complete computational model implemented with the Python programming language and some functions in the Numpy module are presented and discussed.

12.2 MATHEMATICAL MODELING

The three types of methods that are used for modeling are:

1. Graphical

2. Numerical

3. Analytical

Graphical methods apply visualization of the data to help understand the data. Various types of graphs can be used; the most common one is the line graph.

Numerical methods directly manipulate the data of the problem to compute various quantities of interest, such as the average change of the population size in a year.

Analytical methods use various forms of relations and equations to allow computation of the various quantities of interest. For example, an equation can be derived that defines how to compute the population size for any given year. Each method has its advantages and limitations. The three methods complement each other and are normally used in modeling.

12.2.1 Difference Equations

A data list that contains values ordered in some manner is known as a *sequence*. Typically, a sequence is used to represent ordered values of a property of interest in some real problem. Each of these values corresponds to a recorded measure at a specific point in time. In the example of population change over a period of five years, the ordered list will contain the value of the population size for every year. This type of data is discrete data and the expression for the ordered list can be written as:

$$\langle p_1, p_2, p_3, p_4, p_5 \rangle.$$

In this expression, p_1 is the value of the population of year 1, p_2 is the value of the population for year 2, p_5 is the value of the population for year 5, and so on. In this case, the length of the list is 5 because it has only five values, or terms.

Another example is the study of changes in electric energy price in a year in Georgia, given the average monthly price. Table 12.1 shows the values of average retail price of electricity for the state of Georgia.[1] The data given corresponds to the price of electric power that has been recorded every month for the last 12 months. This is another example of *discrete data*. The data list is expressed mathematically as:

$$\langle e_1, e_2, e_3, e_4, e_5, e_6, e_7, e_8, e_9, e_{10}, e_{11}, e_{12} \rangle.$$

Given the data that has been recorded about a problem and that has been recorded in a list, simple assumptions can be made. One basic assumption is that the quantities in the list increase at a constant rate. This means the increment is assumed to be fixed. If the property is denoted by x, the increment is denoted by Δx, and the value of a term measured at a particular point in time is equal to the value of the preceding term and the increment added to it. This can be expressed as:

$$x_n = x_{n-1} + \Delta x. \tag{12.1}$$

Another assumption is that the values of x are actually always increasing,

---

[1]U.S. Energy Information Administration—Independent Statistics and Analysis. http://www.eia.gov/

and not decreasing. This means that the increment is greater than zero, denoted by $\Delta x \geq 0$. At any point in time, the increment of x can be computed as the difference of two consecutive measures of x and has the value given by the expression:

$$\Delta x = x_n - x_{n-1}. \tag{12.2}$$

These last two mathematical expressions, Equation 12.1 and Equation 12.2, are known as *difference equations* and are fundamental for the formulation of simple mathematical models. We can now derive a simple mathematical model for the monthly average price of electric energy, given the collection of monthly recorded energy price in cents per kW-h of the last 12 months. This model is formulated as:

$$e_n = e_{n-1} + \Delta e.$$

The initial value of energy price, prior to the first month of consumption, is denoted by e_0, and it normally corresponds to the energy price of a month from the previous year.

12.2.2 Functional Equations

A functional equation has the general form: $y = f(x)$, where x is the *independent variable*, because for every value of x, the function gives a corresponding value for y. In this case, y is a function of x.

An equation that gives the value of x_n at a particular point in time denoted by n, without using the previous value, x_{n-1}, is known as a *functional equation*. In this case the functional equation can also be expressed as: $x = f(n)$. From the data given about a problem and from the difference equation(s), a functional equation can be derived. Using analytical methods, the following mathematical expression can be derived and is an example of a functional equation.

$$x_n = (n-1)\,\Delta x + x_1. \tag{12.3}$$

This equation gives the value of the element x_n as a function of n. In other words, the value x_n can be computed given the value of n. The value of Δx has already been computed. The initial value of variable x is denoted by x_1 and is given in the problem by the first element in the sequence x.

12.3 MODELS WITH ARITHMETIC GROWTH

Arithmetic growth models are the simplest type of mathematical models. These models have a linear relationship in the variable because the values of the variable increase by equal amounts over equal time intervals. Using x

as the variable, the increase is represented by Δx, and the difference equation defined in Equation 12.1:

$$x_n = x_{n-1} + \Delta x.$$

Examples are time-dependent models, in which a selected property is represented by a variable that changes over time.

From the given data of a real problem, to decide if the model of the problem will exhibit arithmetic growth, the given data must be processed by computing the differences of all *consecutive* values in the data. The differences thus consist of another list of values. If the differences are all equal, then the model exhibits arithmetic growth, and therefore it is a *linear model*.

Equation 12.1 and the simplifying assumption of constant growth can be applied to a wide variety of real problems, such as: population growth, monthly price changes of electric energy, yearly oil consumption, and spread of disease.

Using graphical methods, a line chart or a bar chart can be constructed to produce a visual representation of the changes in time of the variable x. Using numerical methods, given the initial value x_0 and once the increase Δx in the property x has been derived, successive values of x can be calculated using Equation 12.1. As mentioned before, with analytical methods, a functional equation can be derived that would allow the direct calculation of variable x_n at any of the n points in time that are included in the data list given. This equation is defined in Equation 12.3:

$$x_n = x_1 + \Delta x \, (n - 1).$$

12.4 USING THE PYTHON LANGUAGE AND NUMPY

The *NumPy* library includes functions, two of which that are needed to implement computational models with arithmetic growth. These functions are *diff* and *linspace*. The first function, *diff*, computes the differences of a vector (sequence) of data values given.

Calling function *diff* requires one argument, which is the specified vector. A second argument is optional and it specifies the order of the differences of the array. The function produces another array that has the values of the differences of the values in the given vector.

```
>>> import numpy as np
>>> e = np.array ([1.25, 2.15, 4.55, 3.2, 1.05, 2.45, 3.85,
   1.15, 2.75, 3.55])
>>> e
array([ 1.25,   2.15,   4.55,   3.2 ,   1.05,   2.45,   3.85,
   1.15,   2.75,   3.55])
>>> de = np.diff(e)
```

```
>>> de
array([ 0.9 ,   2.4 ,  -1.35,  -2.15,   1.4 ,   1.4 ,  -2.7 ,
   1.6 ,   0.8 ])
```

Table 12.1 Average price of electricity (cents per kW-h) in 2010.

Month	Jan	Feb	Mar	Apr	May	Jun	Jul	Aug
Price	10.22	10.36	10.49	10.60	10.68	10.80	10.88	10.94
	Month	Sep	Oct	Nov	Dec			
	Price	11.05	11.15	11.26	11.40			

Listing 12.1 shows the Python program that computes the differences and the approximate price of electricity for 12 months. The array for the monthly price is denoted by e, and the array for months is denoted by m. The array of differences is denoted by de and is computed in line 17 of the program. The data is taken from Table 12.1.

In the program, the number of measurements is the total number of months, denoted by N and has a value of 12. To compute the average value of the increments in price of electric energy in the year we can use the general expression for calculating average:

$$\Delta e = \frac{1}{n} \sum_{i=1}^{i=n} de_i.$$

Computing the average is implemented in Python with the *mean* NumPy function that computes the average of the values in a vector. Using d to denote the average increment (Δe), the following statement illustrates the call to function *mean* and is included in line 20 of the Python program.

```
deltax = np.mean(de)
```

Listing 12.1: Program for computing the differences in price of electricity.

```
1 # Program:   priceelect.py
2 # Python program for computing monthly price for electric
  energy
3 # J. M. Garrido, August 2014.
4
5 import numpy as np
6
```

```
 7 N = 12;    # 12 months
 8 e = np.array([10.22, 10.36, 10.49, 10.60, 10.68, 10.80,
   10.88, 10.94, 11.05, 11.15, 11.26, 11.40])
 9 mm = np.arange(N)  # month array
10 m = mm + 1
11 print "Monthly price of electricity\n"
12
13 # Array Monthly price for electric energy
14 print e
15 #  differences in sequence e
16 print "\nDifferences of the given data\n"
17 de = np.diff(e)
18 print de
19 #  average of increments
20 deltax = np.mean(de)
21 print "\nAverage difference: ", deltax
22 # Calculating price of electric energy
23 ce = mm * deltax + e[0]
24 print "\nCalculated prices of electricity: \n"
25 print ce
26 print "\nData for Plotting\n"
27 for j in range(N):
28      print m[j], e[j], ce[j]
```

With the value of the average increment of the price of electricity, the functional equation of the model is applied to compute the price of electric energy for any month. The new array, *ce*, is defined and contains all the values computed using the functional equation and the average value of the increments. This is shown in line 23 of the program. The following listing shows the command that starts the Python interpreter and the results produced.

```
$ python priceelect.py
Monthly price of electricity

[ 10.22  10.36  10.49  10.6   10.68  10.8   10.88  10.94  11.05
  11.15  11.26  11.4 ]

Differences of the given data

[ 0.14  0.13  0.11  0.08  0.12  0.08  0.06  0.11  0.1   0.11
  0.14]

Average difference:  0.107272727273

Calculated prices of electricity:
```

```
[ 10.22        10.32727273   10.43454545   10.54181818
  10.64909091   10.75636364   10.86363636   10.97090909
  11.07818182   11.18545455   11.29272727   11.4        ]
```

Data for Plotting

```
1 10.22 10.22
2 10.36 10.3272727273
3 10.49 10.4345454545
4 10.6 10.5418181818
5 10.68 10.6490909091
6 10.8 10.7563636364
7 10.88 10.8636363636
8 10.94 10.9709090909
9 11.05 11.0781818182
10 11.15 11.1854545455
11 11.26 11.2927272727
12 11.4 11.4
```

12.5   PRODUCING THE CHARTS OF THE MODEL

Two lists of values for the price of electric energy are available as arrays. The first one is the data given with the problem and is denoted by e; the second list was computed in the program using the functional equation for all 12 months, and the list is denoted by ce. The array with data representing the months of the year is denoted by m.

Gnuplot is a software tool and is used to produce charts with these three data lists or arrays. Two line charts are generated on the same plot, one line chart of array e with array m, the second chart using array ce and array m. The Gnuplot commands that create and draw the charts are stored in the script file: `priceelect.cgp` and are shown as follows:

```
set title "Plot of Montly Price of Electricity vs time"
set xlabel "Time (secs)"
set ylabel "Monthly price of electricity"
set size 1.0, 1.0
set samples 12
plot "priceelect.gpl" u 1:2 with linespoints, "priceelect.gpl"
    u 1:3 with linespoints
```

Figure 12.1 shows the line chart with the original data given in Table 12.1 and the line with computed values of the monthly price of electricity.

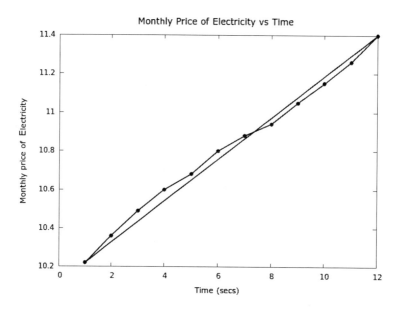

Figure 12.1 Given and computed values of monthly price of electricity.

12.6 VALIDATION OF A MODEL

Validation of a model is the analysis that compares the values computed with the model with the actual values given. For example, starting with the first value of the monthly consumption of electric energy, the model is used to compute the rest of the monthly values of consumption. These values can then be compared to the values given.

If the corresponding values are close enough, the model is considered a reasonable *approximation* to the real system.

12.7 FILE I/O

Computational models typically deal with a large amount of data, which is conveniently stored in data files. The programs discussed and shown previously read and write data to the console, which on a personal computer is the keyboard and the screen. This is also known as standard input and output.

Python provides basic functions and methods necessary to manipulate files. The programs do most of the file manipulation using file objects.

12.7.1 Types of Files

There are two general types of file: text or binary. A text file is typically structured as a sequence of text lines, each being a sequence of characters.

A text line is terminated by a special control character, the EOL (End of

Line) character. The most common line terminator is the \n , or the newline character. The backslash character is used to specify control characters and indicates that the next character will be treated as a newline.

A binary file is basically any file that is not a text file. The main advantage of text files is that no data conversion is necessary.

12.7.2 Opening and Closing Text Files

A file has to be opened before data can be read or written to it. Opening a file creates a file object and is carried out by calling Python's built-in *open* function, which returns a file object. The first argument in the call is a string with the file name, the second argument is the access mode for the file, the third argument is optional and indicates the buffering as an integer. The following table indicates the access modes allowed for files in Python.

Mode	Description
r	Opens a file for reading only. The file pointer is placed at the beginning of the file. This is the default mode.
rb	Opens a file for reading only in binary format. The file pointer is placed at the beginning of the file. This is the default mode.
r+	Opens a file for both reading and writing. The file pointer will be at the beginning of the file.
rb+	Opens a file for both reading and writing in binary format. The file pointer will be at the beginning of the file.
w	Opens a file for writing only. Overwrites the file if the file exists. If the file does not exist, creates a new file for writing.
wb	Opens a file for writing only in binary format. Overwrites the file if the file exists. If the file does not exist, creates a new file for writing.
w+	Opens a file for both writing and reading. Overwrites the existing file if the file exists. If the file does not exist, creates a new file for reading and writing.
wb+	Opens a file for both writing and reading in binary format. Overwrites the existing file if the file exists. If the file does not exist, creates a new file for reading and writing.

Mode	Description
a	Opens a file for appending. The file pointer is at the end of the file if the file exists. That is, the file is in the append mode. If the file does not exist, it creates a new file for writing.
ab	Opens a file for appending in binary format. The file pointer is at the end of the file if the file exists. That is, the file is in the append mode. If the file does not exist, it creates a new file for writing.
a+	Opens a file for both appending and reading. The file pointer is at the end of the file if the file exists. The file opens in the append mode. If the file does not exist, it creates a new file for reading and writing.
ab+	Opens a file for both appending and reading in binary format. The file pointer is at the end of the file if the file exists. The file opens in the append mode. If the file does not exist, it creates a new file for reading and writing.

The following example opens two files: *lengthpart.dat* is opened for writing and the file object created is *outfile*. The second file, *mydata.dat*, is opened for reading and the file object created is *infile*.

```
outfile = open("lengthpart.dat", "w")
infile = open ("mydata.dat", "r")
```

After a file has been used in a program, it should be closed. Method *close* of a file object flushes any unwritten data and closes the file object, after which no more writing can be performed. The following example closes the two files that were opened previously.

```
outfile.close()
infile.close()
```

12.7.3 Writing Data to a File

Writing data to a file involves calling method *write* of the file object and it writes a string to the open file. Method *write* does not add a newline character (\n) to the end of the string.

The following example is a short Python program that opens file *mtest.dat* for writing. In line 3, the string "Testing output data" is written to the file. Line 4 defines variable *v1* with a value of 56. This value has to be converted to a string before it is written to the file. Function *str* is called with variable *v1* as the argument. The new string variable *v1str* is created. The statement in line 6 writes the string variable to the file. In a similar manner, *v2* is a floating-point variable and its value is also converted to a string then written to the file in lines 8–9. The file is closed in line 12.

```
 1 # writing to a data file
 2 mfile = open("mtest.dat", 'w')
 3 mfile.write("Testing output data\n")
 4 v1 = 56
 5 v1str = str(v1)
 6 mfile.write(v1str)
 7 v2 = 10.45
 8 v2str = str(v2)
 9 mfile.write("\n")
10 mfile.write(v2str + "\n")
11 mfile.write(v1str + " " + v2str +"\n")
12 mfile.close()
```

The following lines are produced when the Python interpreter runs the program.

```
Testing output data
56
10.45
56 10.45
```

12.7.4   Reading Data from a File

As mentioned previously, a text file stores data in string form. Reading data from a text file involves reading lines of text from a file. Method *read* reads a string from an open file. This method starts reads from the beginning of the file and tries to read as much as possible, or until the end of file. An optional argument can be used that indicates the number of bytes to read.

Method *readline* reads a single line from the file and returns a string containing characters up to \n. This method is useful for reading data from a file line by line rather than reading the entire file in at once.

```
infile = open('mydata.txt', 'r')
# Read a line from the open file
line = infile.readline()
```

Because several data values can be contained in a line, the *split* string method can be used to separate substrings in the line that separated by a space character. This method returns a list of substrings. An optional argument is the actual separator character to apply. The following example illustrates the use of the *split* string method. It is first called with no arguments, so the default separator is one or more space characters and the list created is *cols*. The second time it is called with a comma as the argument to be used as the substring separator.

```
>>> line = "1234 23.59 part 27134"
>>> cols = line.split()
>>> cols
['1234', '23.59', 'part', '27134']
>>> line2 = "1234, 23.59, part 27134"
>>> cols2 = line2.split(",")
>>> cols2
['1234', ' 23.59', ' part 27134']
```

To read multiple lines from a file, a loop can be used over the file object. This is memory efficient, fast, and leads to simple code. For numerical variables, either integer or floating-point, their values are converted from string to the numeric type using functions *float* and *int*. The following short Python program illustrates the general technique of reading lines of data, separating the various data fields from the line, and converting each data field to the required type.

```
1  # Open input file program: gen_filein.py
2  infile = open('mydata.dat', 'r')
3
4  # Read and ignore a header line
5  headstr = infile.readline()
6
7  # Loop over lines and extract variables
8  for line in infile:
9      line = line.strip()      # remove '\n' char
10     columns = line.split()   # split line into columns
11     print columns
12     var1 = columns[0]
13     var2 = float(columns[1])
14     j = int(columns[2])
15     print var1, var2, j
```

The following data file **mydata.dat** is used for testing the program.

```
Testing data
sequence1 34.56 88
sequence2 10.45 79
sequence3 85.56 45
```

Starting the Python interpreter to run the program, produces the following output:

```
$ python gen_filein.py
['sequence1', '34.56', '88']
sequence1 34.56 88
['sequence2', '10.45', '79']
sequence2 10.45 79
['sequence3', '85.56', '45']
sequence3 85.56 45
```

12.8 SUMMARY

Several basic concepts of mathematical models are presented. A computational model is a computer implementation of a mathematical model. Considering simplifying assumptions and using abstraction are important steps in formulating a mathematical model. This involves a transition from the real world to the abstract world.

Simple mathematical techniques such as difference equations and functional equations are used. With arithmetic growth models, the values of the differences of the data is constant. The difference and functional equations for these models are linear. Numpy is used to create efficient arrays and perform various computations with arrays. The programming techniques discussed compute the first differences of the values in an array. If the differences are constant or are almost all equal, then the model represented by the values of the original array is a linear model. The data of linear models follows arithmetic growth.

Computational models typically deal with a large amount of data, which is conveniently stored in data files. Python provides basic functions and methods necessary to manipulate files. The programs do most of the file manipulation using file objects.

Key Terms

arithmetic growth	mathematical methods	graphical methods
numerical methods	analytical methods	discrete data
data list	ordered list	sequence
differences	average difference	linear models
difference equations	functional equations	model validation
data files	reading files	writing files

12.9 EXERCISES

12.1 Construct a line chart of the data list in Table 12.1 and discuss how the price of electric energy changes in a specified period.

12.2 Construct a bar chart of the data list in Table 12.1 and discuss how the price of electric energy changes in a specified period.

12.3 Formulate a mathematical model based on a difference equation of the data in Table 12.1.

12.4 Develop a Python program that uses the data in Table 12.1 to compute the average increase in price of electric energy per month from Equation 12.1 and/or Equation 12.2. Start with the second month, and calculate the price for the rest of the months. Discuss the difference between the data in the table and the corresponding values calculated.

12.5 Develop a Python program that uses the data from a different problem to compute the differences and decide whether the model of the problem exhibits arithmetic growth.

## CHAPTER 13

Computational Models with Quadratic Growth

13.1   INTRODUCTION

A computational model with quadratic growth is one in which the differences in the data are not constant but growing in some regular manner. Recall that with arithmetic growth these differences are constant. This chapter presents an introduction to computational models in which the differences in the data follow a pattern of arithmetic growth, or in simpler terms, the differences increase or decrease linearly.

This chapter explains the computation of difference and functional equations; their use in the mathematical model is illustrated with a few examples. A complete computational model implemented with the Python programming language is presented and discussed.

13.2   DIFFERENCES OF THE DATA

The data values in a quadratic growth model do not increase (or decrease) by a constant amount. With *quadratic growth*, the differences of the data change linearly. The differences of the differences, known as the *second differences*, are constant.

The data in a problem is used to set up an ordered list of values, or sequence. This type of list is denoted as follows:

$$\langle p_1, p_2, p_3, p_4, p_5 \rangle.$$

The following example presents a data list given in a generic problem and Figure 13.1 shows the graph of these values.

$$S = \langle 6.5, 10.5, 17.5, 27.5, 40.5, 56.5, 75.5, 97.5, 122.5 \rangle$$

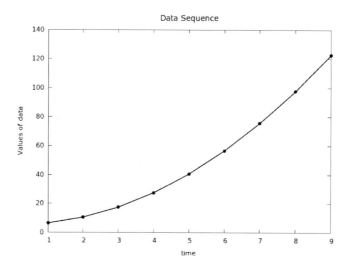

Figure 13.1 Plot of data in the sequence.

The differences of these values is another sequence, D, that can be derived and represents the *differences* of the values in sequence S. The values in D are the increases (or decreases) of the values in the first sequence, S.

$$D = \langle\, 4.0, 7.0, 10.0, 13.0, 16.0, 19.0, 22.0, 25.0 \,\rangle$$

The increases of the values of the differences, D, appear to change linearly and follow an arithmetic growth pattern; this is an important property of quadratic growth models. Figure 13.2 shows the graph with the data of the first differences.

Listing 13.1 shows the source code of a Python program that creates the arrays for the data sequences in the problem, their differences, and their second differences. This program is stored in file: `differences.py`. Note that the first differences are computed in line 18 and the *second differences* are computed in line 21. As mentioned in the previous chapter, the differences are computed by calling function `diff()`, which is in the *NumPy* package.

Listing 13.1: Program that computes the first and second differences.

```
1 # Program:    differences.py
2 # Python program for computing first and second differences
3 # of problem data
4 # J. M. Garrido, August 2014.
5 import numpy as np
6
7 N = 9   # number of data points
8 # Array with original data sequence
```

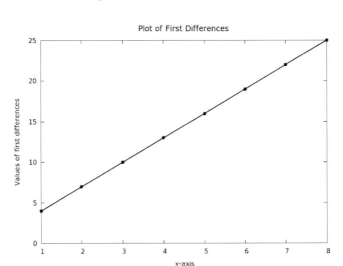

Figure 13.2 Plot of first differences.

```
 9 s = np.array([6.5, 10.5, 17.5, 27.5, 40.5, 56.5, 75.5, 97.5,
    122.5])
10 x = np.arange(N)
11 x = x + 1
12
13 #  Array with problem data
14 print "Data sequence \n"
15 print s
16 # Compute first differences in sequence e
17 print "\nFirst Differences of the given data\n"
18 d = np.diff(s)
19 print d
20 # compute differences of the differences
21 d2 = np.diff(d)
22 print "\nSecond differences\n"
23 print d2
24 print "\nData for Plotting\n"
25 for j in range(N):
26     print x[j], s[j]
27 print "\nData of differences for plotting\n"
28 for j in range(N-1):
29     print x[j], d[j]
```

The following listing shows the Python interpreter processing the program.

The output produced by the execution of the program shows the data given by the problem, the first differences, and the second differences.

```
$ python differences.py
Data sequence

[6.5    10.5   17.5   27.5   40.5   56.5   75.5   97.5  122.5]

First Differences of the given data

[4.    7.   10.   13.   16.   19.   22.   25.]

Second differences

[ 3.   3.   3.   3.   3.   3.   3.]

Data for Plotting

1 6.5
2 10.5
3 17.5
4 27.5
5 40.5
6 56.5
7 75.5
8 97.5
9 122.5

Data of differences for plotting

1 4.0
2 7.0
3 10.0
4 13.0
5 16.0
6 19.0
7 22.0
8 25.0
```

In the example, the second differences are constant; all have value 3.0. This is another important property of quadratic growth models.

13.3 DIFFERENCE EQUATIONS

An ordered data list or sequence is used to represent ordered values of a property of interest in some real problem. Each of these values can correspond to a recorded measure at a specific point in time. The expression for the values in a sequence, S, with n values can be written as:

$$S = \langle s_1, s_2, s_3, s_4, s_5 \ldots s_n \rangle.$$

In a similar manner, the expression for the values in the differences, D, with m values and with $m = n - 1$, can be written as:

$$D = \langle d_1, d_2, d_3, d_4, d_5 \ldots d_m \rangle.$$

Because the values in the sequence of second differences are all the same, the value is denoted by dd, and computed simply as $dd = d_n - d_{n-1}$.

To formulate the difference equation, the value of a term, s_{n+1} in the sequence is computed with the value of the preceding term s_n plus the first term of the differences d_1, plus the single value dd of the second differences added to it times $n - 1$. This equation, which is the difference equation for quadratic growth models, can be expressed as:

$$s_{n+1} = s_n + d_1 + dd\,(n - 1). \tag{13.1}$$

13.4 FUNCTIONAL EQUATIONS

An equation that gives the value of a term x_n without using consecutive values (the previous value, x_{n-1} or the next value, x_{n+1}) , is known as a *functional equation*. The difference equation for quadratic growth models, Equation 13.1, can be rewritten as:

$$s_n = s_{n-1} + d_1 + dd\,(n - 2). \tag{13.2}$$

Equation 13.2 can be manipulated by substituting s_{n-1} for its difference equation, and continuing this procedure until s_1. In this manner, a functional equation can be derived. The following mathematical expression is a general functional equation for quadratic growth models.

$$s_n = s_1 + d_1\,(n - 1) + dd\,(n - 2)n/2. \tag{13.3}$$

Equation 13.3 gives the value s_n as a function of n for a quadratic growth model. The value of the first term of the original sequence is denoted by s_1, the value of the first term of the differences is denoted by d_1, and the single value, dd, is the second difference.

13.5 EXAMPLES OF QUADRATIC MODELS

Equation 13.1 and Equation 13.3 represent mathematical models of quadratic growth models. These can be applied to a wide variety of real problems, such as computer networks, airline routes, roads and highways, and telephone networks. In these models, the first differences increase in a linear manner (arithmetic growth), as the two examples discussed in previous sections of this chapter. Other models with quadratic growth involve addition of ordered values from several sequences that exhibit arithmetic growth.

13.5.1 Growth of Number of Patients

Statistical data maintained by a county with several hospitals include the number of patients every year. Table 13.1 gives the data of the number of patients in the years from 1995 to 2002, and their increases. Figure 13.3 shows the graph for the number of patients in the hospital by year from 1995 through 2002. It can be observed that the number of patients from 1995 through 2002 does not follow a straight line; it is not linear.

Table 13.1 Number of patients for years 1995–2002.

Year	1995	1996	1997	1998	1999	2000
Patients	5,500	8,500	13,500	20,500	29,500	40,500
Increase	0	3,000	5,000	7,000	9,000	11,000

Year	2001	2002
Patients	53,5000	68,500
Increase	13,000	15,000

Table 13.1 shows that the number of patients increases every year by a constant number. The differences in the number of patients from year to year increase in a regular pattern, in a linear manner. This implies that the increases of the number of patients follow an arithmetic growth.

Computing the first and second differences of the data for this problem is similar to the one already discussed.

13.5.2 Growth of Computer Networks

The following example represents a simple network that connects computers directly to each other and a number of links are necessary for the direct connection between computers. To connect two computers, 1 single link is needed. To connect 3 computers, 3 links are needed. To connect 4 computers, 6 links are needed.

It can be noted that as the number of computers increases, the number

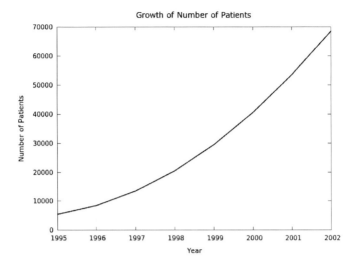

Figure 13.3 Number of patients for 1995–2002.

of links increases in some pattern. To connect 5 computers, 4 new links are needed to connect the new computer to the 4 computers already connected. This gives a total of 10 links. To connect 6 computers, 5 new links are needed to connect the new computer to the 5 computers that are already connected, for a total of 15 links.

Let L_n denote the number of links needed to connect n computers. The difference equation for the number of links can be expressed as:

$$L_n = L_{n-1} + (n - 1).$$

This equation has the same form as the general difference equation for quadratic growth, Equation 13.2. The parameters are set as: $d_1 = 0$ and $dd = 1$.

Using the expression for the difference equation for L_n, the following Python program is used to construct the data sequence for L for n varying from 1 to 50. The source code is stored in the file links.py and is shown in Listing 13.2.

When running the program, all the terms in the sequence L are computed and the results are used to produce a chart using GnuPlot. Figure 13.4 shows the graph of the number of links needed to connect n computers.

Listing 13.2: Python source program that computes the number of links.

```
1 # Program:   links.py
2 # Python program that computes the number of links needed
3 # to connect n computers
4 # J. M. Garrido, August 26, 2014.
```

```
 5 import numpy as np
 6 M = 50  # limit on number of computers
 7 print "Links of computer network\n"
 8 n = np.arange(M)
 9 n = n + 1
10 llinks = np.zeros(M)
11 # compute the links as n varies from 2 to m
12 for j in range(M):
13     llinks[j] = llinks[j-1] + (j-1)
14 # data for plot
15 print "Number of computers  number of links\n"
16 for j in range(M):
17     print n[j], llinks[j]
```

```
$ python links.py
Links of computer network

Number of computers  number of links

1 -1.0 2 -1.0 3 0.0
4 2.0 5 5.0 6 9.0
7 14.0 8 20.0 9 27.0
10 35.0 11 44.0 12 54.0
13 65.0 14 77.0 15 90.0
16 104.0 17 119.0 18 135.0
19 152.0 20 170.0 21 189.0
22 209.0 23 230.0 24 252.0
25 275.0 26 299.0 27 324.0
28 350.0 29 377.0 30 405.0
31 434.0 32 464.0 33 495.0
34 527.0 35 560.0 36 594.0
37 629.0 38 665.0 39 702.0
40 740.0 41 779.0 42 819.0
43 860.0 44 902.0 45 945.0
46 989.0 47 1034.0 48 1080.0
49 1127.0 50 1175.0
```

From the general functional equation, Equation 13.3, the functional equation for the network problem discussed can be expressed as

$$L_{n+1} = L_1 + (n + 1)n/2.$$

Notice that L_1 is always zero ($L_1 = 0$) because no link is necessary when there is only one computer ($n = 1$).

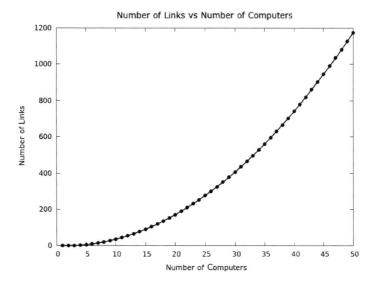

Figure 13.4 Number of links to connect n computers.

$$L_n = n(n-1)/2$$

13.5.3 Models with Sums of Arithmetic Growth

A variety of models have data about a property that follows an arithmetic growth pattern. The summation or running totals of the data of this property is also important. The following example will illustrate this concept.

A county maintains data about the cable installations for multi-purpose services, such as TV, phones, Internet access, and others. The data include new cable installations (in thousands) per year and the total number of cable installations per year. This data is shown in Table 13.2.

Table 13.2 Number of cable installations for years 1995–2002.

Year	1995	1996	1997	1998	1999	2000	2001	2002
New	1.5	1.9	2.3	2.7	3.1	3.5	3.9	4.3
Sum	1.5	3.4	5.7	8.4	11.5	15.0	18.9	25.2

For this type of data, the general principle is that the new cable installations follow an arithmetic growth pattern, and the summations of the cable installations follow a quadratic growth pattern. Two sequences are needed: a for the new cable installations, and b for the sums of cable installations.

$$a = \langle\, a_1, a_2, a_3, a_4, a_5 \ldots a_n \,\rangle \quad b = \langle\, b_1, b_2, b_3, b_4, b_5 \ldots b_n \,\rangle$$

To develop a difference equation and a functional equation of the sums of cable installations, a relation of the two sequences, a and b, needs to be expressed.

It can be observed from the data in Table 13.2 that the data in sequence a follows an arithmetic growth pattern. The difference equation for a can be expressed as:

$$a_n = a_{n-1} + \Delta a, \quad \Delta a = 400, \quad n = 2 \ldots m.$$

The functional equation for a can be expressed as:

$$a_n = 1500 + 400\,(n - 1).$$

It can also be observed from the data in Table 13.2 that the data in sequence b is related to the data in sequence a. The difference equation for sequence b is expressed as:

$$b_n = b_{n-1} + a_n, \quad n = 2 \ldots m.$$

Substituting a_n for the expression in its functional equation and using the general functional equation for quadratic growth, Equation 13.3:

$$b_n = b_1 + d_1\,(n - 1) + dd\,(n - 2)n/2.$$

Finally, the functional equation for the sequence b is expressed as:

$$b_n = 1500 + 1900n + 400(n - 1)n/2.$$

13.6 SUMMARY

This chapter presented some basic concepts of quadratic growth in mathematical models. Simple mathematical techniques, such as difference equations and functional equations, are used in the study of models with quadratic growth. In these models the increases follow an arithmetic growth pattern; the second differences are constants. The most important difference with models with arithmetic growth is that the increases cannot be represented by a straight line, as with arithmetic models. The functional equation of quadratic growth is basically a second-degree equation, also known as a quadratic equation.

Key Terms

quadratic growth	non-linear representation	differences
second differences	network problems	summation
quadratic equation	roots	coefficients

13.7 EXERCISES

13.1 On a typical spring day, the temperature varies according to the data recorded in Table 13.3. Develop a Python program to compute first and second differences. Formulate the difference equations and functional equations for the temperature. Is this discrete or continuous data? Discuss.

Table 13.3 Temperature changes in a 12-hour period.

Time	7	8	9	10	11	12	1	2	3	4	5	6
Temp. (F^0)	51	56	60	65	73	78	85	86	84	81	80	70

13.2 Develop a Python program to compute first and second differences. Produce a line chart of the data list in Table 13.1. Discuss how the number of patients changes in a specified period.

13.3 Formulate a mathematical model based on a difference equation of a modified computer network problem similar to one discussed in this chapter. There are three servers and several client computers connected via communication links. All connections between pairs of clients require two links. The connection between servers also requires two links. Use the concepts and principles explained in this chapter to derive an equation for the number of links.

13.4 Formulate a mathematical model based on a functional equation of a modified computer network problem similar to one discussed in this chapter. There are three servers and several client computers connected via communication links. All connections between pairs of clients require two links. The connection between servers also requires two links. Use the concepts and principles explained in this chapter to derive an equation for the number of links.

13.5 Formulate a mathematical model based on a functional equation of a modified computer network problem similar to one discussed in this chapter. There are K servers and several client computers connected via communication links. All connections between pairs of clients require two links. The connection of a server to the other $K - 1$ servers requires a single link. Use the concepts and principles explained in this chapter to derive an equation for the number of links.

## CHAPTER 14

Models with Geometric Growth

14.1 INTRODUCTION

This chapter presents an introduction to computational models in which the data follow a pattern of geometric growth. In such models, the data exhibit growth in such a way that in equal intervals of time, the data increase by an equal percentage or factor.

The difference and functional equations in models with geometric growth are explained; their use in mathematical modeling is illustrated with a few examples. Several computational models implemented with Python and the *NumPy* library are presented and discussed.

14.2 BASIC CONCEPTS

The data in the sequence represent some relevant property of the model and are expressed as a variable s. An individual value of variable s is known as a *term* in the sequence and is denoted by s_n. A sequence with m data values or terms, is written as follows:

$$\langle s_1, s_2, s_3, s_4, s_5, \ldots, s_m \rangle.$$

In models with geometric growth, the data increase (or decrease) by an equal percentage or *growth factor* in equal intervals of time. The difference equation that represents the pattern of geometric growth has the general form:

$$s_{n+1} = c \, s_n. \tag{14.1}$$

In Equation 14.1, the parameter c is constant and represents the *growth factor*, and n identifies an individual value such that $n \leq m$. With geometric growth, the data increases or decreases by a fixed factor in equal intervals.

14.2.1 Increasing Data with Geometric Growth

The data in a sequence will successively increase in value when the value of the growth factor is greater than 1. For example, consider a data sequence that exhibits geometric growth with a growth factor of 1.45 and a starting value of 50.0. The sequence with 8 terms is:

$$\langle\, 50.0, 72.5, 105.125, 152.43, 221.02, 320.48, 464.70, 673.82\,\rangle.$$

Figure 14.1 shows a graph of the data with geometric growth. Note that the data increases rapidly.

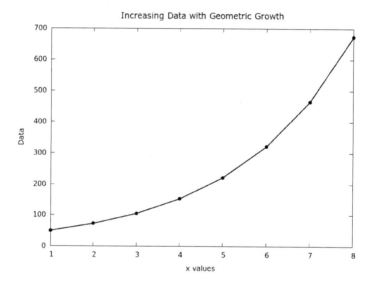

Figure 14.1 Data with geometric growth.

14.2.2 Decreasing Data with Geometric Growth

The data in a sequence will successively decrease in value when the value of the growth factor is less than 1. For example, consider a data sequence that exhibits geometric growth with a growth factor of 0.65 and a starting value of 850.0. The sequence with 10 terms is:

$$\langle\, 850.0, 552.5, 359.125, 233.43, 151.73, 98.62, 64.10, 41.66, 27.08, 17.60\,\rangle.$$

Figure 14.2 shows a graph of the data with geometric growth. Note that the data decreases rapidly because the growth factor is less than 1.0.

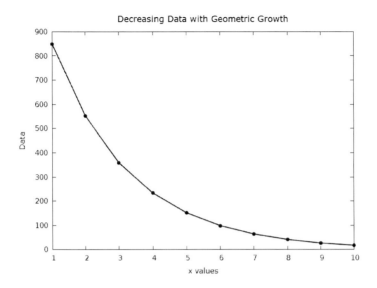

Figure 14.2 Data decreasing with geometric growth.

14.2.3   Case Study 1

The population of a small town is recorded every year; the increases per year are shown in Table 14.1, which gives the data about the population during the years from 1995 to 2003. The table also shows the population growth factor.

Table 14.1 Population of a small town during 1995–2003 (in thousands).

Year	1995	1996	1997	1998	1999	2000	2001	2002	2003
Pop.	81	90	130	175	206	255	288	394	520
Fac.	—	1.111	1.444	1.346	1.177	1.237	1.129	1.368	1.319

Note that although the growth factors are not equal, the data can be considered to grow in a geometric pattern. The values of the growth factor shown in the table are sufficiently close and the average growth factor calculated is 1.2667.

After a brief analysis of the data in the problem, the following tasks are to be computed: (1) create the data lists (arrays) of the sequence s with the values of the original data in Table 14.1; (2) compute the average growth factor from the data in the table; (3) compute the values of a second data list y using 1.267 as the average growth factor and the difference equation $y_{n+1} = 1.267\, y_n$, and (4) plot the graphs.

Listing 14.1 shows the Python program that performs these tasks and is stored in file: `popstown.py`. The factors of the population data are computed in line 19 with a call to function *factors* and the average growth factor is computed in line 22 of the program. The computed data is calculated in lines 25–27.

Listing 14.1: Program that computes the population average growth factor.

```
1 # program: popstown.py
2 # This program computes the growth factor per year
  of the population.
3 #   Uses NumPy
4 #   J Garrido 09-2-2014
5
6 import numpy as np
7
8 def factors(marray):  # Compute factors in marray
9     n = marray.size
10    mf = np.zeros(n-1)  # array of factors
11    for j in range(n-1):
12         mf[j] = marray[j+1]/marray[j]
13    return mf
14
15 N = 9
16 s = np.array([81.0, 90.0, 130.0, 175.0, 206.0, 255.0, 288.0,
   394.0, 520.0])
17 x = np.array([1995, 1996, 1997, 1998, 1999, 2000, 2001,
   2002, 2003])
18 print "Program to compute average growth factor"
19 f = factors(s)   # compute array of factors
20 print "\nFactors in array s: \n"
21 print f
22 meanv = np.mean(f)
23 print "\nMean factor: ", meanv
24 cs = np.zeros(N)
25 cs[0] = s[0]
26 for j in range(N-1):
27     cs[j+1] = meanv * cs[j];
28 print "\nGiven and Computed values of population"
29 for j in range(N):
30     print x[j], s[j], cs[j]
```

When the Python interpreter processes the program, the following listing is produced.

```
Program to compute average growth factor
```

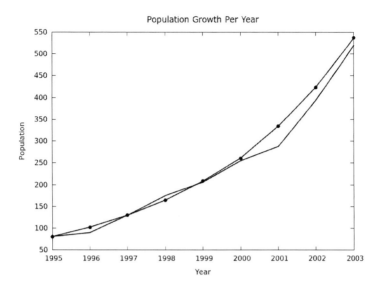

Figure 14.3 Population of a small town for 1995–2003.

```
Factors in array s:

[ 1.11111111  1.44444444  1.34615385  1.17714286  1.23786408
    1.12941176  1.36805556  1.31979695]

Mean factor:   1.26674757639

Given and Computed values of population
1995 81.0 81.0
1996 90.0 102.606553687
1997 130.0 129.976603205
1998 175.0 164.647547097
1999 206.0 208.566881243
2000 255.0 264.201591329
2001 288.0 334.676725494
2002 394.0 423.950930893
2003 520.0 537.038814216
```

Figure 14.3 shows a graph with two curves; one with the population in the town by year from 1995 through 2003 taken directly from Table 14.1. The other curve shown in the graph of Figure 14.3 is the computed data applying Equation 14.1 with 1.267 as the growth factor, using the difference equation $s_{n+1} = 1.267\ s_n$.

A very similar python program is stored in file **popstown2.py**, which reads the population data from a text file and produces the same results.

14.2.4 Case Study 2

In a water treatment process, every application of solvents removes 65% of impurities from the water to make it more acceptable for human consumption. This treatment has to be performed several times until the level of purity of the water is adequate for human consumption. Assume that when the water has less than 0.6 parts per gallon of impurities, it is adequate for human consumption.

In this problem, the data given is the contents of impurities in parts per gallon of water. The initial data is 405 parts per gallon of impurities and the growth factor is 0.35.

Listing 14.2 shows a Python program that computes the impurities of water after each application of the solvent. The program declares the data lists (arrays) of the sequence *s* with the data of the contents of impurities in parts per gallon of water. The output data produced by executing the program and GnuPlot is used to plot the graphs. The program is stored in file: `watertr.py`.

Listing 14.2: Program that computes the impurities of water.

```
1 #  program: watertr.py
2 #  This program computes the impurities in water
     given the growth factor.
3 #  Uses NumPy
4 #  J Garrido 9-2-2014
5 import numpy as np
6
7 N = 10     # Number of applications
8 f = 0.35   # constant factor
9 s = np.zeros(N)    # create array with zeros
10 s[0] = 405.0       # initial impurity of water
11 print "Program to compute growth of impurities in water\n"
12 napp = np.arange(N)
13 napp = napp + 1   # application number
14 print "Initial impurity in water: ", s[0]
15 print "Number of applications: ", N
16 print "Factor: ", f
17 for j in np.arange(N-1):   # compute impurities
18        s[j+1] = f * s[j]
19 print "\nImpurities after each application \n"
20 for j in np.arange(N):
21        print napp[j], s[j]
```

When the Python interpreter processes the program, the following output listing is produced. Figure 14.4 shows the graph of the impurities in parts per gallon of water for several applications of solvents.

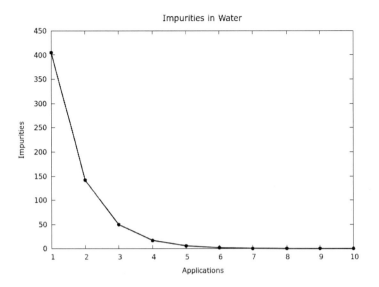

Figure 14.4 Impurities in water (parts/gallon).

```
Program to compute growth of impurities in water

Initial impurity in water:   405.0
Number of applications:   10
Factor:   0.35

Impurities after each application

1 405.0
2 141.75
3 49.6125
4 17.364375
5 6.07753125
6 2.1271359375
7 0.744497578125
8 0.260574152344
9 0.0912009533203
10 0.0319203336621
```

14.3   FUNCTIONAL EQUATIONS IN GEOMETRIC GROWTH

From the difference equation for models with geometric growth, Equation 14.1, the first few terms of sequence s can be written as:

$$s_2 = cs_1$$
$$s_3 = cs_2 = c(cs_1)$$
$$s_4 = cs_3 = c(c(cs_1))$$
$$s_5 = cs_4 = c(c(c(cs_1)))$$
$$s_6 = cs_5 = cc(c(c(cs_1)))$$
$$\ldots$$
$$s_n = c^{n-1}s_1.$$

Equation 14.1 was referenced by substituting s_{n-1} for its difference equation, and continuing this procedure up to s_n. In this manner, a functional equation can be derived. Recall that a functional equation gives the value of a term s_n without using the previous value, s_{n-1}. The following mathematical expression is a general functional equation for geometric growth models.

$$s_n = s_1 \, c^{n-1} \tag{14.2}$$

Equation 14.2 gives the value s_n as a function of n for a geometric growth model, with $n \geq 1$. Note that this functional equation includes the fixed value s_1, which is the value of the first term of the data sequence.

A functional equation such as Equation 14.2 is an example of an exponential function because the independent variable, n, is the exponent. This type of growth in the data is also known as *exponential growth.*

Functional equations can be used to answer additional questions about a model. For example: what will the population be 12 years from now? What amount of impurities are left in the water after 8 repetitions of the application of solvents?

When the growth factor does not correspond to the desired unit of time, then instead of n, a more appropriate variable can be used. For example, in the first population data in Case Study 1, Section 14.2.3, the variable n represents number of years. To deal with months instead of years, a small substitution in the functional equation is needed. Variable t will represent time, and the starting point of the data is at $t = 0$ with an initial value of y_0. This gives meaning to the concept of a continuous model. Because one year has 12 months and using the same growth factor c as before, the following is a modified functional equation and can be applied when dealing with months.

$$y(t) = y_0 \, c^{t/12} \tag{14.3}$$

14.4 SUMMARY

This chapter presented some basic concepts of geometric growth in mathematical models. The data in these models increase or decrease in a constant growth factor for equal intervals. Simple mathematical techniques such as difference equations and functional equations are used in the study of models with geometric growth. The functional equation of geometric growth is basically an exponential function, so this type of growth is also known as exponential growth.

Some important applications involving computational models with geometric growth are pollution control, human drug treatment, population growth, radioactive decay, and heat transfer.

Key Terms

geometric growth	exponential growth	average growth
growth factor	exponent	continuous data
exponentiation rules	logarithms	exponent base

14.5 EXERCISES

14.1 In the population problem Case Study 1 Section 14.2.3, use the average growth factor already calculated and compute the population up to year 14. For this, modify the corresponding Python program and run again. Draw the graphs with GnuPlot.

14.2 In the population problem Case Study 1 Section 14.2.3, use the average growth factor already calculated and compute the population up to year 20. For this, modify the corresponding Python program and run again. Draw the graphs with GnuPlot.

14.3 In the population problem Case Study 1 Section 14.2.3, estimate the population for year 18 and month 4. Use the average growth factor already calculated and the modified functional equation, Equation 14.3. Develop a Python program to solve this problem.

14.4 In the population problem Case Study 1 Section 14.2.3, estimate the population for month 50. Use the average growth factor already calculated and the modified functional equation, Equation 14.3. Develop a Python program to solve this problem.

14.5 In the population problem Case Study 1 Section 14.2.3, compute the year when the population reaches $750,000$. Use the average growth factor and develop a Python program to solve this problem.

14.6 In the population problem Case Study 1 Section 14.2.3, compute the month when the population reaches $875,000$. Use the average growth factor already calculated and the modified functional equation, Equation 14.3. Develop a Python program to solve this problem.

14.7 In the modified water treatment problem Case Study 2 Section 14.2.4, an application of solvents removes 57% of impurities in the water. Compute the levels of impurities after several repetitions of the application of solvents. Use this growth factor and develop a Python program to solve this problem.

14.8 In the original water treatment problem Case Study 2 Section 14.2.4, compute the levels of impurities after 8 repetitions of the application of solvents. Use this growth factor and develop a Python program to solve this problem.

14.9 In the modified water treatment problem, Exercise 14.8, compute the levels of impurities after 8 repetitions of the application of solvents. Use the growth factor and develop a Python program to solve this problem.

14.10 In the original water treatment problem Case Study 2 Section 14.2.4, compute the number of repetitions of the application of solvents that are necessary to reach 0.5 parts per gallon of impurities. Use the growth factor and develop Python program to solve this problem.

## CHAPTER 15

Computational Models with Polynomial Growth

15.1 INTRODUCTION

Linear and quadratic equations are special cases of polynomial functions. These equations are of higher order than quadratic equations and more general mathematical methods are used to solve them.

This chapter presents general concepts and techniques to evaluate and solve polynomial functions with emphasis on numerical and graphical methods. Solutions implemented with the Python programming language and the *polynomial* module of the *Numpy* package are presented and discussed.

15.2 GENERAL FORMS OF POLYNOMIAL FUNCTIONS

Linear and quadratic equations are special cases of polynomial functions. The degree of a polynomial function is the highest degree among those in its terms. A linear function such as $y = 3x + 8$, is a polynomial equation of degree 1 and a quadratic equation such as $y = 4.8x^2 + 3x + 7$, is a polynomial function of degree 2; thus the term "second-degree equation."

A function such as $y = 2x^4 + 5x^3 - 3x^2 + 7x - 10.5$, is a polynomial function of degree 4 because 4 is the highest exponent of the independent variable x. A polynomial function has the general form:

$$y = p_1 x^n + p_2 x^{n-1} + p_3 x^{n-2} + \ldots p_{k-1} x + p_k.$$

This function is a polynomial equation of degree n, and $p_1, p_2, p_3, \ldots p_k$ are the coefficients of the equation and are constant values. Another general form of a polynomial function *P(x)* is:

$$P(x) = c_0 + c_1 x + c_2 x^2 + \cdots + c_{n-1} x^{n-1} + c_n x^n.$$

In this second form of a polynomial, $c_0, c_1, c_2, c_3, \ldots c_n$ are the coefficients of the equation and are constant values. In addition to the algebraic form of a polynomial function, the graphical form is also important; a polynomial function is represented by a graph. As mentioned previously, *Gnuplot* is used for plotting data.

15.3   THE *polynomial* MODULE OF THE *Numpy* PACKAGE

The *polynomial* module of the *Numpy* package provides several functions for manipulating polynomial series. A polynomial function is represented by a vector of *coefficients* in ascending order. The following is a list of the relevant functions for dealing with polynomials.

`polyval(x,c[,tensor])`	Evaluate a polynomial at points x.
`polyval2d(x,y,c)`	Evaluate a 2-D polynomial at points (x, y).
`polyval3d(x,y,z,c)`	Evaluate a 3-D polynomial at points (x, y, z).
`polygrid2d(x,y,c)`	Evaluate a 2-D polynomial on the Cartesian product of x and y.
`polygrid3d(x,y,z,c)`	Evaluate a 3-D polynomial on the Cartesian product of x, y and z.
`polyroots(c)`	Compute the roots of a polynomial.
`polyfromroots(roots)`	Generate a monic polynomial with given roots.
`polyfit(x,y,deg [,rcond,full,w])`	Least-squares fit of a polynomial to data.
`polyvander(x,deg)`	Vandermonde matrix of given degree.
`polyvander2d(x,y,deg)`	Pseudo-Vandermonde matrix of given degrees.
`polyvander3d(x,y,z, deg)`	Pseudo-Vandermonde matrix of given degrees.
`polyder(c[,m,scl, axis])`	Differentiate a polynomial.
`polyint(c[,m,k,lbnd, scl,axis])`	Integrate a polynomial.
`polyadd(c1,c2)`	Add one polynomial to another.
`polysub(c1,c2)`	Subtract one polynomial from another.
`polymul(c1,c2)`	Multiply one polynomial by another.
`polymulx(c)`	Multiply a polynomial by x.
`polydiv(c1,c2)`	Divide one polynomial by another.
`polypow(c,pow[, maxpower])`	Raise a polynomial to a power.
`polyline(off,scl)`	Returns an array representing a linear polynomial.

15.4 EVALUATION OF POLYNOMIAL FUNCTIONS

With a computer program, a relatively large number of values of x can be used to evaluate the polynomial function for every value of x. The set of values of x that are used to evaluate a polynomial function are taken from an interval $a \leq x \leq b$, where a is the lower bound of the interval and b is the upper bound. The interval is known as the *domain* of the polynomial function. In a similar manner, the interval of the values of the function y is known as the *range* of the polynomial function.

A polynomial function is evaluated by using various values of the independent variable x and computing the value of the dependent variable y. In a general mathematical sense, a polynomial function defines y as a function of x. With the appropriate expression, several values of the polynomial can be computed and graphs can be produced.

The functions in the *polynomial* module of the Numpy package evaluate a polynomial, $P(x)$, using Horner's method for stability. Using the general form of the polynomial of degree n,

$$P(x) = c_0 + c_1 x + c_2 x^2 + \cdots + c_{n-1} x^{n-1} + c_n x^n.$$

Function *polyval* evaluates polynomials with real coefficients for the real variable x. The function returns the values of $P(x)$ for the given values in list x. The following assignment statement has a call to function *polyval* that evaluates a polynomial at the values in list x with the coefficients in list c .

```
y = polyval (x, c)
```

The function computes the values of the polynomial (values of y) for all the given values of x. The following example evaluates the polynomial function $y = 2x^3 - 3x^2 - 36x + 14$ and the values in the two arrays x and y are computed by the program in file **polyval1.py**. Only 20 values of x are evaluated in this example. The following listing is the output produced by the program.

```
$ python polyval1.py
Coefficient list
[ 14. -36.  -3.   2.]
Evaluating a polynomial
-6.0 -310.0
-5.31578947368 -179.827525878
-4.63157894737 -82.3265782184
-3.94736842105 -13.6534480245
-3.26315789474 30.0355736988
-2.57894736842 52.5841959469
-1.89473684211 57.8361277154
-1.21052631579 49.6350779997
```

```
-0.526315789474 31.8247557953
0.157894736842 8.24887009768
0.842105263158 -17.2488700977
1.52631578947 -40.8247557953
2.21052631579 -58.6350779997
2.89473684211 -66.8361277154
3.57894736842 -61.5841959469
4.26315789474 -39.0355736988
4.94736842105 4.65344802449
5.63157894737 73.3265782184
6.31578947368 170.827525878
7.0 301.0
```

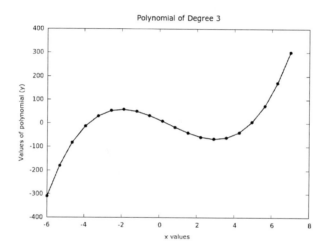

Figure 15.1 Graph of the equation $y = 14 - 36x - 3x^2 + 2x^3$.

Listing 15.1 shows the source code of the Python program in file polyval1.py. The program evaluates the polynomial and computes the values in vector y. Line 18 defines the array with the polynomial coefficients: a, b, c, and d. Line 10 defines array x with 20 different values from the interval $-6.0 \leq x \leq 7.0$. Line 14 calls function *polyval* that computes the value of the function for every value in array x.

Listing 15.1: A program that computes the values of a polynomial.

```
1 # This program evaluates a polynomial given
2 #   the coefficients in list c
3 # at points in list x
4 # J M Garrido, August 2, 2014
5 import numpy as np
```

```
 6 from numpy.polynomial.polynomial import polyval
 7 M = 20     # number of data points
 8 xi = -6.0  # first value of x
 9 xf = 7.0   # final value
10 x = np.linspace(xi, xf, M)
11 c = np.array([14.0, -36.0, -3.0, 2.0])
12 print "Coefficient list"
13 print c
14 y = polyval(x, c)
15 print "Evaluating a polynomial"
16 for j in range(M):
17     print x[j], y[j]
```

A graph is easily produced with Gnuplot using the values in arrays of x and y computed. Figure 15.1 shows the graph of the polynomial function $y = 14 - 36x - 3x^2 + 2x^3$ with the data computed previously. The graph is produced with the script files polyval1.cgp.

To evaluate the polynomial equation $y = -2 + 3x^5$, the corresponding coefficient vector is $[-2.0, 0, 0, 0, 0, 3.0]$. The program calls function *polyval* to evaluate the polynomial in the interval $-2.5 \leq x \leq 2.5$. The program is basically the same as the previous example. The following listing shows the Python interpreter running the program.

```
$ python polyval2.py
Coefficient list
[-2.  0.  0.  0.  0.  3.]
Evaluating polynomial
-2.5 -294.96875
-2.23684210526 -169.995597296
-1.97368421053 -91.8482429544
-1.71052631579 -45.9308953696
-1.44736842105 -21.0553407421
-1.18421052632 -8.98659937214
-0.921052631579 -3.98858195139
-0.657894736842 -2.36974585578
-0.394736842105 -2.02875143775
-0.131578947368 -2.00011831867
0.131578947368 -1.99988168133
0.394736842105 -1.97124856225
0.657894736842 -1.63025414422
0.921052631579 -0.011418048612
1.18421052632 4.98659937214
1.44736842105 17.0553407421
1.71052631579 41.9308953696
```

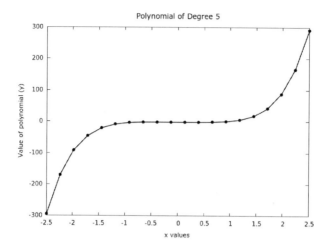

Figure 15.2 Graph of the equation $y = -2 + 3x^5$.

```
1.97368421053 87.8482429544
2.23684210526 165.995597296
2.5 290.96875
```

In a similar manner, executing the command file `polyval2.cgp` with Gnuplot produces a graph of polynomial equation $y = -2 + 3x^5$. Figure 15.2 shows the graph of this polynomial function.

15.5 SOLVING POLYNOMIAL FUNCTIONS

The solution to a quadratic equation, which is a second-degree equation, is relatively straightforward. The solution to this equation involves finding two values of x that give y value zero. These two values of x are known as the *roots* of the function. For higher-order polynomial functions, the degree of the polynomial determines the number of roots of the function. A polynomial function of degree 7 will have 7 roots.

It is not generally feasible to find the roots of polynomial equations of degree 4 or higher by analytical methods.

Function *polyroots* of the *polynomial* module computes the complex roots of a polynomial function. This function takes the coefficients vector of the polynomial function as the argument, and returns a vector with the roots. This vector is known as the *roots vector*. The function implements an algorithm that uses an iterative method to find the approximate locations of roots of polynomials. The only parameter of this function is the array of coefficients of the polynomial function.

Listing 15.2 shows the Python source code of a program that computes the roots of the polynomial function $3+3x-14x^2-17x^3+23x^4$. The program is stored in file `polyrootsp.py`.

Listing 15.2: A program that computes the roots of a polynomial function.

```
 1 # This program computes the roots of a polynomial
 2 #  given the coefficients in list c
 3 # J M Garrido, Aug 2, 2014. Program: polyrootsp.py
 4 import numpy as np
 5 import numpy.polynomial.polynomial as poly
 6 c = np.array([3.0, 3.0, -14.0, -17.0, 23.0])
 7 print "Solving a polynomial"
 8 print "Coefficient list"
 9 print c
10 r = poly.polyroots(c)
11 print "Roots of the polynomial"
12 print r
```

The following output was produced after starting the Python interpreter and running the program. The solution vector, r, has the values of the (complex) roots of the polynomial function.

```
$ python polyrootsp.py
Solving a polynomial
Coefficient list
[  3.    3. -14. -17.   23.]
Roots of the polynomial
[-0.43665695-0.19425443j -0.43665695+0.19425443j
  0.52528824+0.j  1.08715609+0.j ]
```

For the polynomial function: $y = 2x^3 - 3x^2 - 36x + 14$, the following output was produced after starting the Python interpreter and running the program. The solution vector, r, has the values of the (real) roots of the polynomial function. The individual values of vector r are $r_0 = -3.7688$, $r_1 = 0.3799$, and $r_2 = 4.8889$.

```
$ python polyroots2.py
Solving a polynomial
Coefficient list
[ 14 -36  -3   2]
Roots of the polynomial
[-3.76883165  0.37990763  4.88892403]
```

The solution to the polynomial equation $y = -2 + 3x^5$ is computed in a similar manner. Four of the roots computed are complex values, which are known as *complex roots*.

```
$ python polyroots3.py
Solving a polynomial
Coefficient list
[-2.  0.  0.  0.  0.  3.]
Roots of the polynomial
[-0.74600097-0.54200143j -0.74600097+0.54200143j
  0.28494702-0.87697674j  0.28494702+0.87697674j
  0.92210791+0.j         ]
```

15.6 SUMMARY

This chapter presented some basic techniques for solving polynomial equations that are very useful in computational modeling. Linear and quadratic equations are special cases of polynomial equations. The concepts discussed apply to polynomial functions of any degree. The main emphases of the chapter are evaluation of a polynomial function, the graphs of these functions, and solving the functions by computing the roots of the polynomial functions. The list of functions in the *polynomial* module of the *Numpy* package was taken from the *scipy*[1] web page.

Key Terms

polynomial functions	polynomial evaluation	roots
coefficient vector	root vector	evaluation interval
function domain	function range	variable interval

15.7 EXERCISES

15.1 Develop a Python program to evaluate the polynomial function $y = x^4 + 4x^2 + 7$. Find an appropriate interval of x for which the function evaluation is done and plot the graph.

15.2 Develop a Python program to evaluate the polynomial function $y = 3x^5 + 6$. Find an appropriate interval of x for which the function evaluation is done and plot the graph.

15.3 Develop a Python program to evaluate the polynomial function $y = 2x^6 - 1.5x^5 + 5x^4 - 6.5x^3 + 6x^2 - 3x + 4.5$. Find an appropriate interval of x for which the function evaluation is done and plot the relevant data.

---

[1]docs.scipy.org/doc/numpy/reference/routines.polynomials.polynomial.html.

15.4 Develop a Python program to solve the polynomial function $y = x^4 + 4x^2 + 7$.

15.5 Develop a Python program to solve the polynomial function $y = 3x^5 + 6$.

15.6 Develop a Python program to solve the polynomial function $y = 2x^6 - 1.5x^5 + 5x^4 - 6.5x^3 + 6x^2 - 3x + 4.5$.

## CHAPTER 16

Empirical Models with Interpolation and Curve Fitting

16.1 INTRODUCTION

The mathematical model in a computational model can be expressed as a set of polynomial functional equations of any order. If only raw data is available, estimates of the values of the function can be computed for other values of the independent variable, within the bounds of the available set of values of the independent variable. Computing these estimates is carried out using *interpolation* techniques.

If the mathematical model in the form a polynomial function is needed to represent the raw data, then *curve fitting*, also known as *regression*, techniques are used. In the general case, the coefficients of the polynomial can be computed (estimated) for a polynomial function of any degree. This chapter discusses two general numerical techniques that help estimate data values: interpolation and curve fitting. Solutions are implemented with the Scipy, Numpy, and its *polynomial* module used in the Python programming language.

16.2 INTERPOLATION

The given or raw data in a problem usually provides only a limited number of values of data points (x, y). These are normally values of a function y for the corresponding values of variable x. These intermediate values of x and y are not part of the original data. Two well-known interpolation techniques are:

- linear interpolation

- cubic spline interpolation

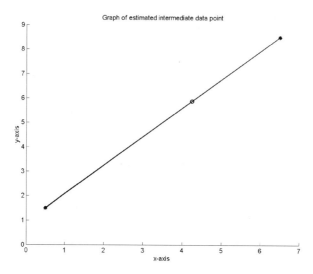

Figure 16.1 Linear interpolation of an intermediate data point.

16.2.1 Linear Interpolation

In the linear interpolation technique, the assumption is that the intermediate data point (value of the function y for a value of x), between two known data points: (x_1, y_1) and (x_2, y_2), can be estimated by a straight line between the known data points. In other words, the intermediate data point (x_i, y_i) falls on the straight line between the known points (x_1, y_1) and (x_2, y_2).

Assume that there are two data points: $(0.5, 1.5)$ and $(6.5, 8.5)$. An intermediate data point between the given points is to be estimated for $x = 4.25$. Applying a linear interpolation technique, the estimated value computed for y is 5.875. The new intermediate data point is therefore $(4.25, 5.875)$. Figure 16.1 illustrates the technique of estimating an intermediate data point on a straight line between two given data points.

There are several functions in Scipy and Numpy that perform linear interpolation given three vectors: x, y, and $xint$. The first two vectors store the values of the given data points. The third vector, $xint$, stores the new or intermediate values of x for which estimates are to be computed and these are new or intermediate data points. In the previous example, vector $xint$ has only a single value, 4.25, and the value of $yint$ computed is 5.875.

The Python program in Listing 16.1 computes the values of the intermediate data points and stores these in vector $yint$. Line 17 in the listing calls function *linterp* to compute the estimated intermediate points using linear interpolation. The program is stored in file `linterp1.py` and the program with additional statement that plot the graph is stored in `linterp1_plot.py`.

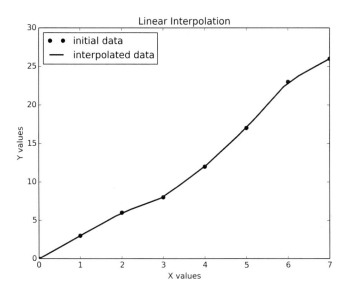

Figure 16.2 Graph of linear interpolation of data points.

Listing 16.1: Program that computes linear interpolation of data points.

```
1 # Program that performs linear interpolation
2 import numpy as np
3
4 y = [0.0, 3.0, 6.0, 8.0, 12.0, 17.0, 23.0, 26]
5 y = np.array(y)
6 print "Values of array y:"
7 print y
8 xi = 0.0
9 xf = 7.0
10 M = np.size(y)
11 x = x = np.linspace(xi, xf, M)
12 print "Values of array x:"
13 print x
14 # generate an array of 20 intermediate points
15 N = 20
16 xint = np.linspace(xi, xf, N)
17 yint = np.interp(xint, x, y)
18 print "Interpotated values of x and y:"
19 for j in range(N):
20     print xint[j], yint[j]
```

In the problem, there are eight given data points in arrays x and y. Array x has equally spaced values of x starting at 0 and increasing by 1. Linear interpolation is used to estimate intermediate data points for every value of x spaced 0.25, starting at 0 and up to 7. The values given in the problem are stored in the three given vectors, x, y, and $xint$, and the values in vector $yint$ are computed using linear interpolation. The values given and the values computed using linear interpolation are shown as follows:

```
$ python interp1.py
Values of array y:
[ 0.   3.   6.   8.  12.  17.  23.  26.]
Values of array x:
[ 0.  1.  2.  3.  4.  5.  6.  7.]
Interpolated values of x and y:
0.0 0.0
0.368421052632 1.10526315789
0.736842105263 2.21052631579
1.10526315789 3.31578947368
1.47368421053 4.42105263158
1.84210526316 5.52631578947
2.21052631579 6.42105263158
2.57894736842 7.15789473684
2.94736842105 7.89473684211
3.31578947368 9.26315789474
3.68421052632 10.7368421053
4.05263157895 12.2631578947
4.42105263158 14.1052631579
4.78947368421 15.9473684211
5.15789473684 17.9473684211
5.52631578947 20.1578947368
5.89473684211 22.3684210526
6.26315789474 23.7894736842
6.63157894737 24.8947368421
7.0 26.0
```

Figure 16.2 shows a plot of the data given in the problem and the interpolated data using the linear interpolation of several intermediate data points given two arrays of the original data points x and y.

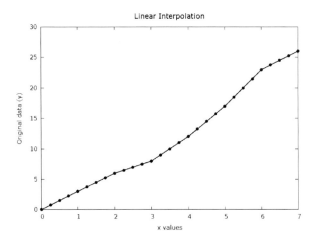

Figure 16.3 Graph of linear interpolation of many data points.

16.2.2 Non-Linear Interpolation

Non-linear interpolation can generate better estimates for intermediate data points than linear interpolation. The Python Scipy package includes functions that implement the *cubic spline* interpolation technique, and smoother curves can be generated using this technique.

Using the Scipy *interpolation* module, two steps are required: (1) a spline representation of the curve is computed, and (2) the spline is evaluated at the desired points. Function *splrep* finds the spline representation of a curve with a direct method. With the computed spline representation, function *splev* computes the intermediate values desired. More detailed documentation is found in the *scipy*[1] web page.

The following example applies the *cubic spline* interpolation technique to data provided by the problem, using an array of values of nx for intermediate data points. The Python program in file `csinterp.py` is shown in Listing 16.2. The statement in line 12 calls function *splrep* to compute spline representation of the curve given by x and y. Function *splev* is called in line 15 to compute the interpolation using the cubic spline technique. The program that includes additional statements that plot the curves is stored in file `csinterp1_plot.py`.

Listing 16.2: Program to compute cubic spline interpolation of data.

```
1 # Cubic spline interpolation
2 #File: csinterp.py   Sep 2, 2014
3 import numpy as np
4 from scipy import interpolate
```

[1]http://docs.scipy.org/doc/scipy/reference/interpolate.html.

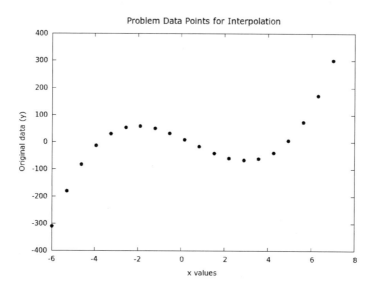

Figure 16.4 Graph of given data points.

```
 5
 6 y = [-310.0, -179.8, -82.3, -13.6, 30.0, 52.6,57.8, 49.6,
   31.8, 8.2, -17.2,-40.8, -58.6, -66.8, -61.5,-39.0, 4.6,
   73.3, 170.8, 301.0]
 7 y = np.array(y)
 8 N = np.size(y)
 9 xi = -6.0
10 xf = 7.0
11 x = np.linspace(xi, xf, N)
12 sprep = interpolate.splrep(x, y, s=0) # spline of y
13 M = int(1.5 * N)   # more points
14 xint = np.linspace(xi, xf, M)
15 yint = interpolate.splev(xint, sprep, der=0) # interp
16 print "Cubic Spline interpolation:"
17 for j in range(M):
18      print xint[j], yint[j]
```

Figure 16.4 shows the graph of original data points that define a curve. Figure 16.5 shows the graph of the original data points and the computed intermediate data points. Executing the program produces the output shown in the following listing.

```
$ python csinterp.py
Cubic Spline interpolation:
```

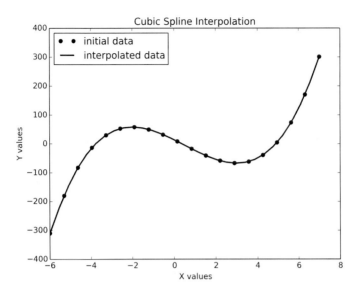

Figure 16.5 Cubic spline interpolation of data points.

```
-6.0 -310.0
-5.55172413793 -220.800476594
-5.10344827586 -146.22867376
-4.65517241379 -85.1612803181
-4.20689655172 -36.4975182072
-3.75862068966 0.761529182443
-3.31034482759 27.7118155661
-2.86206896552 45.5635333789
-2.41379310345 55.3026524832
-1.96551724138 57.9519490911
-1.51724137931 54.684292212
-1.06896551724 46.582496444
-0.620689655172 34.6890729458
-0.172413793103 20.0598682544
0.275862068966 3.84627383363
0.724137931034 -12.8459940601
1.1724137931 -29.0608305789
1.62068965517 -43.6881162489
2.06896551724 -55.5794898433
2.51724137931 -63.6961481782
2.96551724138 -66.9422754367
3.41379310345 -64.2137664572
3.86206896552 -54.4926582486
```

```
4.31034482759 -36.7167186662
4.75862068966 -9.76909231321
5.20689655172 27.5012343371
5.65517241379 76.1610779244
6.10344827586 137.227806359
6.55172413793 211.801416278
7.0 301.0
```

16.3   CURVE FITTING

Curve fitting techniques attempt to find the best polynomial expression that represents a given sequence of data points. The most widely used curve fitting technique is the *least squares* technique. Recall that a polynomial function has the general form:

$$y = p_1 x^n + p_2 x^{n-1} + p_3 x^{n-2} + \ldots p_{k-1} x + p_k.$$

The parameters $p_1, p_2, p_3, \ldots p_k$ are the coefficients of the equation, which are constants. If a polynomial function of degree 1 is fitted to the given data, the technique is known as *linear regression*, and a straight line is fitted to the data points. If the degree of the polynomial is greater than 1, then a curve, instead of a straight line, is fitted to the given data points.

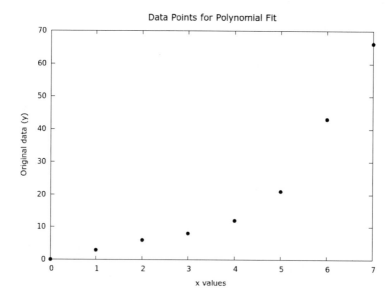

Figure 16.6   Graph of a given set of data points.

The Numpy package in the Python programming language provides function *polyfit* to compute the coefficients (p) of the polynomial function of degree n. The arguments for the function calls are arrays x and y that define the data points, and the value of the desired degree of the polynomial. The *polyfit* function computes an array that consists of the values of the coefficients of a polynomial function of degree n that best fit the given data points in vectors x and y. This array has values that correspond to the coefficients with the highest power first (descending order).

16.3.1 Linear Polynomial Function

The following example illustrates the fitting of a polynomial of degree 1 to a given set of data points. Figure 16.6 shows a given set of data points for which polynomial fit is to be carried out.

Listing 16.4 includes the Python program that computes the coefficients of a polynomial of degree 1 by calling function *polyfit* in line 20. With the coefficients computed, function *polyval* is called in line 25 to evaluate the polynomial with additional data points. The source code of the program code is stored in file **polyf1.py**. Using the *matplotlib* package, the computed data points are plotted and Figure 16.7 shows the graph of the fitted line.

Listing 16.4: Program that fits a linear polynomial to data points.

```
1 # This program fits a polynomial of degree 1
2 # to a set of data points. Then it evaluates a polynomial
3 # given the coefficients in list c
4 # at points in list x. File: polyf1.py
5 # J M Garrido, August 12, 2014
6 import numpy as np
7 from numpy.polynomial.polynomial import polyval
8
9 M = 30       # number of data points to compute
10 xi = 0.0    # first value of x
11 xf = 7.0    # final value of x
12 y = [0.0, 3.0, 6.0, 8.0, 12.0, 21.0, 43.0, 66.0]
13 y = np.array(y)
14 sn = np.size(y) # number of points
15 x = np.linspace(xi, xf, sn)
16 deg = 1   # degree of polynomial function
17 print "Values of X and Y:"
18 for j in range (sn):
19     print x[j], y[j]
20 c = np.polyfit(x, y, deg)
21 print "Coefficient list:"
22 print c
23 # Evaluate polynomial
```

```
24 xc = np.linspace(xi, xf, M) # new x points
25 yc = np.polyval(c, xc) # new y points
26 print "Evaluation of polynomial"
27 for j in range(M):
28       print xc[j], yc[j]
```

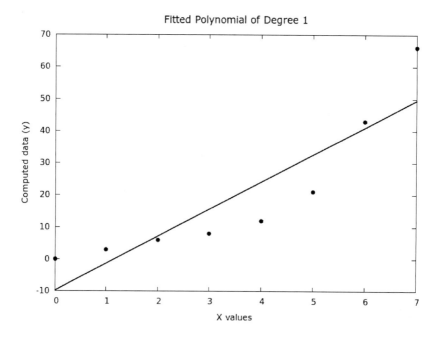

Figure 16.7 Polynomial of degree 1 fit from given data points.

Executing the Python interpreter with the `polyf1.py` program, the output produced are shown in the following listing.

```
$ python polyf1_plot.py
Values of X and Y:
0.0 0.0
1.0 3.0
2.0 6.0
3.0 8.0
4.0 12.0
5.0 21.0
6.0 43.0
7.0 66.0
Coefficient list:
[ 8.46428571 -9.75      ]
```

```
Evaluation of polynomial
0.0 -9.75
0.241379310345 -7.70689655172
0.48275862069 -5.66379310345
0.724137931034 -3.62068965517
0.965517241379 -1.5775862069
1.20689655172 0.465517241379
1.44827586207 2.50862068966
1.68965517241 4.55172413793
1.93103448276 6.59482758621
2.1724137931 8.63793103448
2.41379310345 10.6810344828
2.65517241379 12.724137931
2.89655172414 14.7672413793
3.13793103448 16.8103448276
3.37931034483 18.8534482759
3.62068965517 20.8965517241
3.86206896552 22.9396551724
4.10344827586 24.9827586207
4.34482758621 27.025862069
4.58620689655 29.0689655172
4.8275862069 31.1120689655
5.06896551724 33.1551724138
5.31034482759 35.1982758621
5.55172413793 37.2413793103
5.79310344828 39.2844827586
6.03448275862 41.3275862069
6.27586206897 43.3706896552
6.51724137931 45.4137931034
6.75862068966 47.4568965517
7.0 49.5
```

16.3.2   Fitting Non-Linear Polynomial Functions

Polynomial functions with degree 2 and higher are considered non-linear functions. Executing the program to compute the coefficients of a polynomial of degree 3 by calling function *polyfit* using 3 as the value of the third argument (degree of the polynomial), produces the following output.

```
$ python polyf2.py
Values of X and Y:
0.0 0.0
1.0 3.0
2.0 6.0
```

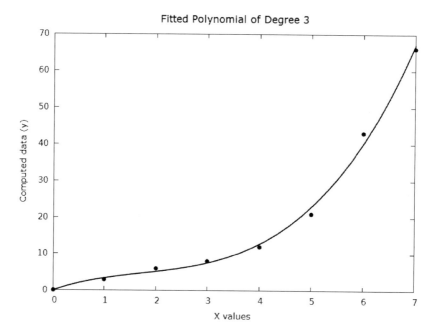

Figure 16.8 Curve fitting of a polynomial of degree 3.

```
3.0 8.0
4.0 12.0
5.0 21.0
6.0 43.0
7.0 66.0
Coefficient list:
[ 0.36616162 -1.8982684    4.90873016  0.03030303]
Evaluation of polynomial
0.0 0.030303030303
0.241379310345 1.10971786834
0.48275862069 1.99882895543
0.724137931034 2.7285338522
0.965517241379 3.32973011927
1.20689655172 3.83331531726
1.44827586207 4.27018700681
1.68965517241 4.67124274853
1.93103448276 5.06738010305
2.1724137931 5.48949663099
2.41379310345 5.96848989298
2.65517241379 6.53525744965
2.89655172414 7.2206968616
3.13793103448 8.05570568947
```

```
3.37931034483 9.07118149389
3.62068965517 10.2980218355
3.86206896552 11.7671242749
4.10344827586 13.5093863726
4.34482758621 15.5557056895
4.58620689655 17.936979786
4.8275862069 20.6841062228
5.06896551724 23.8279825604
5.31034482759 27.3995063597
5.55172413793 31.4295751811
5.79310344828 35.9490865852
6.03448275862 40.9889381328
6.27586206897 46.5800273844
6.51724137931 52.7532519007
6.75862068966 59.5395092422
7.0 66.9696969697
```

Figure 16.8 shows the graph of a fitted curve that corresponds to a polynomial of degree 3 with the same data points used with the previous example. This polynomial of degree 3 is a much better fit to the data points given.

16.4 MODELING THE HEAT CAPACITY OF CARBON DIOXIDE

The amount of energy required to heat a gas depends on its temperature. The heat capacity of a gas is often modeled with polynomial equations. Specific heat of carbon dioxide gas (CO_2) at temperatures ranging from 175 to 6000 K is taken from the web page:

http://www.engineeringtoolbox.com/carbon-dioxide-d_974.html

Given the values of temperature and heat for the CO_2 gas, the problem is to construct a non-linear model that shows the variation of the heat measured in the gas as temperature increases.

The data in the following table is stored in a data file temp_heat_co2.dat. The Python program in file temp_heat_co2.py reads the data from the data file, performs curve fitting to derive a polynomial of degree 3, prints the value of the coefficients of the polynomial, computes values of the heat using the polynomial function, and plots the given values and the computed values of heat with changes of temperature. The listing after the following table shows the results of running the program.

Running the Python program is repeated to derive polynomials of degrees 5 and 6. Figures 16.9 and 16.10 show the plots of the data given and the computed data using the polynomial functions of degrees 3 and 6, respectively.

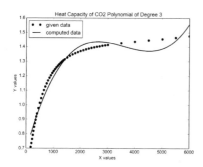

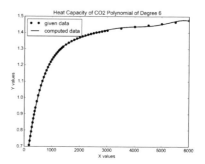

Figure 16.9 Heat Cap CO_2 polynomial deg 3.

Figure 16.10 Heat Cap CO_2 polynomial deg 6.

Carbon Dioxide Gas (CO_2)	
Temperature (K)	Specific Heat (kJ/kgK)
175	0.709
200	0.735
225	0.763
250	0.791
275	0.819
300	0.846
325	0.871
350	0.895
375	0.918
400	0.939
450	0.978
500	1.014
550	1.046
600	1.075
650	1.102
700	1.126
750	1.148
800	1.168
850	1.187
900	1.204
950	1.220
1000	1.234
1050	1.247
1100	1.259
1150	1.270
1200	1.280

Carbon Dioxide Gas (CO_2)	
Temperature (K)	Specific Heat (kJ/kgK)
1250	1.290
1300	1.298
1350	1.306
1400	1.313
1500	1.326
1600	1.338
1700	1.348
1800	1.356
1900	1.364
2000	1.371
2100	1.377
2200	1.383
2300	1.388
2400	1.393
2500	1.397
2600	1.401
2700	1.404
2800	1.408
2900	1.411
3000	1.414
3500	1.427
4000	1.437
4500	1.446
5000	1.455
5500	1.465
6000	1.476

```
$python temp_heat_co2.py
Coefficient list with polynomial of degree:  3
[  1.97002031e-11  -2.12104223e-07   7.09887769e-04
6.77678246e-01]
Evaluation of polynomial
175.0 0.79551849498
289.215686275 0.865723896188
403.431372549 0.930841363004
517.647058824 0.991047011593
631.862745098 1.04651695812
746.078431373 1.09742731874
860.294117647 1.14395420962
974.509803922 1.18627374692
1088.7254902 1.22456204681
1202.94117647 1.25899522545

. . . . .
5200.49019608 1.40385267251
5314.70588235 1.41679485038
5428.92156863 1.43239820502
5543.1372549 1.45083885258
5657.35294118 1.47229290924
5771.56862745 1.49693649114
5885.78431373 1.52494571447
6000.0 1.55649669537
```

16.5   SUMMARY

This chapter presented two important techniques that deal with raw data: interpolation and curve fitting. Interpolation is used to compute estimates of the value of the function for intermediate values of the independent variable, within the bounds of an available set of values of the independent variable. Curve fitting or *regression* techniques are used when a polynomial function is needed that would represent the raw data. Several functions in the Numpy and Scipy packages are used.

Key Terms

intermediate data	estimated data	raw data
linear interpolation	non-linear interpolation	curve fitting
cubic spline interpolation	coefficients	regression

16.6 EXERCISES

16.1 On a typical spring day, the temperature varies according to the data recorded in Table 16.1. Develop a Python program and apply linear interpolation to compute estimates of intermediate values of the temperature. Produce a plot of the given and estimated data.

Table 16.1 Temperature changes in a 12-hour period.

Time	7	8	9	10	11	12	1	2	3	4	5	6
Temp. (F^0)	51	56	60	65	73	78	85	86	84	81	80	70

16.2 On a typical spring day, the temperature varies according to the data recorded in Table 16.1. Develop a Python program and apply cubic spline interpolation to compute estimates of intermediate values of the temperature. Produce a plot of the given and estimated data.

16.3 On a typical spring day, the temperature varies according to the data recorded in Table 16.1. Apply curve fitting to derive the polynomial function of degree 2 that represents the given data. Develop a Python program and use the polynomial function to compute estimates of intermediate values of the temperature. Produce a plot of the given and estimated data.

16.4 On a typical spring day, the temperature varies according to the data recorded in Table 16.1. Apply curve fitting to derive the polynomial function of degree 3 that represents the given data. Develop a Python program and use the polynomial function to compute estimates of intermediate values of the temperature. Produce a plot of the given and estimated data.

16.5 On a typical spring day, the temperature varies according to the data recorded in Table 16.1. Apply curve fitting to derive the polynomial function of degree 4 that represents the given data. Develop a Python program and use the polynomial function to compute estimates of intermediate values of the temperature. Produce a plot of the given and estimated data.

16.6 Develop a Python program that uses the data in Table 16.2 and applies linear interpolation to compute estimates of intermediate values of the number of patients. Produce a plot of the given and estimated data.

16.7 Develop a Python program and use the data in Table 16.2 and apply non-linear interpolation to compute estimates of intermediate values of the number of patients. Produce a plot of the given and estimated data.

Table 16.2 Number of patients for years 1995–2002.

Year	1995	1996	1997	1998	1999	2000
Patients	5,500	8,500	13,500	20,500	29,500	40,500
Increase	0	3,000	5,000	7,000	9,000	11,000

Year	2001	2002
Patients	53,5000	68,500
Increase	13,000	15,000

16.8 Develop a Python program and use the data in Table 16.2 and apply curve fitting to derive a polynomial function of degree 2. Use this function to compute estimates of intermediate values of the number of patients. Produce a plot of the given and estimated data.

16.9 Develop a Python program and use the data in Table 16.2 and apply curve fitting to derive a polynomial function of degree 3. Use this function to compute estimates of intermediate values of the number of patients. Produce a plot of the given and estimated data.

## CHAPTER 17

Using Arrays with Numpy

17.1 INTRODUCTION

This chapter presents an overview of single-dimensional arrays, implementation techniques with programming in Python and Numpy, and a summary of computing with vectors. In general, an array is a term used in programming and is defined as a data structure that is a collection of values and these values are organized in several ways. In programming, a one-dimensional array is often known as a *vector*. The following arrays: X, Y, and Z have their data arranged in different manners. Array X is a one-dimensional array with n elements and it is considered a *row vector* because its elements $x_1, x_2, \ldots, x_n$ are arranged in a single row.

$$X = [x_1\ x_2\ x_3\ \cdots\ x_n] \qquad Z = \begin{bmatrix} z_1 \\ z_2 \\ z_3 \\ \vdots \\ z_m \end{bmatrix}$$

Array Z is also a one-dimensional array; it has m elements organized as a *column vector* because its elements: $z_1, z_2, \ldots, z_m$ are arranged in a single column.

The following array, Y, is a two-dimensional array organized as an $m \times n$ matrix; its elements are arranged in m rows and n columns. The first row of Y consists of elements: $y_{11}, y_{12}, \ldots, y_{1n}$. Its second row consists of elements: $y_{21}, y_{22}, \ldots, y_{2n}$. The last row of Y consists of elements: $y_{m1}, y_{m2}, \ldots, y_{mn}$.

$$Y = \begin{bmatrix} y_{11} & y_{12} & \cdots & y_{1n} \\ y_{21} & y_{22} & \cdots & y_{2n} \\ \vdots & \vdots & \ddots & \vdots \\ y_{m1} & y_{m2} & \cdots & y_{mn} \end{bmatrix}$$

17.2 VECTORS AND OPERATIONS

A vector is a mathematical entity that has magnitude and direction. In physics, it is used to represent characteristics such as the velocity, acceleration, or momentum of a physical object. A vector v can be represented by an n-tuple of real numbers:

$$v = (v_1, v_2, \ldots, v_n)$$

Several operations with vectors are performed with a vector and a scalar or with two vectors.

17.2.1 Addition of a Scalar and a Vector

To add a scalar to a vector involves adding the scalar value to every element of the vector. In the following example, the scalar α is added to the elements of vector Z, element by element.

$$Z = \begin{bmatrix} z_1 \\ z_2 \\ z_3 \\ \vdots \\ z_m \end{bmatrix} \qquad Z + \alpha = \begin{bmatrix} z_1 + \alpha \\ z_2 + \alpha \\ z_3 + \alpha \\ \vdots \\ z_m + \alpha \end{bmatrix}$$

17.2.2 Vector Addition

Vector addition of two vectors that are n-tuple involves adding the corresponding elements of each vector. The following example illustrates the addition of two vectors, Y and Z.

$$Y = \begin{bmatrix} y_1 \\ y_2 \\ y_3 \\ \vdots \\ z_m \end{bmatrix} \qquad Z = \begin{bmatrix} z_1 \\ z_2 \\ z_3 \\ \vdots \\ z_m \end{bmatrix} \qquad Y + Z = \begin{bmatrix} y_1 + z_1 \\ y_2 + z_2 \\ y_3 + z_3 \\ \vdots \\ y_m + z_m \end{bmatrix}$$

17.2.3 Multiplication of a Vector and a Scalar

Scalar multiplication is performed by multiplying the scalar with every element of the specified vector. In the following example, scalar α is multiplied by every element z_i of vector Z.

$$Z = \begin{bmatrix} z_1 \\ z_2 \\ z_3 \\ \vdots \\ z_m \end{bmatrix} \qquad Z \times \alpha = \begin{bmatrix} z_1 \times \alpha \\ z_2 \times \alpha \\ z_3 \times \alpha \\ \vdots \\ z_m \times \alpha \end{bmatrix}$$

17.2.4 Dot Product of Two Vectors

Given vectors $v = (v_1, v_2, \ldots, v_n)$ and $w = (w_1, w_2, \ldots, w_n)$, the dot product $v \cdot w$ is a scalar defined by:

$$v \cdot w = \sum_{i=1}^{n} v_i w_i = v_1 w_1 + v_2 w_2 + \ldots + v_n w_n.$$

Therefore, the dot product of two vectors in an n-dimensional real space is the sum of the product of the vectors' components.

When the elements of the vectors are complex, then the dot product of two vectors is defined by the following relation. Note that $\bar{v}_i$ is the complex conjugate of v_i.

$$v \cdot w = \sum_{i=1}^{n} \bar{v}_i w_i = \bar{v}_1 w_1 + \bar{v}_2 w_2 + \ldots + \bar{v}_n w_n$$

17.2.5 Length (Norm) of a Vector

Given a vector $v = (v_1, v_2, \ldots, v_n)$ of dimension n, the Euclidean norm of the vector denoted by $\|v\|_2$, is the length of v and is defined by the square root of the dot product of the vector:

$$\|v\|_2 = \sqrt{v \cdot v} = \sqrt{v_1^2 + v_2^2 + \cdots + v_n^2}$$

In the case that vector v is a 2-dimensional vector, the Euclidean norm of the vector is the value of the hypotenuse of a right-angled triangle. When vector v is a 1-dimensional vector, then $\|v\|_2 = |v_1|$, the absolute value of the only component v_1.

17.3 VECTOR PROPERTIES AND CHARACTERISTICS

A vector $v = (v_1, v_2, \ldots, v_n)$ in $\mathcal{R}^n$ (an n-dimensional real space) can be specified as a column or row vector. When v is an n column vector, its transpose v^T is an n row vector.

17.3.1 Orthogonal Vectors

Vectors v and w are said to be orthogonal if their dot product is zero. The angle θ between vectors v and w is defined by:

$$cos(\theta) = \frac{v \cdot w}{\|v\|_2 \ \|w\|_2}.$$

where θ is the angle from v to w, and non-zero vectors are orthogonal if and only if they are perpendicular to each other, i.e., when $cos(\theta) = 0$ and θ is equal to $\pi/2$ or 90 degrees. Orthogonal vectors v and w are called *orthonormal* if they are of length one, i.e., $v \cdot v = 1$, and $w \cdot w = 1$.

17.3.2 Linear Dependence

A set k of vectors $\{x_1, x_2, \ldots, x_k\}$ is linearly dependent if at least one of the vectors can be expressed as a linear combination of the others. Assuming there exists a set of scalars $\{\alpha_1, \alpha_2, \ldots, \alpha_k\}$, vector x_k is defined as follows:

$$x_k = \alpha_1 x_1 + \alpha_2 x_2 + \ldots + \alpha_{k-1} x_{k-1}.$$

If a vector w depends linearly on vectors $\{x_1, x_2, \ldots, x_k\}$, this is expressed as follows:

$$w = \alpha_1 x_1 + \alpha_2 x_2 + \ldots + \alpha_k x_k.$$

17.4 USING ARRAYS IN PYTHON WITH NUMPY

Arrays are created and manipulated in Python and Numpy by calling the various library functions. Before using an array, it needs to be created. Numpy function *array* creates an array given the values of the elements. When an array is no longer needed in the program, it can be destroyed by using the *del* Python command.

Numpy function *zeros* creates an array with the specified number of elements, all initialized to zero. Similarly, function *ones* creates an array with its elements initialized to value 1.0. Note that the default type of these arrays is *float*. Function *arange* creates an array of integers starting at value 0 and increasing up to $n - 1$.

The following short Python program illustrates the various Numpy functions used to create arrays. The program is stored in file `test_arrays.py`.

```python
import numpy as np
print "Creating arrays"
x = np.array([4.5, 2.55, 12.0 -9.785])
print "Array x: ", x
```

```
y = np.zeros(12)
print "Array y: ", y
z = np.ones((3, 4)) # 3 rows, 4 cols
print "Array z: "
print z
n = np.arange(12)
print "Array n: ", n
del x   # delete array x
```

Executing the Python interpreter and running the program yields the following output. Note that array z is a two-dimensional array with three rows and four columns with all its elements initialized to 1.0

```
$ python test_arrays.py
Creating arrays
Array x:  [ 4.5     2.55    2.215]
Array y:  [ 0.  0.  0.  0.  0.  0.  0.  0.  0.  0.  0.  0.]
Array z:
[[ 1.  1.  1.  1.]
 [ 1.  1.  1.  1.]
 [ 1.  1.  1.  1.]]
Array n: [ 0  1  2  3  4  5  6  7  8  9 10 11]
```

A vector is manipulated by accessing its individual elements and changing and/or retrieving the value of the elements using indexing.

Listing 17.1 shows the source code of a Python program stored in file test2_arrays.py that uses Numpy functions to create and manipulate a vector. Line 4 calls function *zeros* to create vector p with N elements. In lines 7–8, elements with index j (from 0 to $k - 1$) of vector p are set to value 5.25. In line 11, vector p is destroyed.

Listing 17.1: Program that creates and manipulates a vector.

```
 1 import numpy as np
 2 print "Creating and manipulate an array"
 3 N = 8   # number of elements
 4 p = np.zeros(N)
 5 print "Array p: ", p
 6 k = 5
 7 for j in range(k):
 8      p[j] = 5.25
 9 print "Array p: "
10 print p
11 del p   # delete array p
```

Executing the Python interpreter and running the program yields the following output. Note that only the first k elements of array p are set to value 5.25.

```
$ python test2_arrays.py
Creating and manipulate an array
Array p:  [ 0.  0.  0.  0.  0.  0.  0.  0.]
Array p:
[ 5.25  5.25  5.25  5.25  5.25  0.   0.   0.  ]
```

Slicing can be used to indicate more selective indexing with the : operator. For example, x[0:4] references the elements of array x but only the elements with indices from 0 up to 3. Most of the time it is more efficient (time-wise) to use slicing than using a loop. The following program is a variation of the previous one, but instead of the loop, now slicing is used in line 7 to assign values to select elements of array p. The results are the same as the previous program.

```
1 import numpy as np
2 print "Creating and manipulate an array"
3 N = 8  # number of elements
4 p = np.zeros(N)
5 print "Array p: ", p
6 k = 5
7 p[0:k] = 5.25
8 print "Array p: "
9 print p
```

Arrays can be stacked into a single array by calling Numpy function *hstack*. Arrays can also be split into separate arrays by calling function *hsplit*. The following program creates two arrays p and q in lines 3 and 6, then it stacks them into array *newa* in line 7. Array *newa* is split into three arrays with equal shape in line 10. Arrays x, y, and z are used to reference the three arrays created in lines 12–14. Array *newa* is split after column 4 and up to column 6, basically creating three arrays in line 17.

```
1 import numpy as np
2 print "Stacking and splitting array"
3 p = np.array([1.5, 2.5, 3.5, 4.5, 5.5, 6.5, 7.5, 8.5])
4 print "Array p: "
5 print p
6 q = np.array([2.35, 5.75, 7.75, 3.15])
7 newa = np.hstack((p, q))
8 print "newa: "
```

```
 9 print newa
10 r = np.hsplit(newa,3) # three equally shaped arrays
11 print "Array r:"
12 x = r[0]
13 y = r[1]
14 z = r[2]
15 print "Array x: ", x
16 # after col 4 up to col 6
17 newb = np.hsplit (newa,(4,6))
18 print "Array newb:"
19 print newb[0]
20 print newb[1]
21 print newb[2]
```

```
$ python test2c_arrays.py
Stacking and splitting array
Array p:
[1.5, 2.5, 3.5, 4.5, 5.5, 6.5, 7.5, 8.5]
newa:
[ 1.5    2.5    3.5    4.5    5.5    6.5    7.5    8.5    2.35  5.75
   7.75  3.15]
Array r:
Array x:  [ 1.5  2.5  3.5  4.5]
Array newb:
[ 1.5   2.5   3.5   4.5]
[ 5.5   6.5]
[ 7.5    8.5    2.35  5.75  7.75  3.15]
```

17.5 SIMPLE VECTOR OPERATIONS

Operations on vectors are performed on an individual vector, with a vector and a scalar, or with two vectors.

17.5.1 Arithmetic Operations

To add a scalar to a vector involves adding the scalar value to every element of the vector. This operation adds the specified constant value to the elements of the vector specified.

The following Python program illustrates the arithmetic operations on vectors and scalars. The program is stored in file test3_arrays.py. In line 4, vector p is created and its elements initialized with zero values. In line 8, a scalar value xv is added to vector p. In line 11, the constant 2 (another scalar) is subtracted from vector p and assigned to array q. In line 13, vector q is multiplied by the scalar 3 and assigned to vector $q2$. In line 15, vector $q2$ is divided by the scalar 2.5.

```
 1 import numpy as np
 2 print "Arithmetic operations with a scalar"
 3 N = 8   # number of elements
 4 p = np.zeros(N)
 5 print "Array p: "
 6 print p
 7 xv = 3.75
 8 p = p + xv
 9 print "Array p: "
10 print p
11 q = p - 2.0
12 print "Array q: ", q
13 q2 = q * 3
14 print "Array q2: ", q2
15 q3 = q2 / 2.5
16 print "Array q3: ", q3
```

Executing the Python interpreter with program `test3_arrays.py` yields the following output.

```
$ python test3_arrays.py
Arithmetic operations with a scalar
Array p:
[ 0.   0.   0.   0.   0.   0.   0.   0.]
Array p:
[ 3.75   3.75   3.75   3.75   3.75   3.75   3.75   3.75]
Array q:   [ 1.75   1.75   1.75   1.75   1.75   1.75   1.75   1.75]
Array q2:   [ 5.25   5.25   5.25   5.25   5.25   5.25   5.25   5.25]
Array q3:   [ 2.1   2.1   2.1   2.1   2.1   2.1   2.1   2.1]
```

To add two vectors involves adding the corresponding elements of each vector, and a new vector is created. This addition operation on vectors is only possible if the row vectors (or column vectors) are of the same size. In a similar manner, subtracting two vectors of the same size can be performed.

In the following program, vectors p and q are created with size 8. The operation in line 8 adds the elements of vector q and the elements of vector p and the elements are assigned to vector $q3$, the new vector created. The operation in line 10 subtracts the elements of vector p from the elements of vector q and the new vector created is $q4$. This program is stored in file `test4_arrays.py`.

```
 1 import numpy as np
 2 print "Vector arithmetic operations"
 3 p = np.zeros(8)
 4 p = p + 3.5
 5 print "Array p: ", p
 6 q = p * 3
 7 print "Array q: ", q
 8 q3 = q + p
 9 print "Array q3: ", q3
10 q4 = q - p
11 print "Array q4: ", q4
```

Executing the Python interpreter with program test4_arrays.py yields the following output.

```
$python test4_arrays.py
Vector arithmetic operations
Array p:  [ 3.5   3.5   3.5   3.5   3.5   3.5   3.5   3.5]
Array q:  [ 10.5  10.5  10.5  10.5  10.5  10.5  10.5  10.5]
Array q3: [ 14.  14.  14.  14.  14.  14.  14.  14.]
Array q4: [ 7.   7.   7.   7.   7.   7.   7.   7.]
```

17.5.2 Element Multiplication and Division Operations

Applying element-by-element multiplication, the corresponding elements of two vectors are multiplied. This operation is applied to two vectors of equal size. This operation multiplies the elements of the specified vector by the elements of the second specified vector. Using element-by-element division, the corresponding elements of two vectors are divided. This operation is applied to two vectors of equal size.

The following program shows element-wise multiplication and division with vectors. In line 8, vectors *p* and *q* are multiplied and the results are stored in vector *p3*. In line 10, vector *q* is divided by vector *p* and the results are stored in vector *q4*. The program is stored in file test5_arrays.py.

```
 1 import numpy as np
 2 print "Element multiplication and division operations"
 3 p = np.zeros(8)
 4 p = p + 3.5
 5 print "Array p: ", p
 6 q = p * 3
 7 print "Array q: ", q
 8 q3 = q * p
```

```
 9 print "Array q3: ", q3
10 q4 = q / p
11 print "Array q4: ", q4
```

Executing the Python interpreter with program `test5_arrays.py` yields the following output.

```
$ python test5_arrays.py
Element multiplication and division operations
Array p:  [ 3.5  3.5  3.5  3.5  3.5  3.5  3.5  3.5]
Array q:  [ 10.5  10.5  10.5  10.5  10.5  10.5  10.5  10.5]
Array q3:  [ 36.75  36.75  36.75  36.75  36.75  36.75
    36.75  36.75]
Array q4:  [ 3.  3.  3.  3.  3.  3.  3.  3.]
```

17.5.3   Vector Multiplication

Multiplication of vectors is carried out by using the Numpy function *dot* or *vdot*. This operation is known as the *dot multiplication* of vectors. The following program stored in file **test5b_arrays** illustrates these operations. Lines 8 and 10 apply dot multiplication of vectors p and q.

```
 1 import numpy as np
 2 print "Vector dot multiplication"
 3 p = np.zeros(8)
 4 p = p + 3.5
 5 print "Array p: ", p
 6 q = p * 3
 7 print "Array q: ", q
 8 q4 = np.dot(p,q)
 9 print "Dot product of p, q: ", q4
10 q4v = np.vdot(p,q)
11 print "Dot product of p, q", q4v
```

Executing the Python interpreter with program `test5b_arrays.py` yields the following output.

```
$ python test5b_arrays.py
Vector dot multiplication
Array p:  [ 3.5  3.5  3.5  3.5  3.5  3.5  3.5  3.5]
Array q:  [ 10.5  10.5  10.5  10.5  10.5  10.5  10.5  10.5]
Dot product of p, q:  294.0
Dot product of p, q 294.0
```

17.5.4 Additional Vector Operations

In addition to slicing, selective indexing can also be carried out using arrays of indices. The following program stored in file **test9_arrays.py** creates an array of indices in line 6 and applies it to array p to select the corresponding elements in line 7. These selected elements are assigned the value 2.0 in line 9.

```
 1 import numpy as np
 2 print "Array of indices"
 3 p = np.array([1.5, 2.5, 3.5, 4.5, 5.5, 6.5, 7.5, 8.5])
 4 print "Array p: "
 5 print p
 6 pindx = np.array([1, 4, 6, 7])
 7 q = p[pindx]
 8 print "q: ", q
 9 p[pindx] = 2.0
10 print "Array p:"
11 print p
```

Executing the Python interpreter with program **test9_arrays.py** yields the following output.

```
$ python test9_arrays.py
Array of indices
Array p:
[ 1.5  2.5  3.5  4.5  5.5  6.5  7.5  8.5]
q:  [ 2.5  5.5  7.5  8.5]
Array p:
[ 1.5  2.   3.5  4.5  2.   6.5  2.   2. ]
```

Various additional operations can be applied to vectors. For vector assignment, the resulting vector is only a view of the first vector. The two vectors refer to the same list of values. For the copy operation, method *copy* is called, the two vectors must have the same length, and the element values are all copied to the second vector.

The vector comparison operation is applied to all vectors as a whole by calling Numpy function *array_equal*. The result is a single Boolean value. The element comparison of two vectors for equality is performed element by element by calling Numpy function *equal* and the operation creates a new vector of the same size with values *True* or *False*. Similarly, when comparing for greater-than operation, the Numpy function *greater* is called.

The following program illustrates the use of these operations. Line 10 is an assignment of vector q to vector qq. The two vectors now refer to the same list of values. In line 12, the elements of vector p are copied to vector qqq by calling method *copy* and *qqq* becomes a new vector. In line 14, Numpy function *array_equal* compares vectors *p* and *qqq*. This produces a single Boolean result. In line 16, the Numpy function *equal* is called to compare the elements of vector *p* and *qq*, and the result is another vector with the Boolean result of the comparison for each corresponding pair of element values. In a similar manner, the Numpy function *greater* is called to compare the elements of vectors *q* and *p* in line 18.

```
1 import numpy as np
2 print "Vector assignment, comparisons operations"
3 p = np.zeros(8)
4 p = p + 3.5
5 p[2] = 1.75
6 p[6] = 12.35
7 print "Array p: ", p
8 q = p * 3
9 print "Array q: ", q
10 qq = q
11 print "Array qq: ", qq
12 qqq = p.copy()
13 print "Array qqq: ", qqq
14 yyn = np.array_equal(p, qqq)
15 print "Vector equality: ", yyn
16 yn = np.equal(p, qq)  # p == qq
17 print "Element equality: ", yn
18 byn = np.greater(q, p) # q > p
19 print "Greater: ", byn
```

Executing the Python interpreter with program `test6_arrays.py` yields the following output.

```
$ python test6_arrays.py
Vector assignment, comparisons operations
Array p:  [  3.5     3.5     1.75    3.5     3.5     3.5    12.35
    3.5 ]
Array q:  [ 10.5    10.5     5.25   10.5    10.5    10.5    37.05
   10.5 ]
Array qq:  [ 10.5    10.5     5.25   10.5    10.5    10.5    37.05
   10.5 ]
Array qqq:  [  3.5     3.5     1.75    3.5     3.5     3.5    12.35
    3.5 ]
Vector equality:  True
Element equality:  [False False False False False False
  False False]
Greater:  [ True   True   True   True   True   True   True   True]
```

A Boolean expression can be used in indexing, and this is very convenient to select the elements for which the Boolean expression is true. In the following program, a Boolean expression is applied to all elements of array *p* in line 6. The array of truth values created, *pindx*, is used to index array *p*.

```
1 import numpy as np
2 print "Indexing with boolean expression"
3 p = np.array([1.5, 2.5, 3.5, 4.5, 5.5, 6.5, 3.25, 1.65])
4 print "Array p: "
5 print p
6 pindx = p <= 2.25
7 print "Array pindx: ", pindx
8 q = p[pindx]
9 print "q: ", q
```

Executing the Python interpreter with program `tmat11.py` yields the following output.

```
$  python tmat11.py
Indexing with boolean expression
Array p:
[ 1.5   2.5   3.5   4.5   5.5   6.5   3.25  1.65]
Array pindx:  [ True False False False False False False  True]
q:  [ 1.5   1.65]
```

Function *max* gets the maximum value stored in the specified vector. The following function call gets the maximum value in vector *pv* and assigns this value to *x*.

```
x = numpy.max(pv)
```

In addition to the maximum value in a vector, the index of the element with that value may be desired. Calling function *argmax* returns the index value of the element with the maximum value in a specified vector. In the following function call, the index value of the element with the maximum value in vector *pv* is returned and assigned to integer variable *idx*.

```
idx = numpy.argmax (pv)
```

In a similar manner, function *min* gets the minimum value stored in a vector. Function *argmin* returns the index value of the minimum value in the specified vector.

Listing 17.2 shows a Python source listing of a program that includes the application of the operations on vectors that have been discussed and some additional ones. The program is stored in file `vectorops.py`.

Listing 17.2: Program that shows various operations on vectors.

```
1 import numpy as np
2 print "Vector maximum, minimum, sum, mean, other
    operations"
3 p = np.zeros(8)
4 p = p + 3.5
5 p[2] = 1.75
6 p[6] = 12.35
7 print "Array p: ", p
8 q = p * 3
9 print "Array q: ", q
10 x = np.max(p)
11 print "Max in vector p: ", x
12 idx = np.argmax(p)
13 print "Index of max in p: ", idx
14 y = np.min(q)
15 print "Min in vector q: ", y
16 ydx = np.argmin(q)
17 print "Index of min in q: ", ydx
18 mym = np.mean(p)
19 print "Mean of p: ", mym
20 xm = np.median(p)
21 print "Median of p: ", xm
22 mystd = p.std()
23 print "STD of p: ", mystd
24 mysum = p.sum()
25 print "Sum of p: ", mysum
```

Executing the Python interpreter with program `vectorops.py` yields the following output.

```
$ python vectorops.py
Vector maximum, minimum, sum, mean, other operations
Array p:  [  3.5     3.5     1.75    3.5     3.5     3.5    12.35
    3.5 ]
Array q:  [ 10.5    10.5     5.25   10.5    10.5    10.5    37.05
   10.5 ]
Max in vector p:  12.35
Index of max in p:  6
Min in vector q:  5.25
Index of min in q:  2
Mean of p:  4.3875
Median of p:  3.5
STD of p:  3.06357124122
Sum of p:  35.1
```

17.6 SUMMARY

Arrays are data structures that store collections of data. To refer to an individual element, an integer value, known as the index, is used to indicate the relative position of the element in the array. Python and Numpy manipulate arrays as vectors and matrices. Many operations and functions are defined for creating and manipulating vectors and matrices.

Key Terms

arrays	elements	index
vectors	array elements	matrices
column vector	row vector	two-dimensional array
vector operations	vector functions	complex vectors

17.7 EXERCISES

17.1 Develop a Python program that computes the values in a vector V that are the sines of the elements of vector T. The program must assign to T a vector with 75 elements running from 0 to 2π. Plot the elements of V as a function of the elements of T; use GnuPlot or *matplotlib*.

17.2 Develop a Python program that finds the index of the first negative number in a vector. If there are no negative numbers, it should set the result to -1.

17.3 The Fibonacci series is defined by $F_n = F_{n-1} + F_{n-2}$. Develop a Python program that computes a vector with the first n elements of a Fibonacci series. A second vector should also be computed with the ratios of consecutive Fibonacci numbers. The program must plot this second vector using GnuPlot or *matplotlib*.

17.4 Develop a Python program that reads the values of a vector P and computes the element values of vector Q with the cubes of the positive values in vector P. For every element in P that is negative, the corresponding element in Q should be set to zero.

17.5 Develop a Python program that reads the values of a vector P and assigns the element values of vector Q with every other element in vector P.

17.6 Develop a Python program that reads the values of a matrix M of m rows and n columns. The program must create a column vector for every column in matrix M, and a row vector for every row in matrix M.

17.7 A trapezoid is a four-sided region with two opposite sides parallel. The area of a trapezoid is the average length of the parallel sides, times the distance between them. The area of the trapezoid with width $\Delta x = x_2 - x_1$, is computed with Δx times the average height $(y_2 + y_1)/2$.

$$A_t = \Delta x \frac{y_1 + y_2}{2}$$

For the interval of $[a, b]$ on variable x, it is divided into $n - 1$ equal segments of length Δx. Any value of y_k is defined as $y_k = f(x_k)$. The trapezoid sum to compute the area under the curve for the interval of $[a, b]$, is defined by the summation of the areas of the individual trapezoids and is expressed as follows.

$$A = \sum_{k=2}^{k=n} [\Delta x \frac{1}{2}(f(x_{k-1}) + f(x_k))]$$

The larger the number of trapezoids, the better the approximation of the area. The area from a to b, with segments $a = x_1 < x_2 < \ldots < x_n = b$ is given by:

$$A = \frac{b - a}{2n}[f(x_1) + 2f(x_2) + \ldots + 2f(x_{n-1}) + f(x_n)].$$

Develop a Python function that computes an estimate of the area under a curve given by a sequence of values y with a corresponding sequence of values x. The function must be called from *main* and the program must use GnuPlot or *matplotlib*.

## CHAPTER 18

Models with Matrices and Linear Equations

18.1 INTRODUCTION

As mentioned in previous chapters, in scientific computing, data is conveniently organized as collections of values known as *vectors* and *matrices* and are used to implement data lists and sequences of values. This chapter presents basic concepts of matrices, operations on matrices, and programming computing with matrices in Python and Numpy, and a summary of systems of linear equations that are solved using Numpy.

18.2 MATRICES

In general, a matrix is a two-dimensional array of data values and is organized in rows and columns. The following array, Y, is a *two-dimensional array* organized as an $m \times n$ matrix; its elements are arranged in m rows and n columns.

$$Y = \begin{bmatrix} y_{11} & y_{12} & \cdots & y_{1n} \\ y_{21} & y_{22} & \cdots & y_{2n} \\ \vdots & \vdots & \ddots & \vdots \\ y_{m1} & y_{m2} & \cdots & y_{mn} \end{bmatrix}$$

The first row of Y consists of elements $y_{11}, y_{12}, \ldots, y_{1n}$. The second row consists of elements $y_{21}, y_{22}, \ldots, y_{2n}$. The last row of Y consists of elements $y_{m1}, y_{m2}, \ldots, y_{mn}$. In a similar manner, the elements of each column can be identified.

18.2.1 Basic Concepts

A matrix is defined by specifying the rows and columns of the array. An m by n matrix has m rows and n columns. A *square* matrix has the same number of rows and columns, n rows and n columns, which is denoted as $n \times n$. The following example shows a 2×3 matrix, which has two rows and three columns:

$$\begin{bmatrix} 0.5000 & 2.3500 & 8.2500 \\ 1.8000 & 7.2300 & 4.4000 \end{bmatrix}$$

A matrix of dimension $m \times 1$ is known as a *column vector* and a matrix of dimension $1 \times n$ is known as a *row vector*. A vector is considered a special case of a matrix with one row or one column. A row vector of size n is typically a matrix with one row and n columns. A column vector of size m is a matrix with m rows and one column.

The elements of a matrix are denoted with the matrix name in lower-case and two indices, one corresponding to the row of the element and the other index corresponding to the column of the element. For matrix Y, the element at row i and column j is denoted by y_{ij} or by $y_{i,j}$.

The *main diagonal* of a matrix consists of those elements on the diagonal line from the top left and down to the bottom right of the matrix. These elements have the same value of the two indices. The diagonal elements of matrix Y are denoted by y_{ii} or by y_{jj}. For a square matrix, this applies for all values of i or all values of j. For a square matrix X (an $n \times n$ matrix), the elements of the main diagonal are:

$$x_{1,1}, \; x_{2,2}, \; x_{3,3}, \; x_{4,4}, \; \ldots, \; x_{n,n}.$$

An *identity matrix* of size n, denoted by I_n is a square matrix that has all the diagonal elements with value 1, and all other elements (off-diagonal) with value 0. The following is an identity matrix of size 3 (of order n):

$$I = \begin{bmatrix} 1 & 0 & 0 \\ 0 & 1 & 0 \\ 0 & 0 & 1 \end{bmatrix}$$

18.2.2 Arithmetic Operations

The arithmetic matrix operations are similar to the ones discussed previously for vectors. The multiplication of a matrix Y by a *scalar* λ calculates the multiplication of every element of matrix Y by the scalar λ. The result defines a new matrix.

$$Y = \begin{bmatrix} y_{11} & y_{12} & \cdots & y_{1n} \\ y_{21} & y_{22} & \cdots & y_{2n} \\ \vdots & \vdots & \ddots & \vdots \\ y_{m1} & y_{m2} & \cdots & y_{mn} \end{bmatrix} \qquad \lambda Y = \begin{bmatrix} \lambda y_{11} & \lambda y_{12} & \cdots & \lambda y_{1n} \\ \lambda y_{21} & \lambda y_{22} & \cdots & \lambda y_{2n} \\ \vdots & \vdots & \ddots & \vdots \\ \lambda y_{m1} & \lambda y_{m2} & \cdots & \lambda y_{mn} \end{bmatrix}$$

In a similar manner, the addition of a scalar λ to matrix Y, calculates the addition of every element of the matrix with the scalar λ. The subtraction of a scalar λ from a matrix Y, denoted by $Y - \lambda$, computes the subtraction of the scalar λ from every element of matrix Y.

The *matrix addition* is denoted by $Y+Z$ of two $m \times n$ matrices Y and Z, and calculates the addition of every element of matrix Y with the corresponding element of matrix Z. The result defines a new matrix. This operation requires that the two matrices have the same number of rows and columns.

$$Y + Z = \begin{bmatrix} y_{11} + z_{11} & y_{12} + z_{12} & \cdots & y_{1n} + z_{1n} \\ y_{21} + z_{21} & y_{22} + z_{22} & \cdots & y_{2n} + z_{2n} \\ \vdots & \vdots & \ddots & \vdots \\ y_{m1} + z_{m1} & y_{m2} + z_{m2} & \cdots & y_{mn} + z_{mn} \end{bmatrix}$$

Similarly, the subtraction of two matrices Y and Z, denoted by $Y - Z$, subtracts every element of matrix Z from the corresponding element of matrix Y. The result defines a new matrix.

The *element-wise multiplication* (also known as the Hadamard product or Schur product) of two matrices, multiplies every element of a matrix X by the corresponding element of the second matrix Y and is denoted by $X \circ Y$. The result defines a new matrix, Z. This operation requires that the two matrices have the same number of rows and columns. The following is the general form of the element-wise multiplication of matrix X multiplied by matrix Y.

$$Z = X \circ Y = \begin{bmatrix} x_{11} \times y_{11} & x_{12} \times y_{12} & \cdots & x_{1n} \times y_{1n} \\ x_{21} \times y_{21} & x_{22} \times y_{22} & \cdots & x_{2n} \times y_{2n} \\ \vdots & \vdots & \ddots & \vdots \\ x_{m1} \times y_{m1} & x_{m2} \times y_{m2} & \cdots & x_{mn} \times y_{mn} \end{bmatrix}$$

The *determinant* of a matrix A, denoted by $\det A$ or by $| A |$, is a special number that can be computed from the matrix. The determinant is useful to describe various properties of the matrix that are applied in systems of linear equations and in calculus.

The determinant provides important information when the matrix consists of the coefficients of a system of linear equations. If the determinant is nonzero, the system of linear equations has a unique solution. When the determinant is zero, there are either no solutions or many solutions.

One of several techniques to compute the determinant of a matrix is applying Laplace's formula, which expresses the determinant of a matrix in terms of its minors. The minor $M_{i,j}$ is defined to be the determinant of the submatrix that results from matrix A by removing the ith row and the jth column. Note that this technique is not very efficient. The expression $C_{i,j} = (1)^{i+j} M_{i,j}$ is known as a *cofactor*. The determinant of A is computed by

$$\det A = \sum_{j=1}^{n} C_{i,j} \times a_{i,j} \times M_{i,j}.$$

The *matrix multiplication* of an $m \times n$ matrix X and an $n \times p$ matrix Y, denoted by XY, defines another matrix Z of dimension m by p and the operation is denoted by $Z = XY$. In matrix multiplication, the number of rows in the first matrix has to equal the number of columns in the second matrix. An element of the new matrix Z is determined by

$$z_{ij} = \sum_{k=1}^{n} x_{ik} \, y_{kj},$$

that is, element z_{ij} of matrix Z is computed as follows:

$$z_{ij} = x_{i1} \, y_{1j} + x_{i2} \, y_{2j} + \ldots + x_{in} \, y_{nj}.$$

The matrix multiplication is not normally commutative, that is, $XY \neq YX$. The following example defines a 2 by 3 matrix, X, and a 3 by 3 matrix, Y. The matrix multiplication of matrix X and Y defines a new matrix Z.

$$X = \begin{bmatrix} 1 & 2 & 1 \\ 0 & 2 & 1 \end{bmatrix} \quad Y = \begin{bmatrix} 1 & 2 & 0 \\ 0 & 3 & 1 \\ -2 & 1 & 1 \end{bmatrix}$$

$$Z = XY = \begin{bmatrix} -1 & 9 & 3 \\ -2 & 7 & 3 \end{bmatrix}$$

The *transpose* of an $m \times n$ matrix X is an $n \times m$ matrix X^T formed by interchanging the rows and columns of matrix X. For example, for the given matrix X, the transpose (X^T) is:

$$X = \begin{bmatrix} 1 & 2 & 1 \\ 0 & 5 & 3 \end{bmatrix} \quad X^T = \begin{bmatrix} 1 & 0 \\ 2 & 5 \\ 1 & 3 \end{bmatrix}.$$

The *conjugate* of an $m \times n$ complex matrix Z is a matrix $\overline{Z}$ with all its elements conjugate of the corresponding elements of matrix Z. The *conjugate transpose* of an $m \times n$ matrix Z, is an $n \times m$ matrix that results by taking the transpose of matrix Z and then the complex conjugate. The resulting matrix is denoted by Z^H or by Z^* and is also known as the *Hermitian transpose*, or the *adjoint matrix* of matrix Z. For example:

$$Z = \begin{bmatrix} 1.5 + 2.3i & 2.1 - 1.4i & 1 + 0.7i \\ 0 + 3.2i & 5.2 - 1.5i & 3.7 + 3.5i \end{bmatrix} \qquad Z^H = \begin{bmatrix} 1.5 - 2.3i & 0 - 3.2i \\ 2.1 + 1.4i & 5.2 + 1.5i \\ 1 - 0.7i & 3.7 - 3.5i \end{bmatrix}$$

A square matrix Y is *symmetric* if $Y = Y^T$. The following example shows a 3×3 symmetric matrix Y.

$$Y = \begin{bmatrix} 1 & 2 & 7 \\ 2 & 3 & 4 \\ 7 & 4 & 1 \end{bmatrix}$$

A square matrix X is *triangular* if all the elements above or below its diagonal have value zero. A square matrix Y is *upper triangular* if all the elements below its diagonal have value zero. A square matrix Y is *lower triangular* if all the elements above its diagonal have value zero. The following example shows a matrix Y that is upper triangular.

$$Y = \begin{bmatrix} 1 & 2 & 7 \\ 0 & 3 & 4 \\ 0 & 0 & 1 \end{bmatrix}$$

The *rank* of a matrix Y is the number of independent columns in matrix Y, or the number of linearly independent rows of matrix Y. The *inverse* of an $n \times n$ matrix X is another matrix X^{-1} (if it exists) of dimension $n \times n$ such that their matrix multiplication results in an identity matrix or order n. This relation is expressed as:

$$X^{-1}X = XX^{-1} = I_n.$$

A square matrix X is *orthogonal* if for each column x_i of X, $x_i^T x_j = 0$ for any other column x_j of matrix X. If the rows and columns are orthogonal unit vectors (the norm of each column x_i of X has value 1), then X is *orthonormal*. A matrix X is orthogonal if its transpose X^T is equal to its inverse X^{-1}. This implies that for orthogonal matrix X,

$$X^T X = I.$$

Given matrices A, X, and B, a general matrix equation is expressed by $AX = B$. This equation can be solved for X by pre-multiplying both sides of the matrix equation by the inverse of matrix A. The following expression shows this:

$$A^{-1}AX = A^{-1}B.$$

This results in

$$X = A^{-1}B.$$

18.3 MATRIX MANIPULATION WITH NUMPY

Matrices are created and manipulated in Python by calling the various library functions in the Numpy and Scipy packages. Before using a matrix, it needs to be created. Matrices are created in a similar manner than the one used to create vectors.

18.3.1 Creating, Initializing, and Indexing Matrices

The most straightforward way to create a matrix with Numpy is to call function *array* and specify the values in the matrix arranged in rows and columns. Numpy function *ones* creates a matrix of the specified number of rows and columns, and returns a new matrix. Numpy function *zeros* creates a matrix of the specified number of rows and columns, and returns a new matrix. Matrices are stored in row-major order, and the elements of each row form a contiguous block in memory. When a matrix is no longer needed in the program, it can be destroyed by calling the *del* command.

The following program shows how to create and manipulate matrices in Python and is stored in file `tmat1.py`. In line 3, matrix *amat* is created given the values of the elements arranged into three rows and two columns. In line 6, matrix *bmat* is created by calling Numpy function *ones* and specifying three rows and two columns with all elements initialized to 1. In line 9, calling Numpy function *ones* creates matrix *cmat* with two rows and three columns and all elements initialized to 0. An element of matrix *amat* is accessed by specifying index 2 and index 1, which indicate the element in row 2 and column 1. Finally, in line 14, the assignment gets the element in row 0 and column 1 and assigns the value to *q*.

```
1 import numpy as np
2 print "Create and manipulate a matrix"
3 amat = np.array([[1.0, 3.0], [2.0, 5.0], [4.0, 7.0]])
4 print "Matrix amat: "
5 print amat
6 bmat = np.ones([3,2])
7 print "Matrix bmat: "
8 print bmat
```

```
 9 cmat = np.zeros([2,3])
10 print "Matrix cmat:"
11 print cmat
12 p = amat[2,1]
13 print "Value of p: ", p
14 q = amat[0,1]
15 print "Value of q: ", q
```

Executing the Python interpreter and running the program in file tmat1.py, yields the following output. Note that matrices *amat* and *bmat* are two-dimensional arrays with three rows and two columns, and matrix *cmat* is a matrix of two rows and three columns.

```
$ python tmat1.py
Create and manipulate a matrix
Matrix amat:
[[ 1.  3.]
 [ 2.  5.]
 [ 4.  7.]]

Matrix bmat:
[[ 1.  1.]
 [ 1.  1.]
 [ 1.  1.]]

Matrix cmat:
[[ 0.  0.  0.]
 [ 0.  0.  0.]]

Value of p:  7.0
Value of q:  3.0
```

Numpy function *identity* creates an identity matrix and sets it to the specified matrix and number of rows. Therefore, the elements of the main diagonal will have a value of 1, and all other elements will have a value of zero. The following segment of Python code creates identity matrix *imat* of four rows.

```
import numpy as np
print "Create an identity matrix of four rows"
imat = np.identity(4)
print imat
```

Running the code with the Python interpreter produces the following output.

```
Create an identity matrix of four rows
[[ 1.   0.   0.   0.]
 [ 0.   1.   0.   0.]
 [ 0.   0.   1.   0.]
 [ 0.   0.   0.   1.]]
```

18.3.2 Element Addition and Subtraction Operations

The basic operations on matrices are performed with a matrix and a scalar or with two matrices. These are carried out by calling several Numpy or Scipy functions in Python.

As with vectors, adding a scalar to a matrix, involves adding the scalar value to every element of the matrix. With Python and Numpy, the addition operator $(+)$ is applied directly to add a value to a matrix.

In the following portion of Python code, in matrix *amat* the constant value 5.5 is added to the matrix and the result is used to create matrix *cmat*. In a similar manner, subtracting a scalar from a matrix is specified.

```
import numpy as np
print "Simple matrix operations"
amat = np.array([[1.0, 3.0], [2.0, 5.0], [4.0, 7.0]])
print "Matrix amat: "
print amat
print "Scalar addition and subtraction: "
cmat = amat + 5.5
dmat = amat - 3.2
print "Matrix cmat:"
print cmat
print "Matrix dmat:"
print dmat
```

Running the code with the Python interpreter produces the following output.

```
Simple matrix operations
Matrix amat:
[[ 1.   3.]
 [ 2.   5.]
 [ 4.   7.]]
Scalar addition and subtration:
Matrix cmat:
[[  6.5   8.5]
 [  7.5  10.5]
 [  9.5  12.5]]
```

```
Matrix dmat:
[[-2.2 -0.2]
 [-1.2  1.8]
 [ 0.8  3.8]]
```

Matrix addition adds two matrices and involves adding the corresponding elements of each matrix. This addition operation on matrices is only possible if the two matrices are of the same size. *Matrix subtraction* is carried out by subtracting the corresponding elements of each of two matrices. The matrices must be of the same size.

The following program is stored in file **tmat4.py** and creates matrix *amat* and matrix *bmat*, which are both matrices of size 3×2. The statement in line 9 adds the elements of matrix *amat* with the elements of matrix *bmat* and then creates the resulting matrix *cmat*. In line 12, the elements of matrix *bmat* are subtracted from the elements of matrix *amat* and creates the resulting matrix *dmat*.

```
 1 import numpy as np
 2 print "Matrix addition and subtraction"
 3 amat = np.array([[1.0, 3.0], [2.0, 5.0], [4.0, 7.0]])
 4 print "Matrix amat: "
 5 print amat
 6 bmat = np.ones([3,2])
 7 print "\nMatrix bmat: "
 8 print bmat
 9 cmat = amat + bmat # matrix addition
10 print "\nMatrix cmat:"
11 print cmat
12 dmat = amat - bmat
13 print "\nMatrix dmat: "
14 print dmat
```

Executing the Python interpreter and running the program in file tmat4.py, yields the following output. Note that matrices *amat* and *bmat* are two-dimensional arrays with three rows and two columns.

```
$ python tmat4.py
Matrix addition and subtraction
Matrix amat:
[[ 1.  3.]
 [ 2.  5.]
 [ 4.  7.]]

Matrix bmat:
```

```
[[ 1.   1.]
 [ 1.   1.]
 [ 1.   1.]]
```

```
Matrix cmat:
[[ 2.   4.]
 [ 3.   6.]
 [ 5.   8.]]
```

```
Matrix dmat:
[[ 0.   2.]
 [ 1.   4.]
 [ 3.   6.]]
```

18.3.3   Element Multiplication and Division

Matrix scalar multiplication involves multiplying each element of the matrix by the value of the scalar. With Numpy, the multiplication operator ($*$) is directly applied to multiply the specified matrix by the constant factor.

Element-by-element matrix multiplication involves multiplying the corresponding elements of two matrices. With Numpy, the multiplication operator ($*$) is directly applied to multiply two matrices of equal size ($m \times n$). The operation multiplies the elements of the first specified matrix by the elements of the second specified matrix.

The following program is stored in file **tmat5.py** and creates matrix *amat* and matrix *bmat*, which are both matrices of size 3×2. The statement in line 9 multiplies the elements of matrix *amat* by the constant 2 and the resulting matrix is *cmat*. In line 12, the elements of matrix *amat* are multiplied by the elements of matrix *bmat* and creates the resulting matrix *dmat*.

```
1 import numpy as np
2 print "Matrix scalar and elementwise multiplication"
3 amat = np.array([[1.0, 3.0], [2.0, 5.0], [4.0, 7.0]])
4 print "Matrix amat: "
5 print amat
6 bmat = np.array([[2.5, 1.5], [3.0, 1.5], [2.0, 4.25]])
7 print "\nMatrix bmat: "
8 print bmat
9 cmat = amat * 2.0 # matrix scalar multiplication
10 print "\nMatrix cmat:"
11 print cmat
12 dmat = amat * bmat
13 print "\nMatrix dmat: "
14 print dmat
```

Executing the Python interpreter and running the program in file tmat5.py, yields the following output. Note that matrices *amat* and *bmat* are two-dimensional arrays with three rows and two columns.

```
$ python tmat5.py
Matrix scalar and elementwise multiplication
Matrix amat:
[[ 1.   3.]
 [ 2.   5.]
 [ 4.   7.]]

Matrix bmat:
[[ 2.5    1.5 ]
 [ 3.     1.5 ]
 [ 2.     4.25]]

Matrix cmat:
[[  2.    6.]
 [  4.   10.]
 [  8.   14.]]

Matrix dmat:
[[  2.5     4.5 ]
 [  6.      7.5 ]
 [  8.     29.75]]
```

Matrix scalar division involves dividing each element of the matrix by the value of the scalar. With Numpy, the division operator ($/$) is applied directly for dividing the specified matrix by the constant factor.

Element-by-element matrix division involves dividing the corresponding elements of two matrices. With Numpy, the division operator ($/$) is applied directly to two matrices of equal size ($m \times n$) and it divides one matrix by the second matrix. This operation divides the elements of the first specified matrix by the elements of the second specified matrix.

The following program is stored in file **tmat6.py** and creates matrix *amat* and matrix *bmat*, which are both matrices of size 3×2. The statement in line 9 divides the elements of matrix *amat* by the constant 2 and the resulting matrix is *cmat*. In line 12, the elements of matrix *amat* are divided by the elements of matrix *bmat* and creates the resulting matrix *dmat*.

```
1 import numpy as np
2 print "Matrix scalar and elementwise division"
3 amat = np.array([[1.0, 3.0], [2.0, 5.0], [4.0, 7.0]])
4 print "Matrix amat: "
```

```
 5 print amat
 6 bmat = np.array([[2.5, 1.5], [3.0, 1.5], [2.0, 4.25]])
 7 print "\nMatrix bmat: "
 8 print bmat
 9 cmat = amat / 2.0 # matrix scalar multiplication
10 print "\nMatrix cmat:"
11 print cmat
12 dmat = amat / bmat
13 print "\nMatrix dmat: "
14 print dmat
```

Executing the Python interpreter and running the program in file tmat6.py, yields the following output. Note that matrices *amat* and *bmat* are two-dimensional arrays with three rows and two columns.

```
$ python tmat6.py
Matrix scalar and elementwise division
Matrix amat:
[[ 1.   3.]
 [ 2.   5.]
 [ 4.   7.]]

Matrix bmat:
[[ 2.5    1.5 ]
 [ 3.     1.5 ]
 [ 2.     4.25]]

Matrix cmat:
[[ 0.5  1.5]
 [ 1.   2.5]
 [ 2.   3.5]]

Matrix dmat:
[[ 0.4         2.        ]
 [ 0.66666667  3.33333333]
 [ 2.          1.64705882]]
```

18.3.4 Additional Matrix Functions

Various additional operations can be applied to matrices. Recall that simple assignment of matrices results in two references to the same matrix and is considered another *view* of the matrix. Copying one matrix to another matrix is performed by calling the Numpy function *copy*. This function copies the elements of the specified source matrix into the specified destination matrix. The two matrices must have the same length $m \times n$. In the following example, matrix *amat* is copied to matrix *bmat*.

```
bmat = np.copy (amat)
```

The slicing operator (:) allows referencing of a subset of a matrix. The following program is stored in file **tmat7a.py** and creates matrix *amat*, which is a matrix of size 3×2. The statement in line 6 applies slicing to reference column 0 of matrix *amat* and is assigned to vector *bmat*. In line 9, rows 0 and 1 of matrix *amat* are referenced and assigned to *cmat*. In line 11, row 1 of matrix *amat* is referenced and assigned to *dmat*. In line 14, row 2 of matrix *amat* is assigned new values from a specified vector.

```
 1 import numpy as np
 2 print "Applying slicing on a matrix"
 3 amat = np.array([[1.0, 3.0], [2.0, 5.0], [4.0, 7.0]])
 4 print "Matrix amat: ", amat.shape
 5 print amat
 6 bmat = amat[:, 0]         # column 0
 7 print "\nColumn 0 of matrix amat: ", bmat.shape
 8 print bmat
 9 cmat = amat[0:2,1] # rows 0 and 1 of col 1
10 print "\nRows 0 and 1 of col 1 Matrix amat: ", cmat
11 dmat = amat[1,:]    # row 1 all cols
12 print "\nRow 1 of matrix amat: ", dmat
13 # assign new values for row 2
14 amat[2, :] = [21.35, 8.55]
15 print "Updated matrix amat: "
16 print amat
```

Executing the Python interpreter and running the program in file tmat7a.py, yields the following output. Note that matrix *amat* is a two-dimensional array with three rows and two columns.

```
$ python tmat7a.py
Applying slicing on a matrix
Matrix amat:  (3, 2)
[[ 1.   3.]
 [ 2.   5.]
 [ 4.   7.]]

Column 0 of matrix amat:  (3,)
[ 1.   2.   4.]
```

```
Rows 0 and 1 of col 1 Matrix amat:   [ 3.   5.]

Row 1 of matrix amat:   [ 2.   5.]
Updated matrix amat:
[[  1.     3.   ]
 [  2.     5.   ]
 [ 21.35   8.55]]
```

Copying rows and columns of a matrix to a vector or another matrix is performed by slicing with the : operator and using function *copy*. The following program is stored in file tmat7.py and creates matrix *amat*, which is a matrix of size 3×2. In line 6, matrix *amat* is copied to matrix *bmat*. In line 9, rows 0 and 1 in column 0 of matrix *amat* are copied to to matrix *cmat*. In line 11, row 1 of matrix *amat* is copied to to *dmat*.

```
 1 import numpy as np
 2 print "Copy a matrix, row, or column"
 3 amat = np.array([[1.0, 3.0], [2.0, 5.0], [4.0, 7.0]])
 4 print "Matrix amat: "
 5 print amat
 6 bmat = np.copy(amat)        # copy entire matrix
 7 print "\nMatrix bmat: "
 8 print bmat
 9 cmat = np.copy(amat[0:2,0]) # rows 0 to 1 of col 0
10 print "\nMatrix cmat: ", cmat
11 dmat = np.copy(amat[1,:])   # row 1 all cols
12 print "\nMatrix dmat: ", dmat
```

Executing the Python interpreter and running the program in file tmat7.py, yields the following output. Note that matrix *amat* is a two-dimensional array with three rows and two columns.

```
$ python tmat7.py
Copy a matrix, row, or column
Matrix amat:
[[ 1.  3.]
 [ 2.  5.]
 [ 4.  7.]]

Matrix bmat:
[[ 1.  3.]
 [ 2.  5.]
 [ 4.  7.]]
```

```
Matrix cmat:   [ 1.   2.]
```

```
Matrix dmat:   [ 2.   5.]
```

A copy of the first specified row or column is selected; then an in-place copy is performed to the second specified row or column. The following program is stored in file `tmat8c.py` and exchanges rows 0 and 1 of matrix *amat*. Function *exchange_col* is defined in lines 4–7 and is called in line 13 to exchange columns 0 and 1 of matrix *amat*.

```
 1 import numpy as np
 2 def exchange_col (x, ncol1, ncol2):
 3     tmat = np.copy(x[:,ncol1])
 4     x[:,ncol1] = x[:,ncol2]
 5     x[:,ncol2] = tmat
 6
 7 print "Exchange columns of a matrix"
 8 amat = np.array([[1.0, 3.0], [2.0, 5.0], [4.0, 7.0]])
 9 print "Matrix amat: ", amat.shape
10 print amat
11 exchange_col(amat, 0, 1)
12 print "Updated matrix amat: "
13 print amat
```

Executing the Python interpreter and running the program in file tmat8c.py, yields the following output. Note that matrix *amat* is a two-dimensional array with three rows and two columns.

```
$ python tmat8c.py
Exchange columns of a matrix
Matrix amat:   (3, 2)
[[ 1.   3.]
 [ 2.   5.]
 [ 4.   7.]]
Updated matrix amat:
[[ 3.   1.]
 [ 5.   2.]
 [ 7.   4.]]
```

In a similar manner, the exchange of two rows is performed by calling function *exchange_row*, which is defined in program `tmat8r.py`

The comparison of two matrices using relational operations is performed by calling the Numpy functions or directly using the standard operators ==, !=,

>, >=, <, and <=. These are element-wise operations on the matrices. Function *array_equal* returns the single truth value *True* if two arrays have the same shape and elements and *False* otherwise.

The following program is stored in file `tmat9.py` and applies the various relational operators on matrices. In line 9, function *array_equal* is called to compare the shape and elements of matrices *amat* and *bmat*. In the rest of the program the relational operators are directly applied.

```
 1 import numpy as np
 2 print "Relational operations on  matrices"
 3 amat = np.array([[1.0, 3.0], [2.0, 5.0], [4.0, 7.0]])
 4 print "Matrix amat: ", amat.shape
 5 print amat
 6 bmat = np.copy(amat)
 7 print "Matrix bmat: "
 8 print bmat
 9 flageq = np.array_equal(amat, bmat)
10 print "Flag matrix equal shape and elements: ", flageq
11 flageq2 = np.equal(amat, bmat) # or amat == bmat
12 print "Flag equal: ", flageq2
13 flaggt = np.greater(amat, bmat)
14 print "Flag greater: ", flaggt
15 flaggte = amat >= bmat
16 print "Flag greater equal: ", flaggte
17 flaglt = amat < bmat
18 print "Flag less than: ", flaglt
19 flagle = amat <= bmat
20 print "Flag less equal: ", flagle
```

Executing the Python interpreter and running the program in file `tmat9.py`, yields the following output.

```
$ python tmat9.py
Relational operations on  matrices
Matrix amat:   (3, 2)
[[ 1.  3.]
 [ 2.  5.]
 [ 4.  7.]]
Matrix bmat:
[[ 1.  3.]
 [ 2.  5.]
 [ 4.  7.]]
Flag matrix equal shape and elements:  True
Flag equal2:  [[ True  True]
```

```
[ True   True]
[ True   True]]
Flag greater:   [[False False]
 [False False]
 [False False]]
Flag greater equal:   [[ True   True]
 [ True   True]
 [ True   True]]
Flag less than:   [[False False]
 [False False]
 [False False]]
Flag less equal:   [[ True   True]
 [ True   True]
 [ True   True]]
```

Relational operations can also be applied to evaluate a Boolean expression on the elements or to select the elements that result from a relational expression. The following program is stored in file `tmat9b.py` and applies the various relational operators on a matrix. In line 6, the Boolean relational expression `>= 1.25` is applied on the elements of matrix *amat* and results in a matrix of Boolean values. In line 8, the Boolean expression is applied on matrix *amat* to select the elements that make the Boolean expression *True*. Lines 10 and 11, applies the Boolean expression only to column 0 of matrix *amat*.

```
 1 import numpy as np
 2 print "Relational operations on elements with specific
     values"
 3 amat = np.array([[1.0, 3.0], [2.0, 5.0], [4.0, 7.0]])
 4 print "Matrix amat: ", amat.shape
 5 print amat
 6 bmat = amat >= 1.25
 7 print "Elements >= 1.25 in amat: ", bmat
 8 cmat = amat[amat >= 1.25]
 9 print "Elements >= 1.25 in amat: ", cmat
10 dmat = amat[0,:]
11 emat = dmat[dmat >= 1.25]
12 print "Elements >= 1.25 in amat in row 0: ", emat
```

Executing the Python interpreter and running the program in file `tmat9b.py`, yields the following output.

```
$ python tmat9b.py
Relational operations on elements with specific values
Matrix amat:  (3, 2)
```

```
[[ 1.   3.]
 [ 2.   5.]
 [ 4.   7.]]
Elements >= 1.25 in amat:   [[False   True]
 [ True   True]
 [ True   True]]
Elements >= 1.25 in amat:   [ 3.   2.   5.   4.   7.]
Elements >= 1.25 in amat in row 0:   [ 3.]
```

Numpy function *amax* computes the maximum value stored in the specified matrix. The following function call gets the maximum value of all elements in matrix *p* and assigns this value to variable *x*. When the *axis* is specified as the second argument in the function call, it indicates that the maximum value is computed in every column (`axis=0`) or in every row (`axis=1`). Therefore, in the following example, *xc* and *xr* are vectors.

```
x = numpy.amax(p)
xc = numpy.amax(p, axis=0) # max in columns
xr = numpy.amax(p, axis=1) # max in columns
```

In addition to the maximum value in a matrix, the index of the element with that value may be desired in every row and every column. Calling function *argmax* computes the index values of the element with the maximum value in the specified in every column (`axis=0`) matrix and in every row (`axis=1`). In the following function call, the index values of the element with the maximum value in matrix *p* in every column are stored in vector *idxc* and the index values of maximum in every row in vector *idxr*.

```
idxc = numpy.argmax (p, axis=0)
idxr = numpy.argmax (p, axis=1)
```

In a similar manner, function *amin* computes the maximum value stored in the specified matrix. Function *argmin* gets the index values (row and column) of the minimum value in the specified matrix.

Listing 18.3 shows a Python program stored in file `array2d_ops.py` that performs the operations discussed for computing maximum values in a matrix.

```
1 import numpy as np
2 # Maximum values and indices in a matrix
3 # Program: array2d_ops.py
4 print "Sum, maximum values in a matrix"
5 amat = np.array([[3.11, 5.12, 2.13], [1.21, 8.22, 5.23],
      [6.77, 2.88, 7.55]])
```

```
 6 print "Array amat: "
 7 print amat
 8 suma = np.sum(amat)
 9 print "Sum of all elements of amat: ", suma
10 sumac = np.sum(amat, axis=0)
11 print "Sum of columns: ", sumac
12 sumar = np.sum(amat, axis=1)
13 print "Sum of rows: ", sumar
14 maxa = np.amax(amat)
15 print "Maximum of all elements of amat: ", maxa
16 maxac = np.amax(amat, axis=0)
17 print "Maximum of columns: ", maxac
18 maxar = np.amax(amat, axis=1)
19 print "Maximum of rows: ", maxar
20 idxc = np.argmax(amat, axis=0)
21 print "Indices of maximum in columns: ", idxc
22 idxr = np.argmax(amat, axis=1)
23 print "Indices of minimum in rows: ", idxr
```

Executing the Python interpreter and running the program in file array2d_ops.py, yields the following output.

```
$ python array2d_ops.py
Sum, maximum values in a matrix
Array amat:
[[ 3.11  5.12  2.13]
 [ 1.21  8.22  5.23]
 [ 6.77  2.88  7.55]]
Sum of all elements of amat:  42.22
Sum of columns:  [ 11.09  16.22  14.91]
Sum of rows:  [ 10.36  14.66  17.2 ]
Maximum of all elements of amat:  8.22
Maximum of columns:  [ 6.77  8.22  7.55]
Maximum of rows:  [ 5.12  8.22  7.55]
Indices of maximum in columns:  [2 1 2]
Indices of minimum in rows:  [1 1 2]
```

The *linalg* module of the Numpy package provides function *inv* that computes the inverse of a matrix. Function *transpose* rearranges the matrix by permuting the rows and columns of a matrix. The following program stored in file array2d_ops2.py illustrates the use of these functions.

```
1 import numpy as np
2 # Inverse and transpose of a matrix
```

```
 3 # Program: array2d_ops2.py
 4 print "Inverse and transpose of a matrix"
 5 amat = np.array([[3.11, 5.12, 2.13], [1.21, 8.22, 5.23],
      [6.77, 2.88, 7.55]])
 6 print "Array amat: "
 7 print amat
 8 imat = np.linalg.inv(amat)
 9 print "Inverse of matrix amat: "
10 print imat
11 tamat = np.transpose(amat)
12 print "Transpose of matrix amat: "
```

Executing the Python interpreter and running the program in file array2d_ops2.py, yields the following output.

```
$ python array2d_ops2.py
Inverse and transpose of a matrix
Array amat:
[[ 3.11  5.12  2.13]
 [ 1.21  8.22  5.23]
 [ 6.77  2.88  7.55]]
Inverse of matrix amat:
[[ 0.27717054 -0.19179357  0.0546632 ]
 [ 0.15493469  0.053433   -0.0807239 ]
 [-0.30763662  0.15159675  0.11422715]]
Transpose of matrix amat:
[[ 3.11  1.21  6.77]
 [ 5.12  8.22  2.88]
 [ 2.13  5.23  7.55]]
```

Function *det* of the *linalg* module of the Numpy package computes the determinant of a square matrix. Function *dot* performs the matrix multiplication of two matrices. Recall that in this operation, the matrices have to be aligned; the number of columns of the first matrix must be equal to the number of rows in the second matrix. The first matrix is $n \times m$, the second matrix is $m \times k$, and the resulting matrix is $n \times k$.

In the following program stored in file array2d_ops3.py creates a 3×3 matrix, *amat*, in line 5. The determinant of matrix *amat* is computed in line 8. A 3×2 matrix, *bmat*, is created in line 10. The dot product of matrix *amat* and matrix *bmat* is computed in line 12.

```
1 import numpy as np
2 # Matrix determinant and multiplication
3 # Program: array2d_ops2.py
```

```
 4 print "Matrix determinant and multiplication "
 5 amat = np.array([[3.11, 5.12, 2.13], [1.21, 8.22, 5.23],
     [6.77, 2.88, 7.55]])
 6 print "Array amat: "
 7 print amat
 8 adet = np.linalg.det(amat)
 9 print "Determinant of amat: " , adet
10 bmat = np.array([[4.5, 6.5], [8.5, 1.5], [5.25, 7.35]])
11 print bmat
12 dpmat = np.dot(amat, bmat)
13 print "Dot product of amat and bmat: "
14 print dpmat
```

Executing the Python interpreter and running the program in file array2d_ops3.py, yields the following output.

```
$ python array2d_ops3.py
Matrix determinant and multiplication
Array amat:
[[ 3.11  5.12  2.13]
 [ 1.21  8.22  5.23]
 [ 6.77  2.88  7.55]]
Determinant of amat:  169.56564
[[ 4.5   6.5 ]
 [ 8.5   1.5 ]
 [ 5.25  7.35]]
Dot product of amat and bmat:
[[  68.6975   43.5505]
 [ 102.7725   58.6355]
 [  94.5825  103.8175]]
```

18.4   SOLVING SYSTEMS OF LINEAR EQUATIONS

Several methods exist for solving systems of linear equations applying vectors and matrices. Some of these methods are: substitution, cancellation, and matrix manipulation. A system of n linear equations can be expressed by the general equations

$$
\begin{aligned}
a_{11}x_1 &+ a_{12}x_2 + \ldots + a_{1n}x_n &= b_1 \\
a_{21}x_1 &+ a_{22}x_2 + \ldots + a_{2n}x_n &= b_2 \\
&\vdots \qquad \vdots \qquad\qquad \vdots \qquad\ \vdots \\
a_{n1}x_1 &+ a_{n2}x_2 + \ldots + a_{nn}x_n &= b_n.
\end{aligned}
$$

This system of linear equations is more conveniently expressed in matrix form in the following manner:

$$
\begin{bmatrix}
a_{11} & a_{12} & \cdots & a_{1n} \\
a_{21} & a_{22} & \cdots & a_{2n} \\
\vdots & \vdots & \ddots & \vdots \\
a_{m1} & a_{2m} & \cdots & a_{mn}
\end{bmatrix}
\begin{bmatrix}
x_1 \\
x_2 \\
\vdots \\
x_n
\end{bmatrix}
=
\begin{bmatrix}
b_1 \\
b_2 \\
\vdots \\
b_m
\end{bmatrix}. \tag{18.1}
$$

This can also be expressed in a more compact matrix form as $AX = B$. Matrix A is the coefficients matrix (of the variables x_i for $i = 1 \ldots n$). X is the vector of unknowns x_i, and B is the vector of solution. Consider a simple linear problem that consists of a system of three linear equations ($n = 3$):

$$
\begin{array}{rrrr}
5x_1 & +2x_2 & +x_3 & = 25 \\
2x_1 & +x_2 & +3x_3 & = 12 \\
-x_1 & +x_2 & +2x_3 & = 5.
\end{array}
$$

In matrix form, this system of three linear equations can be written in the following manner:

$$
\begin{bmatrix}
5 & 2 & 1 \\
2 & 1 & 3 \\
-1 & 1 & 2
\end{bmatrix}
\begin{bmatrix}
x_1 \\
x_2 \\
x_3
\end{bmatrix}
=
\begin{bmatrix}
25 \\
12 \\
5
\end{bmatrix}. \tag{18.2}
$$

Matrix A is a square ($n \times n$) of coefficients, X is a vector of size n, and B is the solution vector also of size n.

$$
A =
\begin{bmatrix}
5 & 2 & 1 \\
2 & 1 & 3 \\
-1 & 1 & 2
\end{bmatrix}
\quad
X =
\begin{bmatrix}
x_1 \\
x_2 \\
x_3
\end{bmatrix}
\quad
B =
\begin{bmatrix}
25 \\
12 \\
5
\end{bmatrix} \tag{18.3}
$$

Decomposing the coefficient matrix, A is the main technique used to compute the determinant, matrix inversion, and the solution a set of linear equations. Common numerical methods used are:

- Gaussian elimination

- LU decomposition

- SV decomposition

- QR decomposition

The Numpy function *linalg.solve* finds the solution to a linear matrix equation, or system of linear scalar equations, given the coefficient matrix *a* and the vector of dependent variable *b*. The function applies LU decomposition to the specified square matrix. The function returns the vector *x*, which has the same dimension as vector *b*.

The following program, stored in file `linsolve.py`, creates a square matrix *amat* in line 6. The solution vector *b* is defined in line 10. The vector of unknowns *x* is created in line 12 by calling function *linalg.solve*.

```
 1 import numpy as np
 2 # Solution to a set of linear equations
 3 # A X = B
 4 # Program: linsol.py
 5 print "Solving a set of linear equations"
 6 amat = np.array([[3.11, 5.12, 2.13], [1.21, 8.22, 5.23],
   [6.77, 2.88, 7.55]])
 7 print "Matrix a: "
 8 print amat
 9 print "Vector b: "
10 b = np.array([4.5, 6.5, 8.5])
11 print b
12 x = np.linalg.solve(amat, bmat)
13 print "Vector x: "
14 print x
```

Executing the Python interpreter and running the program in file `linsolve.py`, yields the following output.

```
$ python linsolve.py
Solving a set of linear equations
Matrix a:
[[ 3.11  5.12  2.13]
 [ 1.21  8.22  5.23]
 [ 6.77  2.88  7.55]]
Vector b:
[ 4.5  6.5  8.5]
Vector x:
[ 0.46524638  0.35836741  0.57194488]
```

18.5   INDUSTRIAL MIXTURES IN MANUFACTURING

Manufacturing of various products require specified amounts of several materials to produce products of acceptable quality. The optimization of such mixtures is discussed in detail in the chapters on linear optimization.

For example, every unit weight (in grams) of product A requires 0.59 g of material P, 0.06 g of material Q, 0.037 g of material R, and 0.313 g of material S. This can be written in a general expression of the form:

$$A = b_1 + b_2 + b_3 + b_4.$$

In the expression, b_1 is the amount material P, b_2 is the amount of material Q, b_3 is the amount of material R, and b_4 is the amount of material S. A similar expression can be used for the mix required for the manufacturing of product B, so on.

The materials needed in the manufacturing process are acquired as substances that contain various amounts of the materials mentioned previously. For example, the following table provides data on the unit content (1 g) of the substances used to obtain the materials needed for manufacturing of products A and B.

Substance	Material P	Material Q	Material R	Material S
S1	0.67	0.2	0.078	0.135
S2	0.87	0.04	0.0029	0.0871
S3	0.059	0.018	0.059	0.864
S4	0.72	0.02	0.03	0.23

From the data in the table, the expression that is used to compute the amount b_1 of material P using y_1 grams of $S1$, y_2 grams of $S2$, y_3 grams of $S3$, and y_4 grams of $S4$ is:

$$b_1 = 0.67\, y_1 + 0.87\, y_2 + 0.059\, y_3 + 0.72\, y_4.$$

In a similar manner, the expression that is used to compute the amount b_2 of material Q using y_1 grams of $S1$, y_2 grams of $S2$, y_3 grams of $S3$, and y_4 grams of $S4$ is:

$$b_2 = 0.2\, y_1 + 0.04\, y_2 + 0.018\, y_3 + 0.02\, y_4$$

Similar expressions can be written for computing the amount of materials R and S needed for the manufacturing of product A.

The system of 4 linear equations that corresponds to this problem can be expressed by:

$$
\begin{array}{llllll}
0.67\, y_1 & + \, 0.87\, y_2 & + \, 0.059\, y_3 & + \, 0.72\, y_4 & = & b_1 \\
0.2\, y_1 & + \, 0.04\, y_2 & + \, 0.018\, y_3 & + \, 0.02\, y_4 & = & b_2 \\
0.078\, y_1 & + \, 0.0029\, y_2 & + \, 0.059\, y_3 & + \, 0.03\, y_4 & = & b_3 \\
0.135\, y_1 & + \, 0.0871\, y_2 & + \, 0.864\, y_3 & + \, 0.23\, y_4 & = & b_4.
\end{array}
$$

This system of linear equations is more conveniently expressed in matrix form in the following manner:

$$
\begin{bmatrix}
a_{11} & a_{12} & a_{13} & a_{14} \\
a_{21} & a_{22} & a_{23} & a_{24} \\
a_{31} & a_{32} & a_{33} & a_{34} \\
a_{41} & a_{42} & a_{43} & a_{44}
\end{bmatrix}
\begin{bmatrix}
y_1 \\
y_2 \\
y_3 \\
y_4
\end{bmatrix}
=
\begin{bmatrix}
b_1 \\
b_2 \\
b_3 \\
b_4
\end{bmatrix}.
\tag{18.4}
$$

This can also be expressed in a more compact matrix form as: $AY = B$. Matrix A is the coefficients matrix (of the variables y_i for $i = 1 \ldots n$). Y is the vector of unknowns y_i, and B is the vector of solution.

The Python program `mixmanuf.py` solves the problem by solving the system of four equations. This provides the solution by computing the values of vector Y, which are the quantities necessary of substances S1, S2, S3, and S4 that are required to produce 1 gram of product A. The following listing shows running the program with the Python interpreter.

```
$ python mixmanuf.py
Mix for Manufacturing Products
Solving the set of linear equations
Matrix A:
[[ 0.67     0.87     0.059    0.72  ]
 [ 0.2      0.04     0.018    0.02  ]
 [ 0.078    0.0029   0.059    0.03  ]
 [ 0.135    0.0871   0.864    0.23  ]]
Vector B:
[ 0.59    0.06    0.037   0.313]
Vector Y:
[ 0.19145029   0.31650837   0.23670601   0.23944495]
```

18.6   SUMMARY

Computations that involve single numbers are known as scalars. Matrices are data structures that store collections of data in two dimensions: rows and columns. To refer to an individual element, two index values are used: one to indicate the row and the other to indicate the column of the element in the array. With the Python programming language and the Numpy package, several functions are available to create and manipulate matrices. Several case studies are presented that show for each problem, the Python source program and the output produced by the program execution.

Key Terms

arrays	elements	index
matrices	array elements	Numpy arrays
column vector	row vector	double-dimension array
matrix operations	matrix functions	determinant
inverse matrix	system of linear equations	LU decomposition

18.7 EXERCISES

18.1 Develop a Python program that reads the values of a matrix M of m rows and n columns. The program must use Numpy and create a new matrix that has the same number of rows and columns, from the appropriate elements in matrix M. Hint: if $m < n$, then the second matrix would be an $m \times m$ square matrix.

18.2 Develop a computational model that inputs and processes the rainfall data for the last five years. For every year, four quarters of rainfall are provided, measured in inches. Hint: use a matrix to store these values. The attributes are the precipitation (in inches), the year, and the quarter. The program must compute the average, minimum, and maximum rainfall per year and per quarter (for the last five years). Implement with Python using a Numpy array.

18.3 Compute the inverse matrix and the determinant of the following matrix:

$$A = \begin{bmatrix} 3.0 & 5.0 & 2.0 \\ 2.0 & 3.0 & -1.0 \\ 1.0 & -2.0 & -3.0 \end{bmatrix}$$

18.4 A computational model has a mathematical representation as a set of three linear equations. Develop a Python program that computes the solution to the following set of linear equations using LU elimination.

$$\begin{aligned} 3x_1 &+5x_2 &+2x_3 &= 8 \\ 2x_1 &+3x_2 &-x_3 &= 1 \\ x_1 &-2x_2 &-3x_3 &= -1 \end{aligned}$$

18.5 A computational model has a mathematical representation as a set of the following linear equations. Develop a Python program that computes the solution to the following set of linear equations using LU elimination.

$$
\begin{array}{rrrrrrrl}
3x_1 & +4x_2 & +2x_3 & -x_4 & +x_5 & +7x_6 & +x_7 & = 42 \\
2x_1 & -2x_2 & +3x_3 & -4x_4 & +5x_5 & +2x_6 & +8x_7 & = 32 \\
x_1 & +2x_2 & +3x_3 & +x_4 & +2x_5 & +4x_6 & +6x_7 & = 12 \\
5x_1 & +10x_2 & +4x_3 & +3x_4 & +9x_5 & -2x_6 & +x_7 & = -5 \\
3x_1 & +2x_2 & -2x_3 & -4x_4 & -5x_5 & -6x_6 & +7x_7 & = 10 \\
-2x_1 & +9x_2 & +x_3 & +3x_4 & -3x_5 & +5x_6 & +x_7 & = 18 \\
x_1 & -2x_2 & -8x_3 & +4x_4 & +2x_5 & +4x_6 & +5x_7 & = 17
\end{array}
$$

## CHAPTER 19

Introduction to Models of Dynamical Systems

19.1   INTRODUCTION

Computational models of dynamical systems are used to study the behavior of systems over time. The foundations for modeling dynamical systems are based on the mathematical concepts of derivatives, integrals, and differential equations. Models of dynamical systems use difference and differential equations to describe the behavior of the systems they represent. This chapter discusses models of dynamical systems and the computer (numerical) solution to the corresponding types of equations using Python and Numpy.

A continuous model is one in which the changes of state in the model occur continuously with time. Often the state variables in the model are represented as continuous functions of time. For example, a model that represents the temperature in a boiler as part of a power plant can be considered a continuous model because the state variable that represents the temperature of the boiler is implemented as a continuous function of time. These types of models are usually modeled as a set of differential equations.

19.2   AVERAGE AND INSTANTANEOUS RATE OF CHANGE

A mathematical function defines the relation between two (or more) variables and this relation is expressed as: $y = f(x)$. In this expression, variable y is a function of variable x, and x is the *independent variable* because for a given value of x, there is a corresponding value of y.

The average rate of change of a variable, y, with respect to a variable, x, (the independent variable) is defined as the proportion of the change of y, denoted as Δy, over a finite interval of x, denoted as Δx.

The Cartesian plane is used to illustrate the concept of rate of change of y with respect to x. It consists of two directed lines that perpendicularly intersect their respective zero points. The horizontal directed line is called

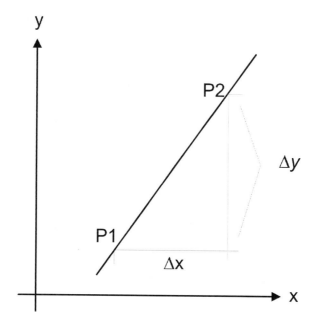

Figure 19.1 The slope of a line.

the *x-axis* and the vertical directed line is called the *y-axis*. The point of intersection of the x-axis and the y-axis is called the *origin* and is denoted by the letter O.

The graphical interpretation of the average rate of change of a variable with respect to another is the *slope* of a line drawn in the Cartesian plane. The vertical axis is usually associated with the values of the dependent variable, y, and the horizontal axis is associated with the values of the independent variable, x.

Figure 19.1 shows a straight line on the Cartesian plane. Two points on the line, P_1 and P_2, are used to compute the slope of the line. Point P_1 is defined by two coordinate values (x_1, y_1) and point P_2 is defined by the coordinate values (x_2, y_2). The horizontal distance between the two points, Δx, is computed by the difference $x_2 - x_1$. The vertical distance between the two points is denoted by Δy and is computed by the difference $y_2 - y_1$.

The *slope* of the line is the inclination of the line and is computed by the expression $\Delta y/\Delta x$, which is the same as the average rate of change of a variable y over an interval Δx. Note that the slope of the line is constant, on any pair of points on the line.

As mentioned previously, if the dependent variable y does not have a linear relationship with the variable x, then the graph that represents the relationship between y and x is a curve instead of a straight line. The average rate of change of a variable y with respect to variable x over an interval Δx, is

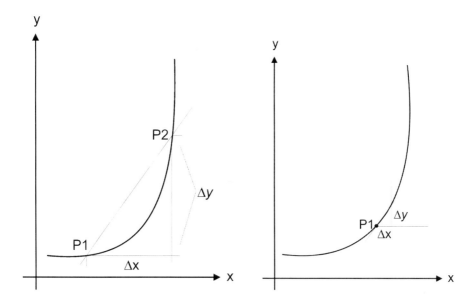

Figure 19.2 Slope of a secant.   Figure 19.3 Slope of a tangent.

computed between two points, P_1 and P_2. The line that connects these two points is called a *secant* of the curve. The average rate on that interval is defined as the slope of that secant. Figure 19.2 shows a secant to the curve at points P_1 and P_2.

The *instantaneous* rate of change of a variable, y, with respect to another variable, x, is the value of the rate of change of y at a particular value of x. This is computed as the slope of a line that is tangent to the curve at a point P.

Figure 19.3 shows a tangent of the curve at point P_1. The instantaneous rate of change at a specified point $P1$ of a curve can be approximated by calculating the slope of a secant and using a very small interval, in different words, choosing Δx very small. This can be accomplished by selecting a second point on the curve closer and closer to point $P1$ (in Figure 19.3), until the secant almost becomes a tangent to the curve at point $P1$.

Examples of rate of change are: the average velocity, $\bar{v}$, computed by $\Delta y/\Delta t$, and the average acceleration $\bar{a}$, computed by $\Delta v/\Delta t$. These are defined over a finite time interval, Δt.

19.3 THE FREE-FALLING OBJECT

A problem is solved by developing a computational model and running the corresponding program. Developing a computational model generally involves applying the software development process discussed previously.

19.3.1 Initial Problem Statement

The problem requires the calculation of the values of the vertical position and the velocity of a free-falling object as time passes. The solution to this problem is the calculation of vertical distance traveled and the velocity as the object approaches the ground. Several relevant questions related to the free-falling object need to be answered. Some of these are:

1. How does the acceleration of gravity affect the motion of the free-falling object?

2. How does the height of the free-falling object change with time, while the object is falling?

3. How does the velocity of the free-falling object change with time, while the object is falling?

4. How long does the free-falling object take to reach ground level, given the initial height, y_0? This question will not be answered here, it is left as an exercise.

19.3.2 Analysis

A brief analysis of the problem involves:

1. Understanding the problem. The main goal of the problem is to develop a model to compute the vertical positions of the object from the point where it was released and the speed accordingly with changes in time.

2. Finding the relevant concepts and principles on the problem being studied. Studying the mathematical expressions for representing the vertical distance traveled and the vertical velocity of the falling object. This knowledge is essential for developing a mathematical model of the problem.

3. Listing the limitations and assumptions about the mathematical relationships found.

19.3.2.1 Assumptions

The main assumption for this problem is that near the surface of the earth, the acceleration, due to the force of gravity, is constant with value 9.8 m/s^2, which is also 32.15 ft/s^2. The second assumption is that the object is released from rest. The third important assumption is that the frictional drag, due to resistance of the air, is not considered.

19.3.2.2 Basic Definitions

The vertical motion of an object is defined in terms of displacement (y), velocity (v), acceleration (g), and time (t).

A time change, denoted by Δt, is a finite interval of time defined by the final time instance minus the initial time instance of the interval of time: $(t_2 - t_1)$. A change of displacement is denoted by Δy, and it represents the difference in the vertical positions of the object in a finite interval: $(y_2 - y_1)$. In a similar manner, a change of velocity is denoted by Δv, and it represents the difference in the velocities in a finite interval: $(v_2 - v_1)$.

The velocity is the *rate of change* of displacement, and the acceleration is the rate of change of velocity. The average velocity, denoted by $\bar{v}$, is the average rate of change of displacement with respect to time on the interval Δt. The average acceleration, denoted by $\bar{a}$, is the average rate of change of the velocity with respect to time on the interval Δt. These are defined by the following mathematical expressions:

$$\bar{v} = \frac{\Delta y}{\Delta t} \qquad \bar{a} = \frac{\Delta v}{\Delta t}$$

19.3.3 Design

The solution to the problem consists of the mathematical formulas expressing the vertical displacement and the velocity of the object in terms of the time since the object was released and began free fall. Note that a general way to compute the average velocity, $\bar{v}$, is from the following expression:

$$\bar{v} = \frac{v_0 + v}{2},$$

with v_0 being the initial velocity and v the final velocity in that interval.

The mathematical model of the solution for a vertical motion of a free-falling object is considered next. Recall that in this model, the air resistance is ignored and the vertical acceleration is the constant $-g$. The vertical position as the object falls is expressed by the equation:

$$y = y_0 + v_0 t - \frac{gt^2}{2}. \tag{19.1}$$

The velocity of the object at any time is given by the equation:

$$v_y = v_0 - gt, \tag{19.2}$$

where y is the vertical position of the object; t is the value of time; v is the vertical velocity of the object; v_0 is the initial vertical velocity of the object; and y_0 is the initial vertical position of the object. Equation 19.1

and Equation 19.2 represent the relationships among the variables: vertical position, vertical velocity, initial velocity, time instant, and initial vertical position of the object.

Note that in this model, the system state changes continuously with time and the problem can be expressed completely by a set of mathematical equations (or expressions).

19.3.4 Implementation

The next step is to implement the mathematical model using a Python program. The computational model has the mathematical expression (formula) for the vertical position, y, and the vertical velocity v_y of the object, and allows arbitrary values given for time t. This really means that the program will use the equations (Equation 19.1 and Equation 19.2) defined previously.

Listing 19.1 shows the source program in Python and stored in file ffallobj3.py. Two constants are first declared in lines 9 and 10. The value for the acceleration of gravity is g, in meters per seconds squared, and the value of the initial height is y_0, in meters. The convention used here is to name the symbolic constants in upper case.

The initial value of the vertical position is read from input in line 12. The values of time t of the falling object is computed in line 14. The values of the vertical position are computed in line 17. The vertical velocity at time t is computed in lines 22–23. The values of the acceleration are computed in lines 30–31.

Listing 19.1: Python program of free-falling object.

```
1 # File: ffallobj3.py
2 # Compute height, velocity, and acceleration
3 # of a free-falling object
4 # Compute vertical velocity and acceleration vs time
5 #   using finite rates of change
6 # J Garrido, Updated 9-1-2014. CS Department, KSU.
7
8 import numpy as np
9 N = 20     # number of data points
10 g = 9.8    # acceleration of gravity m/sec2
11 print "Free-falling object \n"
12 y0 = input("Type initial vertical pos: ")
13 print "\n Time     Vertical Position"
14 tf = np.linspace(0.0, 2.87, N)     # values of time
15 dtf = tf[1] - tf[0]                # delta t
16 # Compute vertical position of falling object
17 hf = y0 - 0.5 * g * pow(tf,2)
18 for j in range(N):
19        print tf[j], hf[j]
```

```
20 # Compute Vertical velocity
21 # using rates of change
22 dhf = np.diff(hf)    # Differences of vertical pos
23 vel = dhf/dtf        # velocity
24 print "\n Time, Vertical velocity"
25 for j in range(N-1):
26      #vel[j+1] = dhf[j]/dtf; // rate of change of hf
27      print tf[j+1], vel[j]
28 print "\n Time  Acceleration of object"
29 # Compute differences of the vertical velocity
30 dvel = np.diff(vel)
31 accel = dvel/dtf
32 for j in range(N-2):
33      print tf[j+2], accel[j]
```

The following lines show executing the Python interpreter and executing program ffallobj3.py.

```
$ python ffallobj3.py
Free-falling object

Type initial vertical pos: 40.0

 Time      Vertical Position
0.0 40.0
0.151052631579 39.8881972022
0.302105263158 39.5527888089
0.453157894737 38.9937748199
0.604210526316 38.2111552355
0.755263157895 37.2049300554
0.906315789474 35.9750992798
1.05736842105 34.5216629086
1.20842105263 32.8446209418
1.35947368421 30.9439733795
1.51052631579 28.8197202216
1.66157894737 26.4718614681
1.81263157895 23.9003971191
1.96368421053 21.1053271745
2.11473684211 18.0866516343
2.26578947368 14.8443704986
2.41684210526 11.3784837673
2.56789473684 7.68899144044
2.71894736842 3.77589351801
2.87 -0.36081
```

```
Time, Vertical velocity
0.151052631579 -0.740157894737
0.302105263158 -2.22047368421
0.453157894737 -3.70078947368
0.604210526316 -5.18110526316
0.755263157895 -6.66142105263
0.906315789474 -8.14173684211
1.05736842105 -9.62205263158
1.20842105263 -11.1023684211
1.35947368421 -12.5826842105
1.51052631579 -14.063
1.66157894737 -15.5433157895
1.81263157895 -17.0236315789
1.96368421053 -18.5039473684
2.11473684211 -19.9842631579
2.26578947368 -21.4645789474
2.41684210526 -22.9448947368
2.56789473684 -24.4252105263
2.71894736842 -25.9055263158
2.87 -27.3858421053
```

To compute the vertical position and velocity of the falling object for several values of time, the program is made to run several times. Table 19.1 shows most of the values used of the height and the vertical velocity computed with the values of time shown. This table represents a simple and short set of results of the original problem.

Table 19.1 Values of height and vertical velocity.

t	0.0	0.5	0.7	1.0	1.2	1.8	2.2	2.5	2.8
y	40.0	38.7	37.6	35.1	32.9	24.12	16.28	9.37	1.58
v_y	0.0	−4.9	−6.86	−9.8	−11.7	−17.6	−21.5	−24.5	−27.4

19.4 DERIVATIVE OF A FUNCTION

A mathematical function defines a relationship between two (or more) variables. A simple relation is expressed as $y = f(x)$. In this expression, variable y is a function of variable x, which is the *independent variable*; for a given value of x there is a corresponding value of y.

The derivative of a function is used to study some relevant properties of the function, for example, the *rate of change*. The derivative of a function

$y = f(x)$ at a particular point is the slope of the tangent line at that point and can be computed with the value of $\Delta y/\Delta x$ when Δx is *infinitely small*. In mathematics, this is the limit when Δx approaches zero or $\Delta x \rightarrow 0$. A description of a graphical interpretation of the slope of a tangent to a function at some specified point appears in Section 19.2. This is the rate of change of y with respect to x. The exact slope, m, of the tangent is expressed as:

$$m = \lim_{\Delta x \to 0} \frac{\Delta y}{\Delta x}.$$

The *derivative* of a variable y with respect to variable x is expressed as:

$$\frac{dy}{dx} = \lim_{\Delta x \to 0} \frac{\Delta y}{\Delta x}.$$

It is assumed that y is a function of x, expressed as $y = f(x)$, and that y is a *continuous* function in an interval of interest. The function $f(x)$ is continuous in an interval if its limit exists for every value of x in the interval.

The derivative of y with respect to x is the instantaneous rate of change of y with respect to x and is denoted as:

$$\frac{dy}{dx} \quad \text{or} \quad y'.$$

The second derivative of y with respect to x is denoted as:

$$y'' \quad \text{or} \quad \frac{d^2 y}{dx^2}.$$

The third and higher-order derivatives are similarly denoted. A derivative of order n is denoted as:

$$y^{(n)} \quad \text{or} \quad \frac{d^n y}{dx^n}.$$

When a variable q is a function of two independent variables, $q = f(x, y)$, the derivative of q has to be specified with respect to x or with respect to y. This concept is known as a *partial derivative*. The partial derivative of q with respect to x is denoted by $\delta q/\delta x$ and the partial derivative of q with respect to y is denoted by $\delta q/\delta y$.

19.4.1 Computing the Derivative with Finite Differences

The derivative of a curve given by $y = f(x)$ at some specified point $x = c$ can be approximated by the use of *finite differences*. Figure 19.4 shows the curve given by $y = f(x)$ at point $x = c$. A finite difference or change of the values of variable y at $x = c$ is denoted by $\Delta y = f(c + h) - f(c)$ and a finite difference or change of the values of variable x is denoted by $\Delta x = h$. An approximation of the derivative of $f(x)$, denoted by $f'(x)$, at $x = c$ can be computed by:

$$f'(x)\big|_{x=c} \approx \frac{f(c+h) - f(c)}{h}.$$

This expression is known as the *forward difference* of $f(x)$ at $x = c$. A similar expression allows computing an approximation to the derivative of $f(x)$ at $x = c$ using the *backward difference*.

$$f'(x)\big|_{x=c} \approx \frac{f(c) - f(c - h)}{h}$$

Another similar expression allows computing an approximation to the derivative of $f(x)$ at $x = c$ using the *central difference*.

$$f'(x)\big|_{x=c} \approx \frac{f(c+h) - f(c - h)}{2h}$$

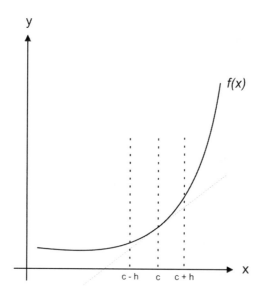

Figure 19.4 The slope of $x = c$.

The value of h can be chosen smaller and smaller to improve the approximation of the derivative of $f(x)$ at $x = c$. Because of roundoff and truncation errors, care must be taken in applying the previous expressions of finite differences. There are finite difference expressions for higher-order derivatives. For example, an estimate of the second derivative of $f(x)$ at $x = c$ can be computed with the central finite difference expression:

$$f''(x)\big|_{x=c} \approx \frac{f(c+h) - 2f(c) + f(c-h)}{h^2}.$$

19.4.2 Computing the First Derivative Using Python

Listing 19.2 shows the source program `deriv1.py` that computes the approximations of the derivative values of $f(x) = x^2$ using forward, backward, and central differences for several values of h.

Listing 19.2: Program that computes the approximation of a derivative.

```
1 # File: deriv1.py
2 # This program estimates the derivative of the function
3 #    f(x) = x^2 at x=2.
4 #    The derivative is computed using forward, backward,
    and central differences.
5 #    J M Garrido. Updated in Python Sep 28, 2014
6
7 def myf (x):
8      return pow (x, 2)
9
10 N = 10    # number of iterations decreasing h
11 h = 0.5   # finite diff interval of x
12 x = 2.0
13 exactv = 2.0 * x   # exact value of derivative
14 print "Exact value of derivative: ", exactv
15 print "h,  forward diff, backward diff, central diff,
   errors"
16 for j in range(N):
17     intf = (myf(x+h) - myf(x)) / h        # forward diff
18     intb = (myf(x) - myf(x-h)) / h        # backward diff
19     intc = (myf(x+h) - myf(x-h))/(2.0*h) # central diff
20     abserrf = (intf - exactv)/intf
21     abserrb = (intb - exactv)/intb
22     abserrc = (intc - exactv)/intc
23     print "h=", h, "forward=",intf, "error=", abserrf
24     print "h=", h, "backward=", intb, "error=",abserrb
25     print "h=", h, "central=", intc, "error=", abserrc
26     h = h / 5.0
```

The following listing shows executing the Python interpreter and running the program `deriv1.py`. The best value of the derivative is 4.0000000361 using a value for h near 0.00016 with forward differences. The best value using central differences appears using $h = 0.1$

```
$ python deriv1.py
Exact value of derivative:  4.0
h,  forward diff, backward diff, central diff, errors
h= 0.5 forward= 4.5 error= 0.111111111111
h= 0.5 backward= 3.5 error= -0.142857142857
h= 0.5 central= 4.0 error= 0.0
h= 0.1 forward= 4.1 error= 0.0243902439024
h= 0.1 backward= 3.9 error= -0.025641025641
h= 0.1 central= 4.0 error= 2.22044604925e-16
h= 0.02 forward= 4.02 error= 0.00497512437811
h= 0.02 backward= 3.98 error= -0.00502512562814
h= 0.02 central= 4.0 error= 8.881784197e-16
h= 0.004 forward= 4.004 error= 0.000999000998975
h= 0.004 backward= 3.996 error= -0.001001001001
h= 0.004 central= 4.0 error= -1.29896093881e-14
h= 0.0008 forward= 4.0008 error= 0.000199960007783
h= 0.0008 backward= 3.9992 error= -0.000200040008145
h= 0.0008 central= 4.0 error= -1.79523063082e-13
h= 0.00016 forward= 4.00016000001 error= 3.9998401598e-05
h= 0.00016 backward= 3.99984 error= -4.0001600292e-05
h= 0.00016 central= 4.0 error= 6.53033183084e-13
h= 3.2e-05 forward= 4.00003200002 error= 7.99994043779e-06
h= 3.2e-05 backward= 3.999968 error= -8.00006296855e-06
h= 3.2e-05 central= 4.00000000001 error= 2.73470135425e-12
h= 6.4e-06 forward= 4.00000640013 error= 1.60003104617e-06
h= 6.4e-06 backward= 3.99999359996 error= -1.6000133499e-06
h= 6.4e-06 central= 4.00000000005 error= 1.14082077117e-11
h= 1.28e-06 forward= 4.0000012809 error= 3.20223912095e-07
h= 1.28e-06 backward= 3.9999987201 error= -3.19975786555e-07
h= 1.28e-06 central= 4.0000000005 error= 1.24165344657e-10
h= 2.56e-07 forward= 4.00000025463 error= 6.36584084889e-08
h= 2.56e-07 backward= 3.99999974462 error= -6.3843767019e-08
h= 2.56e-07 central= 3.99999999963 error= -9.26752008554e-11
```

Function *scipy.misc.derivative* can be used to compute the n-th derivative of a function at a specified point. The function definition has to be provided, as shown in the following Python program stored in file tt deriv2.py. The function and the specified point to compute the derivative is the same as in the previous program.

```
1 # File: deriv2.py
```

```
 2 # This program estimates the derivative of the function
 3 #    f(x) = x^2 at x=2.  The derivative is computed
 4 #    calling scipy.misc.derivative.
 5 #    J M Garrido. Sep 28, 2014
 6
 7 # import numpy as np
 8 import scipy.misc as scder
 9
10 def myf (px):
11     return pow (px, 2)
12
13 N = 20   # number of iterations decreasing h
14 h = 0.5  # finite diff interval of x
15 x = 2.0
16 exactv = 2.0 * x  # exact value of derivative
17 print "Exact value of derivative: ", exactv
18 y = scder.derivative(myf, x, h)
19 print "Derivative, h, error"
20 for j in range(N):
21     y = scder.derivative(myf, x, h)
22     abserrf = (y - exactv)/y
23     print y, h, abserrf
24     h = h / 2.5
```

The following listing shows executing the Python interpreter and running the program deriv2.py.

```
$ python deriv2.py
Exact value of derivative:  4.0
Derivative, h, error
4.0 0.5 0.0
4.0 0.2 2.22044604925e-16
4.0 0.08 8.881784197e-16
4.0 0.032 8.881784197e-16
4.0 0.0128 -6.10622663544e-15
4.0 0.00512 -1.90958360236e-14
4.0 0.002048 -3.00870439673e-14
4.0 0.0008192 -4.36317648678e-14
4.0 0.00032768 -1.79189996175e-13
4.0 0.000131072 -3.48610029732e-13
4.0 5.24288e-05 7.48290318597e-14
4.0 2.097152e-05 1.13375975275e-12
3.99999999995 8.388608e-06 -1.21010979017e-11
4.00000000015 3.3554432e-06 3.75297570784e-11
3.99999999995 1.34217728e-06 -1.21010979017e-11
```

```
4.00000000086 5.36870912e-07 2.15373718813e-10
4.00000000128 2.147483648e-07 3.18771231532e-10
3.99999999455 8.589934592e-08 -1.36143940751e-09
3.99999998292 3.4359738368e-08 -4.26949621925e-09
4.00000000231 1.37438953472e-08 5.77265124257e-10
```

19.5   NUMERICAL INTEGRATION

Several methods exist for the approximation of the integration of functions. The simplest methods are the Trapezoid method and Simpson's method. A more advanced method is that of Gauss Quadrature. The integral of function $f(x)$ is formulated as:

$$I = \int_{x_a}^{x_b} f(x)\, w(x)\, dx.$$

The function $w(x)$ is a weight function of $f(x)$, x_a is the lower bound, and x_b is the upper bound of the integration interval. In the practical cases presented in this chapter, $w(x) \equiv 1$.

The Scipy and Numpy packages provide several functions that implement methods for adaptive and non-adaptive integration of general functions, and some functions with specialized methods for specific cases. In the function call, the user specifies the absolute and relative error bounds required.

In partitioning the integration interval, the adaptive functions tend to adjust according to the behavior of the function $f(x)$.

19.5.1   Area under a Curve

A general method for approximating the area under a curve in the interval $x = x_a$ and $x = x_b$ is the trapezoid method. It consists of dividing the interval $[x_a, x_b]$ into several *trapezoids*, computing the areas of the trapezoids, and adding these areas.

A trapezoid is a geometric figure with four sides and only two parallel opposite sides. The area of a trapezoid with width $\Delta x = x_{i+1} - x_i$, is computed as:

$$q = \Delta x\, \frac{y_i + y_{i+1}}{2}.$$

There are $n - 1$ equal subintervals, Δx, in the interval $[x_a, x_b]$ on variable x and $y_i = f(x_i)$. The sum of the areas of the trapezoids is:

$$A = \sum_{k=1}^{k=n-1} [\Delta x\, \frac{1}{2}\, (y_k + y_{k+1})].$$

The approximation of the area can be improved with smaller values of the width (Δx) of the trapezoids. Figure 19.5 shows a segment of a curve divided into $n-1$ trapezoids. The area from x_a to x_b with $x_a = x_1 < x_2 < \ldots < x_n = x_b$ is:

$$A = \frac{x_b - x_a}{2n} \left[y_1 + 2y_2 + \ldots + 2y_{n-1} + y_n \right].$$

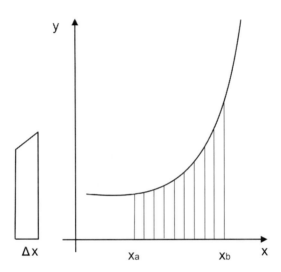

Figure 19.5   The area under a curve.

19.5.2   Using the Trapezoid Method

Numerical integration consists of approximating the integral of a function in a given interval. The area under a curve on an interval is the basic concept used for computing an approximation of the integral of a function. This applies for functions continuous and nonnegative on the given interval. For functions that have negative values in a subinterval, the area of this subinterval is given a negative sign.

The program in Listing 19.3, stored in file `trapz1.py`, calls the Numpy function *trapz* that implements the Trapezoid method to compute an approximate integral of function $f(x) = e^x$ in the interval $(0.0, 1.0)$ of x.

Listing 19.3: Python program that computes an integral of $f(x) = e^x$.

```
1 # File: trapz1.py
2 # This program estimates the integral of the function
3 #    f(x) = e^x on the interval (0, 1.0)
4 #    The program calls trapz function.
```

```
 5 #    It start with a small number of trapezoids and
 6 #    increasing
 7 #    J M Garrido. Updated in Python Sep 28, 2014
 8 import numpy as np
 9 def myf (x):
10     return np.exp (x)
11
12 n = 4      # initial number of trapezoids
13 xa = 0.0  # lower bound interval of x
14 xb = 1.0  # upper bound
15 exactint = 1.718282  # exact value of integral
16 print "Compute the integral of f(x) = e^x "
17 print "Exact value of integral: ", exactint
18 while n <= 16500:
19     x = np.linspace(xa, xb, n)
20     y = myf(x)
21     intv = np.trapz(y, x)  # compute integral
22     errint = intv - exactint
23     print "n=", n, " integral =", intv, " error =", errint
24     n = n * 2
```

The following listing shows the execution of the Python interpreter and running of the program. Note that with $n = 64$ (number of trapezoids), the computed approximation is reasonably close to the exact value of the integral, 1.718282.

```
$ python trapz1.py
Compute the integral of f(x) = e^x
Exact value of integral:  1.718282
n= 4   integral = 1.73416246012   error = 0.0158804601234
n= 8   integral = 1.72120308299   error = 0.00292108298745
n= 16   integral = 1.718918182   error = 0.000636182000449
n= 32   integral = 1.71843082707   error = 0.000148827074113
n= 64   integral = 1.71831790544   error = 3.59054434467e-05
n= 128   integral = 1.71829070626   error = 8.70625707416e-06
n= 256   integral = 1.71828403054   error = 2.0305369488e-06
n= 512   integral = 1.71828237683   error = 3.76826060133e-07
n= 1024   integral = 1.71828196528   error = -3.47170827641e-08
n= 2048   integral = 1.71828186263   error = -1.37368397768e-07
n= 4096   integral = 1.718281837   error = -1.63001986131e-07
n= 8192   integral = 1.71828183059   error = -1.69406733086e-07
n= 16384   integral = 1.71828182899   error = -1.71007461303e-07
```

The program in file **trapz2.py** applies a cumulatively integration of $y(x)$ using the composite trapezoidal rule. Scipy function *romb* applies Romberg integration using samples of a specified function. The program in file `romb1.py` applies this method. The following listing shows the result of computing the same problem.

```
$ python romb1.py
Compute the integral of f(x) = e^x
Exact value of integral:  1.718282
n= 3   integral = 1.85914091423  error = 0.14085891423
n= 5   integral = 1.71886115188  error = 0.000579151876593
n= 9   integral = 1.71828268792  error = 6.87924757159e-07
n= 17  integral = 1.71828182879  error = -1.71205469801e-07
n= 33  integral = 1.71828182846  error = -1.7154092169e-07
n= 65  integral = 1.71828182846  error = -1.71540954552e-07
n= 129 integral = 1.71828182846  error = -1.71540954996e-07
n= 257 integral = 1.71828182846  error = -1.71540955218e-07
n= 513 integral = 1.71828182846  error = -1.71540955218e-07
```

19.5.3 Using Adaptive Quadrature

The Scipy function *quad* applies an adaptive quadrature method for integration. The first parameter is a reference to the function to integrate, the next two parameters are the lower and upper bounds of the integration interval. The *quad* function returns a tuple with the estimated value of the function and an upper bound of the error.

The following program is stored in file `quad1.py` and it computes the integral of the same function in the previous program. It calls function *quad* to compute an estimate of the given function in the specified interval.

```
1 # File: quad1.py
2 # This program estimates the integral of the function
3 #    f(x) = e^x on the interval (0, 1.0)
4 #    The program calls function quad.
5 #    J M Garrido. Sep 28, 2014
6 import numpy as np
7 import scipy.integrate as scint
8 def myf (x):
9     return np.exp (x)
10
11 xa = 0.0  # lower bound interval of x
12 xb = 1.0  # upper bound
13 exactint = 1.718282  # exact value of integral
14 print "Compute the integral of f(x) = e^x "
15 (intv, errub) = scint.quad(myf, xa, xb) # compute integral
16 errint = intv - exactint
17 print "Upper bound on error: ", errub
```

```
18 print "Exact: ", exactint, " integral =", intv,
   " rel error =", errint
```

The following listing shows the execution of the Python interpreter and running of program quad1.py.

```
$ python quad1.py
Compute the integral of f(x) = e^x
Upper bound on error:  1.90767604875e-14
Exact:  1.718282  integral = 1.71828182846
  rel error = -1.71540954774e-07
```

19.6   WORK PRODUCED IN A PISTON WITH AN IDEAL GAS

A piston with gas in its cylinder is a simple thermodynamic system. The process is isothermal if the temperature is kept constant. A gas confined by a piston in a cylinder is not heated or cooled, but the piston is slowly moved so that the gas expands or is compressed.

The temperature is maintained at a constant value by putting the system in contact with a constant-temperature reservoir (the thermodynamic definition of a reservoir is something large enough that it can transfer heat into or out of a system without changing temperature).

The behavior of an ideal gas is the relationship of pressure (P), volume (V), and temperature (T), and can be summarized in the ideal gas law:

$$PV = nRT,$$

where n is the number of moles of gas, and $R = 8.314$ J / (mol K) is known as the universal gas constant.

If the volume increases while the temperature is constant, the pressure must decrease, and if the volume decreases the pressure must increase. The work produced is due to the gas pressure on the piston. The work can also be represented per unit mass of fuel and air. Work is simply a force multiplied by the distance moved in the direction of the force.

For a small displacement, dx, the work is dW.

$$dW = F\,dx = PA\,dx = P\,dV$$

For a finite volume change, work is given by:

$$W = \int_{v_1}^{v_2} P\,dV.$$

The python program wpiston.py computes the work produced in a piston using an ideal gas. The temperature value is 300.0 degrees Kelvin and 1 mole of gas used. The volume changes applied on the piston are: $V1 = 1.0$ to

$V2 = 7.5$ cubic meters. The program calls the Scipy function *quad* that applies an adaptive quadrature method for integration. The following listing shows the results produced after running the program.

```
$python wpiston.py
Compute the work produced on a piston
Volume limits (cubic meters):  1.0 7.5
Upper bound on error:  4.8116672127e-07
Work produced:  5025.57111384  kJ
```

19.7 DIFFERENTIAL EQUATIONS

An ordinary differential equation (ODE) is one that includes derivatives of a function with a single independent variable. For example, the following equation is a differential equation of order one.

$$\frac{dy}{dx} + 20y - 6x = 23.5$$

The order of a differential equation is determined by the highest derivative. The following expression is an example of a differential equation of order two.

$$y'' + 20y'x - 3y'y = 12$$

The general form of a differential equation of order n is expressed in the following manner:

$$\frac{d^n z(t)}{dt^n} + a_1 \frac{d^{n-1} z(t)}{dt^{n-1}} + \cdots + a_{n-1} \frac{dz(t)}{dt} + a_n z(t) = Q. \tag{19.3}$$

The solution to a differential equation is an expression for the function $y = f(x)$. It is convenient when solving a differential equation of a higher order with numerical methods, to reduce the order of the equation to order one.

A differential equation of order n can be reduced to n first-order differential equations. The following variable substitutions are necessary to reduce Equation (19.3) to a system of n first-order equations:

$$x_1 = z(t), \quad x_2 = \frac{dz(t)}{dt}, \quad x_3 = \frac{d^2 z(t)}{dt^2}, \quad \ldots, \quad x_n = \frac{d^{n-1} z(t)}{dt^{n-1}}. \tag{19.4}$$

Substituting the Equations (19.4) in Equation (19.3), the following first-order equations result:

$$\frac{dx_1}{dt} = x_2$$

$$\frac{dx_2}{dt} = x_3$$

$$\frac{dx_3}{dt} = x_4$$

$$\cdots$$

$$\frac{dx_n}{dt} = Q - a_1 x_n - a_2 x_{n-1} - \cdots - a_{n-1} x_2 - a_n x_1. \tag{19.5}$$

19.8 MODELS OF DYNAMICAL SYSTEMS

Models of dynamical systems describe the behavior of systems varying with time. Differential equations are used to model this behavior as it changes continuously with time. A model of a dynamical system has two major components:

1. the state vector that indicates the current state at a particular time instance and

2. a set of (linear) differential equations that describe the continuous change of state.

Figure 19.6 illustrates the high-level view of a dynamical system and includes the variables used in the modeling: $U(t)$ is the vector of input variables, $Y(t)$ is the vector of output variables, and $X(t)$ is the vector of state variables. All these variables are functions of time.

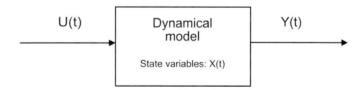

Figure 19.6 High-level view of a dynamical system.

The dynamic behavior of a continuous linear system is described by the following set of linear differential equations.

$$
\begin{aligned}
\dot{x}_1 &= a_{11}x_1 + a_{12}x_2 + \ldots + a_{1n}x_n + b_{11}u_1 + b_{12}u_2 + \cdots + b_{1m}u_m \\
\dot{x}_2 &= a_{21}x_1 + a_{22}x_2 + \ldots + a_{2n}x_n + b_{21}u_1 + b_{22}u_2 + \cdots + b_{2m}u_m \\
&\cdots \\
\dot{x}_n &= a_{n1}x_1 + a_{n2}x_2 + \ldots + a_{nn}x_n + b_{n1}u_1 + b_{n2}u_2 + \cdots + b_{nm}u_m. \tag{19.6}
\end{aligned}
$$

19.8.1 State Equations

Equations (19.6) are known as *state equations*, and are expressed in matrix form as:

$$\dot{X} = AX + BU. \tag{19.7}$$

This state equation uses the following matrix definitions:

$$\dot{X} = \begin{bmatrix} \dot{x}_1 \\ \dot{x}_2 \\ \vdots \\ \dot{x}_n \end{bmatrix} \quad A = \begin{bmatrix} a_{11} & a_{12} & \cdots & a_{1n} \\ a_{21} & a_{22} & \cdots & a_{2n} \\ \vdots & \vdots & \ddots & \vdots \\ a_{m1} & a_{2m} & \cdots & a_{mn} \end{bmatrix} \quad X = \begin{bmatrix} x_1 \\ x_2 \\ \vdots \\ x_n \end{bmatrix}$$

$$B = \begin{bmatrix} b_{11} & b_{12} & \cdots & b_{1m} \\ b_{21} & b_{22} & \cdots & b_{2m} \\ \vdots & \vdots & \ddots & \vdots \\ b_{n1} & b_{n2} & \cdots & b_{nm} \end{bmatrix} \quad U = \begin{bmatrix} u_1 \\ u_2 \\ \vdots \\ u_m \end{bmatrix}. \tag{19.8}$$

In the state equations, $\dot{x}$ denotes dx/dt, A is an $m \times n$ matrix, X is a column vector of size n, $\dot{X}$ is a column vector of size n, B is an $n \times m$ matrix, and U is a column vector of size m.

19.8.2 Output Equations

The output equations of a model of a dynamical system are expressed as follows:

$$
\begin{aligned}
y_1 &= c_{11}x_1 + c_{12}x_2 + \ldots + c_{1n}x_n + d_{11}u_1 + d_{12}u_2 + \cdots + d_{1m}u_m \\
y_2 &= c_{21}x_1 + c_{22}x_2 + \ldots + c_{2n}x_n + c_{21}u_1 + d_{22}u_2 + \cdots + d_{2m}u_m \\
&\cdots \\
y_n &= a_{k1}x_1 + a_{k2}x_2 + \ldots + a_{kn}x_n + d_{k1}u_1 + d_{k2}u_2 + \cdots + d_{km}u_m.
\end{aligned} \tag{19.9}
$$

Equation (19.9) can also be written in matrix form as $Y = CX + DU$, in which C is an $k \times m$ matrix, Y is a column vector of size k, D is an $k \times m$ matrix, and U is a column vector of size m.

In a more compact form, the model of a dynamical system can be expressed with two matrix equations:

$$\dot{X} = AX + BU$$
$$Y = CX + DU. \tag{19.10}$$

19.9 FORMULATING SIMPLE EXAMPLES

This section describes the formulation of problems using state variables with differential equations and applying basic laws of physics.

19.9.1 Free-Falling Object

An object is released from a certain height, h_0, and falls freely until it reaches ground level. The problem is to study the changes in the vertical location of the object and its velocity as time progresses.

Assume that the mass of the object is m, and the only force applied on the object is that due to gravity, g. Let x denote the vertical displacement of the object, that is, its height as a function of time, and v its vertical velocity. Applying Newton's law of force, which relates mass, acceleration, and force, gives the following expression:

$$-mg = m\frac{d^2x}{dt^2}.$$

This differential equation of order 2 can be reduced to two first-order differential equations. Because the velocity is the instantaneous rate of change of the displacement and the acceleration is the instantaneous rate of change of the velocity, the two first-order differential equations are:

$$v = \frac{dx}{dt}$$
$$-mg = m\frac{dv}{dt}.$$

The two state variables are the velocity, v, and the displacement, x. There is only one input variable, u_1, and its value is the constant g. The two state equations in the form of general state equations in (19.6), are the following:

$$\frac{dx}{dt} = 0x + v + 0$$
$$\frac{dv}{dt} = 0x + 0v + -g.$$

Following the general matrix and vector form in Equation (19.8), the state vector X, matrix A, matrix B, and vector U are:

$$X = \begin{bmatrix} x \\ v \end{bmatrix} \qquad A = \begin{bmatrix} 0 & 1 \\ 0 & 0 \end{bmatrix} \qquad B = \begin{bmatrix} 0 \\ -1 \end{bmatrix} \qquad U = [g].$$

The output equations are:

$$
\begin{aligned}
y_1 &= x + 0v \\
y_2 &= 0x + v.
\end{aligned}
$$

Matrix C and matrix D are:

$$C = \begin{bmatrix} 1 & 0 \\ 0 & 1 \end{bmatrix} \qquad D = \begin{bmatrix} 0 \\ 0 \end{bmatrix}.$$

19.9.2 Object on Horizontal Surface

A force, F, is applied to an object on a horizontal surface. The resistance of the surface on the object due to friction is proportional to the velocity of the object, and its value is $-kv$. The constant k is the coefficient of friction of the surface. The horizontal displacement of the object is denoted by x. The problem is to find the instantaneous change in the displacement of the object.

The dynamic behavior of the object is defined by Newton's law, and is expressed as follows:

$$m \frac{d^2x}{dt^2} = F - kv.$$

Because the velocity is the instantaneous change of the displacement, and the acceleration is the instantaneous change in the velocity of the object, the previous equation is expressed as follows:

$$
\begin{aligned}
v &= \frac{dx}{dt} \\
m \frac{dv}{dt} &= F - kv.
\end{aligned}
$$

For this problem, the two state variables are displacement x, and the velocity v, of the object. The input variable, u_1, is the force F. The two state equations are expressed as follows:

$$\frac{dx}{dt} = 0x + v + 0$$

$$\frac{dv}{dt} = 0x - \frac{kv}{m} + F/m.$$

In this problem, the state vector X, matrix A, matrix B, and vector U are:

$$X = \begin{bmatrix} x \\ v \end{bmatrix} \quad A = \begin{bmatrix} 0 & 1 \\ 0 & -k/m \end{bmatrix} \quad B = \begin{bmatrix} 0 \\ 1/m \end{bmatrix} \quad U = [F].$$

The output equations are:

$$y_1 = x + 0v$$

$$y_2 = 0x + v.$$

Matrix C and matrix D are:

$$C = \begin{bmatrix} 1 & 0 \\ 0 & 1 \end{bmatrix} \quad D = \begin{bmatrix} 0 \\ 0 \end{bmatrix}.$$

19.9.3 Object Moving on an Inclined Surface

A force F is applied to an object on an inclined surface. The elevation angle is θ. The frictional force that resists movement is proportional to the velocity of the object. The problem is to derive the change of the displacement and the velocity of the object on the inclined surface.

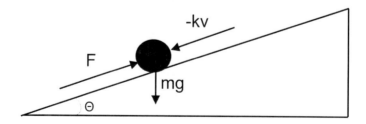

Figure 19.7 Object on inclined surface.

The projection of the force of gravity of the object on the inclined surface is $mg \sin \theta$. As explained in previous problems, applying the law of Newton, the following equations are derived:

$$v = \frac{dx}{dt}$$

$$m\frac{dv}{dt} = F - kv - mg \sin \theta.$$

For this problem, the two state variables are displacement x, and the velocity v, of the object. The input variables are the force F and the acceleration due to the gravity. The two state equations are expressed as follows:

$$\frac{dx}{dt} = 0x + v + 0$$

$$\frac{dv}{dt} = 0x - \frac{kv}{m} + F/m - g \sin \theta.$$

In this problem, the state vector X, matrix A, matrix B, and vector U are:

$$X = \begin{bmatrix} x \\ v \end{bmatrix} \quad A = \begin{bmatrix} 0 & 1 \\ 0 & -k/m \end{bmatrix} \quad B = \begin{bmatrix} 0 & 0 \\ 1/m & -\sin \theta \end{bmatrix} \quad U = \begin{bmatrix} F \\ g \end{bmatrix}.$$

The output equations are:

$$y_1 = x + 0v$$

$$y_2 = 0x + v.$$

Matrix C, vector Y, and matrix D are:

$$C = \begin{bmatrix} 1 & 0 \\ 0 & 1 \end{bmatrix} \quad Y = \begin{bmatrix} x \\ v \end{bmatrix} \quad D = \begin{bmatrix} 0 & 0 \\ 0 & 0 \end{bmatrix}.$$

19.10   SOLUTION OF DIFFERENTIAL EQUATIONS

There are many ordinary differential equations (ODEs) that cannot be solved analytically. Numerical methods are techniques that compute estimates using software implementations. Euler's method is the simplest technique for solving differential equations numerically. With this method, the time step is constant from one iteration to the next. However, this may not be feasible for many functions or may result in an inaccurate solution.

Methods that adjust the time step as the computation proceeds are known as *adaptive* methods. The Dormand–Prince pair of Runge–Kutta is one of the best adaptive methods.

Some general-purpose ODE solvers use a modified Newton iteration method is used to solve the system of non-linear equations. This is generally suitable for stiff problems and requires the Jacobian.

In a *stiff problem*, some methods for solving the equation are numerically unstable, and the step size is forced to be unacceptably small in a region where the solution curve is very smooth.

The general structure of a Python program that numerically solves a set of first-order ordinary differential equations with Scipy has the following sequence of parts in its code:

1. Set up the ODE system with the following components: the programmer-defined function with the right-hand side of the first-order differential equations to solve, the array of initial conditions of differential equations to solve, and an array with the values of time to be used for solving the equations.

2. A call to the Scipy function *odeint*, the ODE solver. The arguments to the function call are the reference to the supplied function that defines the system of differential equations to solve; the array of initial conditions; and the array of time values.

3. Get the solution array and extract from each column the corresponding variable.

4. Optionally, plot every variable with respect to time.

19.10.1 Model with a Single Differential Equation

This case study is a very simple model represented by only a single first-order differential equation. The problem consists of numerically solving the following differential equation:

$$\frac{dy}{dt} = \alpha x \left[1 + \sin(\omega t)\right]. \tag{19.11}$$

This equation is solved with $\alpha = 0.015$ and $\omega = 2\pi/365$, an initial value $y(0) = 2$, and an interval of t from 0.0 to 365.0.

Listing 19.5 shows the Python program that numerically solves the single differential equation and it is stored in file `tode1.py`. The programmer-defined function *dfunc* that specifies the differential equation (19.11) is defined in lines 16–18. Function *odeint* is called in line 33. The function that generates the plot is called in line 37.

Listing 19.5: Python program that solves a single differential equation.

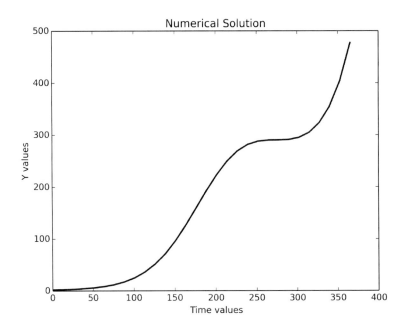

Figure 19.8 Case Study 1: Model with single differential equation.

```
 1 # Program: tode1.py
 2 #   This program solves a single differential equation
 3 #   dxdt = alpha x [1+ sin(w t)]
 4 #   J M Garrido, Sep 2014. CS dept, KSU
 5 #   Uses Scipy odeint ODE function
 6 #   This program solves the equation
 7 #         x'(t) - alpha x(t) (1+sin(omega t)) = 0
 8 import math as m
 9 import numpy as np
10 import matplotlib.pyplot as plt
11 from scipy.integrate import odeint
12
13 alpha = 0.015              # parameter for the diff eq
14 omega = 2.0 * m.pi/365.0 # parameter
15 # dfunct - defines the first order differential equation
16 def dfunc (y, t):
17     f0 = alpha * y[0] * (1+ m.sin (omega*t))
18     return f0
19
20 N = 30   # Number of data points
21 print "This program solves a system with a single
   diff  equation"
```

```
22 tmin = 0.0       # starting t value
23 tmax = 365.0     # final t value
24 dimension = 1    # number of diff eqs
25 yinit = 2.0          # initial value of x
26 print "Input data: "
27 print " alpha = ", alpha, " omega = ", omega
28 print " Time parameters: ", tmin, tmax
29 print " Number of equations: ", dimension
30 print "\n     Time          y          "
31 t  = np.linspace(tmin, tmax, N)    # time grid
32 # solve the ODE
33 ysol = odeint(dfunc, yinit, t)
34 yy = ysol[:,0] # extract column 0
35 for j in range (N):
36     print t[j], yy[j]
37 plt.plot(t, yy, linestyle="-", color="black", linewidth=2.0,
   label="ODE solution")
38 plt.title('Numerical Solution')
39 plt.xlabel('Time values')
40 plt.ylabel('Y values')
41 # Save figure using 300 dots per inch
42 plt.savefig("tode1p.png",dpi=300)
43 plt.show() # Show on screen
```

The following listing shows resulting output when the program tode1.py
runs. The result includes an array with the values of time and an array with
the values of y.

```
$ python tode1.py
This program solves a system with a single diff   equation
Input data:
 alpha =  0.015   omega =  0.017214206321
 Time parameters:  0.0 365.0
 Number of equations:  1

     Time             y
  0.0              2.0
 12.5862068966 2.46529755215
 25.1724137931 3.1622055048
 37.7586206897 4.20891786935
 50.3448275862 5.78679035902
 62.9310344828 8.16884389072
 75.5172413793 11.7536287335
 . . .
```

```
339.827586207 354.637897137
352.413793103 403.318381344
365.0 477.300923692
```

Figure 19.8 shows the graph generated with *matplotlib* of the values of y with time and is a visual representation of the numerical solution of the differential equation (19.11).

19.10.2 Model with a System of Differential Equations

Most practical models are represented by a system of two or more first-order and most often correspond to one or more higher-order differential equations. For example, consider the following system of linear first-order differential equations:

$$\frac{dx_1}{dt} = -x_1 - x_2$$
$$\frac{dx_2}{dt} = x_1 - 2x_2. \tag{19.12}$$

These equations can be expressed in the form of state equations and for this, state vector X and matrix A are:

$$X = \begin{bmatrix} x_1 \\ x_2 \end{bmatrix} \qquad A = \begin{bmatrix} -1 & -1 \\ 1 & -2 \end{bmatrix}. \tag{19.13}$$

For this problem, matrix B, and vector U are empty. The output equations are:

$$y_1 = x_1 + 0x_2$$
$$y_2 = 0x_1 + x_2.$$

Matrix C and vector Y are expressed in the following form:

$$C = \begin{bmatrix} 1 & 0 \\ 0 & 1 \end{bmatrix} \qquad Y = \begin{bmatrix} x_1 \\ x_2 \end{bmatrix}. \tag{19.14}$$

This system of differential equations can be solved numerically using Python and Scipy in a similar manner as with the previous case study. The Python code of the programmer-defined function *dfunc* that specifies the two first-order differential equations (19.12) is shown as follows.

```
def dfunc (y, t):
    # evaluate the right-hand-side at t
    f0 = - y[0] - y[1]
    f1 =   y[0] - 2.0 * y[1]
    return [f0, f1]
```

The initial conditions are set as $x_1 = 1.0$ and $x_2 = 1.0$ at $t = 0$; the time span is set for values of t from 0.0 to 5.0. The program that implements the solution of the model with the two differential equations (19.12) is stored in file tode2.py. The following portion of code shows how the program solves the ODE system.

```
ysol = odeint(dfunc, xinit, t)
x1 = xsol[:, 0]      # extract column 0
x2 = xsol[:, 1]      # column 1
```

Figure 19.9 shows the graph generated by *matplotlib* of the values of variables t, *x1*, and *x2* with time, and represents visually the numerical solution of the differential equations (19.12).

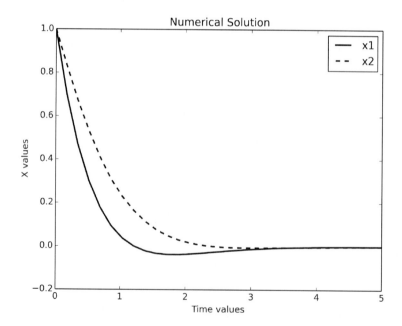

Figure 19.9 Numerical solution to a system of two differential equations.

19.10.3 Model with Drag Force

An object is released from a specified height, h_0, and falls freely until it reaches ground level. The problem is to formulate and solve a model to study the changes in the vertical position of the object and its velocity as time progresses.

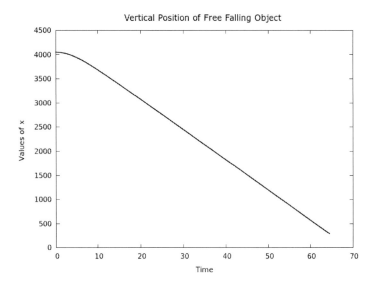

Figure 19.10 Vertical position of free-falling object.

Assume that the mass of the object is m, and there are two forces applied on the object: one due to gravity with acceleration g and the second is the drag force against the direction of movement. Let x denote the vertical displacement of the object, and v its vertical velocity. The drag force is cv^2, in which c is the drag constant. Applying Newton's law of force that relates mass, acceleration, and force, gives the following expression:

$$-mg + cv^2 = m\frac{d^2x}{dt^2}.$$

The vertical velocity is the instantaneous rate of change of the vertical position and the acceleration is the instantaneous rate of change of the vertical velocity. The second-order differential equation can be reduced to two first-order differential equations:

$$v = \frac{dx}{dt}$$
$$-mg + cv^2 = m\frac{dv}{dt}.$$

The two state variables are the vertical position, x, and the vertical velocity, v. The two state equations are the following:

$$\frac{dx}{dt} = v$$

$$\frac{dv}{dt} = (c/m)v^2 - g.$$

The output equations are:

$$y_1 = x$$

$$y_2 = v.$$

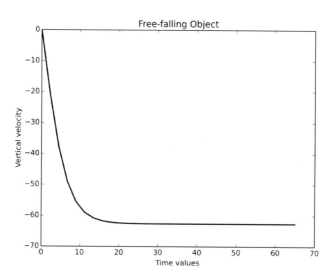

Figure 19.11 Vertical velocity of free-falling object.

This problem is solved with the parameters $m = 80.0$ and $c = 0.2$; the acceleration constant G is always 9.8 (m/s^2). The Python program that computes the solution to the model of the free-falling object is stored in file odefall.py. The Python code of the programmer-defined function *dfunc* that specifies the two first-order differential equations is shown as follows.

```
def dfunc (y, t):
     # evaluate the right-hand-side at t
     f0 = y[1]
     f1 = (c/m) * y[1] * y[1] - G
     return [f0, f1]
```

Figure 19.10 shows the graph of the vertical displacement with time generated by *matplotlib*. Observe that after about 10 seconds, the displacement changes linearly. Figure 19.11 shows the graph of vertical velocity of the object with time. Observe that the velocity of the object increases negatively at an exponential rate until about 20 seconds. After this time instant, the velocity remains constant.

19.10.4 Prey and Predator Model

The prey and predator model helps to study how the population of two animal species changes over time. The prey (rabbits) population is represented by $x_1(t)$. The predator (wolves) population is represented by $x_2(t)$. Without the predator, the prey population will grow as:

$$\frac{dx_1}{dt} = ax_1, \qquad a > 0.$$

Without the prey, the population of the predator will decrease as:

$$\frac{dx_2}{dt} = -bx_2, \qquad b > 0.$$

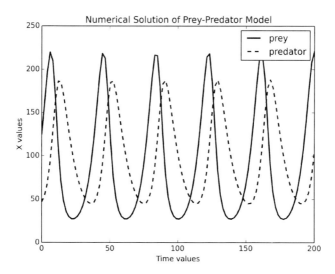

Figure 19.12 Population changes of prey and predator.

When the two species live and interact in the same environment, the prey population will decline when the predator population increases. This interaction produces additional changes in the population of the two species that are given by the following state equations:

$$\frac{dx_1}{dt} = ax_1 - cx_1x_2$$

$$\frac{dx_2}{dt} = -bx_2 + dx_1x_2.$$

In these expressions, a, b, c, and d are constants. The output equations are:

$$y_1 = x_1$$

$$y_2 = x_2.$$

For a numerical solution, assume $a = 0.25$, $b = 0.12$, $c = 0.0025$, and $d = 0.0013$. At the beginning, there are 125 rabbits and 47 wolves, so the initial conditions for the problem are $x_1(0) = 125$ and $x_2(0) = 47$.

The Python function that defines the evaluation of the two differential equations is shown in the following listing and the complete Python program is stored in file **predprey.py**.

```
def dfunc (y, t):
    # evaluate the right-hand-side at t
    f0 = a * y[0] - c * y[0] * y[1]
    f1 = - b * y[1] + d * y[0] * y[1]
    return [f0, f1]
```

Figure 19.12 shows the graph of the population changes of the prey and predator over time. Observe that after about 10 seconds, the displacement changes linearly. Figure 19.13 shows the phase plot of the two population changes. Because this shows a closed curve, it implies that the prey and predator populations follow periodic cycles.

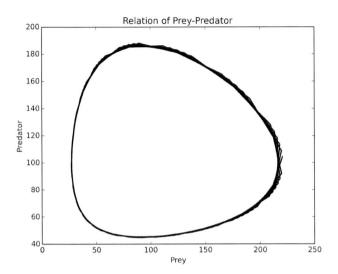

Figure 19.13 Phase plot of the population changes.

19.11 SUMMARY

Function differentiation and integration are used to better understand the behavior of functions that represent relationships in the properties of a system and are important in formulating computational models. Many systems in science and engineering can be modeled with ordinary differential equations. This chapter presented how to implement mathematical models that are formulated with differential equations, and their numerical solutions. Several case studies were presented and discussed with solutions implemented in Python with Numpy and Scipy.

Key Terms

functions	derivative	integrals
rate of change	area under a curve	finite differences
continuous models	dynamical systems	state variables
state equations	output equations	differential equations
initial values	Runge–Kutta methods	ODE solver

19.12 EXERCISES

19.1 Develop a program that computes the derivate at the point $x = 2$ of function $f(x) = (5x - 3)^2 e^x$. Use finite differences.

19.2 Develop a program that uses the Numpy and computes the derivate of function $f(x) = (5x - 3)^2 e^x$ at the point $x = 2$.

19.3 An object falls freely, and the vertical position of the object is given by the expression $y(t) = 16t^2 + 32t + 6$. Compute the velocity and the acceleration at various values of time. To solve this problem, develop a Python program that applies finite differences (central, backward, and forward).

19.4 An object falls freely, and the vertical position of the object is given by the expression $y(t) = 16t^2 + 32t + 6$. Compute the velocity and the acceleration at various values of time. To solve this problem, develop a Python program that uses Numpy (or Scipy) and applies finite differences.

19.5 Develop a program that computes an approximation of the integral of the function $f(x) = x^2$ in the interval $(0.0, 2.0)$ of x. In the program, apply the Trapezoid method. Use a value $n = 6$, then a value $n = 20$.

19.6 Develop a Python program that uses Numpy and computes an approximation of the integral of the function $f(x) = x^2$ in the interval $(0.0, 2.0)$ of x.

19.7 Develop a Python program that implements and solves numerically the following mathematical model, and generate a graph with *matplotlib* using the values of the variables in the numerical solution. Use a time interval from $t = 0$ to $t = 8.0$.

$$\frac{dy}{dt} = t + e^{t/2\pi} \cos(2\pi t)$$

19.8 This problem presents a simplified model of a water heater. The temperature of the water in a tank is increased by the heating elements and the temperature varies with time. The differential equation is

$$cm\frac{dT}{dt} = q - hA(T - t_s),$$

where T is the temperature of the water as a function of time, m is the mass of the water (in Kg), A is the area of the surface of the heater, q is the constant rate at which the elements produce heat, h is the Newton cooling coefficient, and t_s is the temperature of the surroundings (a constant). Develop a program that implements and solves numerically the mathematical model, and generate a graph of the temperature with time. Use the following values: $c = 4175$, $m = 245$, $A = 2.85$, $h = 12.0$, $q = 3600$, $t_s = 15$. The initial condition is $T(0) = 15$ and perform computations over an interval from $t = 0.0$ to $t = 2750.0$.

19.9 Edward N. Lorenz was a pioneer of Chaos Theory; the following simplified differential equations are known as the Lorenz Equations. Develop a program that implements and solves numerically the mathematical model:

$$\frac{dx}{dt} = \sigma(y - x)$$
$$\frac{dy}{dt} = x(\rho - z)$$
$$\frac{dz}{dt} = xy - \beta z.$$

Use the following values for the constants $\sigma = 10$, $\beta = 8/3$, and $\rho = 28$. Use initial conditions $x(0) = 8$, $y(0) = 8$, and $z(0) = 27$ and a time span from $t = 0$ to $t = 20$. Draw a plot of the variables in the numerical solution; first plot z with respect to x, then the three variables with respect to time. Investigate and plot a three-dimensional plot (x, y, z).

19.10 Develop a program that implements and solves numerically the following mathematical model:

$$\frac{d^2u}{dx^2} + e^x \frac{dv}{dx} + 3u = e^{2x}$$
$$\frac{d^2v}{dx^2} + \cos(x)\frac{du}{dx} + u = \sin(x).$$

Use the initial values $u(0) = 1$, $u'(0) = 2$, $v(0) = 3$, and $v'(0) = 4$. Solve over the interval from $x = 0$ to $x = 3$. Generate a plot of the variables in the numerical solution.

19.11 Investigate the SIR model for disease spread, which was proposed by W. O. Kermack and A. G. McKendrick. In this model, three groups of population are considered: $S(t)$ is the population group not yet infected and is susceptible, $I(t)$ is the population group that has been infected and is capable of spreading the disease, and $R(t)$ is the population group that has recovered and is thus immune. The mathematical model consists of the following differential equations:

$$\frac{ds}{dt} = -\alpha si$$
$$\frac{di}{dt} = \alpha si - \beta i$$
$$\frac{dr}{dt} = \beta i.$$

Use the values of the constants $\alpha = 1$ and $\beta = 0.3$. The initial conditions are $s = 0.999$, $i = 0.001$, and $r = 0.0$. Develop a program that implements and numerically solves the mathematical model and plot the values of the variables in the numerical solution.

V

Linear Optimization Models

## CHAPTER 20

Linear Optimization Modeling

20.1 INTRODUCTION

Mathematical optimization consists of finding the best possible values of variables from a given set that can maximize or minimize a real function. An optimization problem can be formulated in such a form that it can be possible to find an optimal solution; this is known as mathematical modeling and is used in almost all areas of science, engineering, business, industry, and defense. The goal of optimization modeling is of formulating a mathematical model of the system and attempting to optimize some property of the model. The actual optimization is carried out by executing the computer implementation of the model. This chapter discusses the general Simplex algorithm and the formulation of simple linear optimization models.

20.2 GENERAL FORM OF A LINEAR OPTIMIZATION MODEL

The following is a general form of a linear optimization model that is basically organized in three parts.

1. The objective function, f, to be maximized or minimized, mathematically expressed as:

$$f(x_1, x_2, \ldots, x_n) = c_1 x_1 + c_2 x_2 + \ldots + c_n x_n. \qquad (20.1)$$

2. The set of m constraints, which is of the form:

$$a_{i,1} x_1 + a_{i,2} x_2 + \ldots + a_{i,n} x_n \leq b_i \quad i = 1, \ldots m. \qquad (20.2)$$

The other form is:

$$a_{i,1}x_1 + a_{i,2}x_2 + \ldots + a_{i,n}x_n \geq b_i \quad i = 1, \ldots m. \quad (20.3)$$

3. The sign restriction for variables: $x_j \geq 0$, or $x_j \leq 0$, or x_j unrestricted in sign, $j = 1, \ldots n$.

Many problems are formulated with a mix of m constraints with $\leq$, $=$, and $\geq$ forms. Note that the objective function, which is expressed mathematically in (20.1), and the constraints, which are expressed mathematically in (20.2) and (20.3), are linear mathematical (algebraic) expressions.

An important assumption included in the general formulation of a linear optimization problem is that the variables, $x_i, i = 1, \ldots n$, take numeric values that are real or fractional. In the case that one or more variables only take integer values, then other techniques and algorithms are used. These methods belong to the class of *Integer Linear Optimization* or *Mixed Integer Optimization*.

20.3 THE SIMPLEX ALGORITHM

The *Simplex algorithm*, due to George B. Dantzig, is used to solve linear optimization problems. It is a tabular solution algorithm and is a powerful computational procedure that provides fast solutions to relatively large-scale applications. There are many software implementations of this algorithm, or variations of it. The basic algorithm is applied to a linear programming problem that is in standard form, in which all constraints are equations and all variables non-negative.

20.3.1 Foundations of the Simplex Algorithm

For a given linear optimization problem, *a point* is the set of values corresponding to one for each decision variable. The *feasible region* for the problem is the set of all points that satisfy the constraints and all sign restrictions. If there are points that are not in the feasible region, they are said to be in an *infeasible region*.

The *optimal solution* to a linear maximization problem is a point in the feasible region with the largest value of the objective function. In a similar manner, the *optimal solution* to a linear minimization problem is a point in the feasible region with the smallest value of the objective function.

There are four cases to consider in a linear optimization problem.

1. A unique optimal solution

2. An infinite number of optimal solutions

3. No feasible solutions

4. An unbounded solution

In a linear maximization problem, a constraint is *binding* at an optimal solution if it holds with equality when the values of the variables are substituted in the constraint.

20.3.2 Problem Formulation in Standard Form

Because the Simplex algorithm requires the problem to be formulated in *standard form*, the general form of the problem must be converted to standard form.

- For each constraint of $\leq$ form, a *slack variable* is defined. For constraint i, slack variable s_i is included. Initially, constraint i has the general form:

$$a_{i,1}x_1 + a_{i,2}x_2 + \ldots + a_{i,n}x_n \leq b_i. \tag{20.4}$$

To convert constraint i of the general form of the expression in (20.4) to an equality, slack variable s_i is added to the constraint, and $s_i \geq 0$. The constraint will now have the form:

$$a_{i,1}x_1 + a_{i,2}x_2 + \ldots + a_{i,n}x_n + s_i = b_i. \tag{20.5}$$

- For each constraint of $\geq$ form, an *excess variable* is defined. For constraint i, excess variable e_i is included. Initially, constraint i has the general form:

$$a_{i,1}x_1 + a_{i,2}x_2 + \ldots + a_{i,n}x_n \geq b_i. \tag{20.6}$$

To convert constraint i of the general form of the expression in (20.6) to an equality, excess variable e_i is subtracted from the constraint, and $e_i \geq 0$. The constraint will now have the form:

$$a_{i,1}x_1 + a_{i,2}x_2 + \ldots + a_{i,n}x_n - e_i = b_i. \tag{20.7}$$

Consider the following formulation of a numerical example:

Maximize: $5x_1 + 3x_2$
Subject to:

$$
\begin{array}{rcrcl}
2x_1 & + & x_2 & \leq & 40 \\
x_1 & + & 2x_2 & \leq & 50 \\
& & x_1 & \geq & 0 \\
& & x_2 & \geq & 0
\end{array}
$$

After rewriting the objective function and adding two slack variables s_1 and s_2 to the problem, the transformed problem formulation in standard form is:

Maximize: $z - 5x_1 - 3x_2 = 0$.
Subject to the following constraints:

$$
\begin{aligned}
2x_1 &+ x_2 &+ s_1 && &= 40 \\
x_1 &+ 2x_2 && &+ s_2 &= 50
\end{aligned}
$$

$$
\begin{aligned}
x_1 &\geq 0 \\
x_2 &\geq 0 \\
s_1 &\leq 0 \\
s_2 &\leq 0
\end{aligned}
$$

20.3.3 Generalized Standard Form

A generalized standard form of a linear optimization problem is:

Maximize (or minimize) $f = c_1 x_1 + c_2 x_2 + \ldots + c_n x_n$

Subject to the following constraints:

$$
\begin{aligned}
a_{1,1}x_1 &+ a_{1,2}x_2 &+ \ldots &+ a_{1,n}x_n &= b_1 \\
a_{2,1}x_1 &+ a_{2,2}x_2 &+ \ldots &+ a_{2,n}x_n &= b_2 \\
&\vdots && \vdots & \vdots \\
a_{m,1}x_1 &+ a_{m,2}x_2 &+ \ldots &+ a_{m,n}x_n &= b_m
\end{aligned}
\tag{20.8}
$$

$$
x_i \geq 0, \quad i = 1, 2, \ldots, n.
$$

The constraints can be written in matrix form as follows:

$$
\begin{bmatrix}
a_{1,1} & a_{1,2} & \cdots & a_{1,n} \\
a_{2,1} & a_{2,2} & \cdots & a_{2,n} \\
\vdots & \vdots & \ddots & \vdots \\
a_{m,1} & a_{2,m} & \cdots & a_{m,n}
\end{bmatrix}
\begin{bmatrix}
x_1 \\
x_2 \\
\vdots \\
x_n
\end{bmatrix}
=
\begin{bmatrix}
b_1 \\
b_2 \\
\vdots \\
b_m
\end{bmatrix}
\tag{20.9}
$$

Equation (22.2) can also be written as $AX = B$, in which A is an $m \times n$ matrix, X is a column vector of size n, and B is a column vector of size m.

20.3.4 Additional Definitions

To derive a basic solution to Equation (22.2), a set m of variables known as the *basic variables* is used to compute a solution. These variables are the ones left after setting the *nonbasic variables*, which is the set of $n - m$ variables chosen and set to zero.

There can be several different basic solutions in a linear optimization problem. There could be one or more sets of m basic variables for which a basic solution cannot be derived.

A *basic feasible solution* to the standard formulation of a linear optimization problem is a basic solution in which the variables are non-negative.

The solution to a linear optimization problem is the best basic feasible solution to $AX = B$ (or Equation (22.2)).

20.4 DESCRIPTION OF THE SIMPLEX ALGORITHM

In addition to transforming the constraints to standard form, the expression of the objective function must be changed to an equation with zero on its right-hand side. The general expression:

$$f = c_1 x_1 + c_2 x_2 + \ldots + c_n x_n$$

is changed to

$$f - c_1 x_1 - c_2 x_2 - \ldots - c_n x_n = 0.$$

This equation becomes row 0 in the complete set of equations of the problem formulation. After this transformation, the Simplex method can be used to solve the linear optimization problem.

20.4.1 General Description of the Simplex Algorithm

The following is a general description of the Simplex algorithm:

1. Find a basic feasible solution to the linear optimization problem; this solution becomes the initial basic feasible solution.

2. If the current basic feasible solution is the optimal solution, stop.

3. Search for an adjacent basic feasible solution that has a greater (or smaller) value of the objective function. An *adjacent basic feasible solution* has $m - 1$ variables in common with the current basic feasible solution. This becomes the current basic feasible solution, continue in step 2.

A linear optimization problem has an *unbounded solution* if the objective function can have arbitrarily large values for a maximization problem, or arbitrarily small values for a minimization problem. This occurs when a variable with a negative coefficient in the objective row (row 0) has a non-positive coefficient in every constraint.

20.4.2 Detailed Description of the Simplex Algorithm

A shorthand form of the set of equations known as the *simplex tableau* is used in the algorithm. Each tableau corresponds to a movement from one basic variable set BVS (extreme or corner point) to another, making sure that the objective function improves at each iteration until the optimal solution is reached. The following sequence of steps describes the application of the simplex solution algorithm:

1. Convert the LP to the following form:

   (a) Convert the minimization problem into a maximization one.

   (b) All variables must be non-negative.

   (c) All RHS values must be non-negative.

   (d) All constraints must be inequalities of the form $\leq$.

2. Convert all constraints to equalities by adding a slack variable for each constraint.

3. Construct the initial simplex tableau with all slack variables in the basic variable set (BVS). The row 0 in the table contains the coefficient of the objective function.

4. Determine whether the current tableau is optimal. That is: If all RHS values are non-negative (called, the feasibility condition) and if all elements of the row 0 are non-positive (called, the optimality condition). If the answers to both questions are Yes, then stop. The current tableau contains an optimal solution. Otherwise, continue.

5. If the current basic variable set (BVS) is not optimal, determine which nonbasic variable should become a basic variable and which basic variable should become a nonbasic variable. To find the new BVS with the better objective function value, perform the following tasks:

   (a) Identify the entering variable: The entering variable is the one with the largest positive coefficient value in row 0. (In case of a tie, the variable that corresponds to the leftmost of the columns is selected).

   (b) Identify the outgoing variable: The outgoing variable is the one with smallest non-negative column ratio (to find the column ratios, divide the RHS column by the entering variable column, wherever possible). In case of a tie, the variable that corresponds to the upmost of the tied rows is selected.

(c) Generate the new tableau: Perform the Gauss–Jordan pivoting operation to convert the entering column to an identity column vector (including the element in row 0).

6. Go to step 4.

At the start of the simplex procedure, the set of basis is constituted by the slack variables. The first BVS has only slack variables in it. The row 0 presents the increase in the value of the objective function that will result if one unit of the variable corresponding to the jth column was brought in the basis. This row is sometimes known as the *indicator row* because it indicates if the optimality condition is satisfied.

Criterion for entering a new variable into the BVS will cause the largest per-unit improvement of the objective function. Criterion for removing a variable from the current BVS maintains feasibility (making sure that the new RHS, after pivoting, remain non-negative). Warning: Whenever during the Simplex iterations you get a negative RHS, it means you have selected a wrong outgoing variable.

Note that there is a solution corresponding to each simplex tableau. The numerical of basic variables are the RHS values, while the other variables (non-basic variables) are always equal to zero. Note also that variables can exit and enter the basis repeatedly during the simplex algorithm.

20.4.3   Degeneracy and Convergence

A linear optimization problem (LP) is *degenerate* if the algorithm loops endlessly, cycling among a set of feasible basic solutions and never gets to the optimal solution. In this case, the algorithm will not converge to an optimal solution. Most software implementations of the Simplex algorithm will check for this type of non-terminating loop.

20.4.4   Two-Phase Method

The Simplex algorithm requires a starting basic feasible solution. The two-phase method can find a starting basic feasible solution whenever it exists. The two-phase simplex method proceeds in two phases, phase I and phase II. Phase I attempts to find an initial basic feasible solution. Once an initial basic feasible solution has been found, phase II is then applied to find an optimal solution to the original objective function.

The simplex method iterates through the set of basic solutions (feasible in phase II) of the LP problem. Each basic solution is characterized by the set of m basic variables $x_{B1}, \ldots, x_{Bm}$. The other n variables are called nonbasic variables and denoted by $x_{N1}, \ldots, x_{Nn}$.

20.5 FORMULATION OF LINEAR OPTIMIZATION MODELS

The formulating of a problem for linear constrained optimization is also known as *linear optimization* modeling or the mathematical modeling of a linear optimization problem (LP). Linear optimization modeling consists of four general steps, and these are as follows:

1. Identify a linear function, known as the *objective function*, to be maximized or minimized. This function is expressed as a linear function of the decision variables.

2. Identify the *decision variables* and assign to them symbolic names, x, y, etc. These decision variables are those whose values are to be computed.

3. Identify the set of *constraints* and express them as linear equations and inequations in terms of the decision variables. These constraints are derived from the given conditions.

4. Include the restrictions on the non-negative values of decision variables.

The objective function, f, to be maximized or minimized is expressed by:

$$f(x_1, x_2, \ldots, x_n) = c_1 x_1 + c_2 x_2 + \ldots + c_n x_n. \tag{20.10}$$

The set of m constraints are expressed in the form:

$$a_{i,1} x_1 + a_{i,2} x_2 + \ldots + a_{i,n} x_n \leq b_i \quad i = 1, \ldots m. \tag{20.11}$$

Or, of the form:

$$a_{i,1} x_1 + a_{i,2} x_2 + \ldots + a_{i,n} x_n \geq b_i \quad i = 1, \ldots m. \tag{20.12}$$

The sign restrictions for variables are denoted by $x_j \geq 0$, or $x_j \leq 0$, or x_j unrestricted in sign, $j = 1, \ldots n$. Many problems are formulated with a mixed of m constraints with $\leq$, $=$, and $\geq$ forms.

20.6 EXAMPLE PROBLEMS

There are many real and practical problems to which the linear optimization modeling may be applied. The following examples, although very simple because they use only two variables, help to illustrate the general method involved in linear optimization modeling.

20.6.1 Case Study 1

An industrial chemical plant produces two products, A and B. The market price for a pound of A is \$12.75, and that of B is \$15.25. Each pound of substance A produced requires 0.25 lbs of material P and 0.125 lbs of material Q. Each pound of substance B produced requires 0.15 lbs of material P and 0.35 lbs of material Q. The amounts of materials available in a week are 21.85 lbs of material P and 29.5 lbs of material Q. Management estimates that at the most, 18.5 pounds of substance A can be sold in a week. The goal of this problem is to compute the amounts of substance A and B to manufacture in order to optimize sales.

20.6.1.1 Understanding the Problem

For easy understanding and for deriving the mathematical formulation of the problem, the data given are represented in a table as follows. As stated previously, the main resource required in the production of the chemical substances A and B are the amounts of material of type P and Q.

Material	Available	Substance of type A	Substance of type B
P	21.85	0.250	0.15
Q	29.50	0.125	0.35

20.6.1.2 Mathematical Formulation

Let x_1 denote the amount of substance (lbs) of type A to be produced, and x_2 denote the amount of substance (lbs) of type B to be produced. The total sales is $12.75x_1 + 15.25x_2$ (to be maximized). The objective function of the linear optimization model formulation of the given problem is:

Maximize $S = 12.75x_1 + 15.25x_2$
Subject to the constraints:

$$
\begin{aligned}
0.25x_1 + 0.15x_2 &\leq 21.85 \\
0.125x_1 + 0.35x_2 &\leq 29.5 \\
x_1 &\leq 18.5 \\
x_1 &\geq 0 \\
x_2 &\geq 0
\end{aligned}
$$

20.6.2 Case Study 2

A manufacturer of toys produces two types of toys: X and Y. In the production of these toys, the main resource required is machine time and three machines are used: M1, M2, and M3. The machine time required to produce a toy of type X is 4.5 hours of machine M1, 6.45 hours of machine M2, and 10.85 hours of machine M3. The machine time required to produce a toy of type Y is 7.25

hours of machine M1, 3.65 hours of machine M2, and 4.85 hours of machine M3. The maximum available machine time for the machines M1, M2, M3 are 415, 292, and 420 hours, respectively. A toy of type X gives a profit of 4.75 dollars, and a toy of type Y gives a profit of 3.55 dollars. Find the number of toys of each type that should be produced to get maximum profit.

20.6.2.1 Understanding the Problem

For easy understanding and for deriving the mathematical formulation of the problem, the data given are represented in a table as follows. As stated previously, the main resource required in the production of toys is machine time of machines M1, M2, and M3.

Machine	Total time available	Req time toy type X	Req time toy type Y
M1	415	4.5	7.25
M2	292	6.45	3.64
M3	420	10.85	4.85

20.6.2.2 Mathematical Formulation

Let x denote the number of toys of type X to be produced, and y denote the number of toys of the type Y to be produced. The total profit is $= 4.75x + 3.55y$ (to be maximized). The objective function of the linear optimization model formulation of the given problem is:

Maximize: $P = 4.75x + 3.55y$
Subject to the constraints:

$$
\begin{aligned}
4.5x + 7.25y &\leq 415 \\
6.45x + 3.65y &\leq 292 \\
10.85x + 4.85y &\leq 420 \\
x &\geq 0 \\
y &\geq 0
\end{aligned}
$$

20.6.3 Case Study 3

A person needs to follow a diet that has at least 5,045 units of carbohydrates, 450.75 units of fat, and 325.15 units of protein. Two types of food are available: P and Q. A unit of food of type P costs 2.55 dollars and a unit of food of type Q costs 3.55 dollars. A unit of food of type P contains 9.75 units of carbohydrates, 18.15 units of fat, and 13.95 units of protein. A unit of food type Q contains 22.95 units of carbohydrates, 12.15 units of fat, and 18.85 units of protein. A mathematical linear model is needed to find the minimum cost for a diet that consists of a mixture of the two types of food and that meets the minimum diet requirements.

20.6.3.1 Understanding the Problem

For easy understanding and for deriving the mathematical formulation of the problem, the data given is represented in a table. For each type of food, the data include the cost per unit of food, and the contents, carbohydrates, fat, and proteins. The data for the diet problem is represented as follows:

Food type	Cost	Carbohydrates	Fat	Protein
P	2.55	9.75	18.15	13.95
Q	3.55	22.95	12.15	18.85

20.6.3.2 Mathematical Formulation

Let x_1 denote the amount of units of food type P and x_2 the units of food type Q contained in the diet. The total cost of the diet is $2.55x_1 + 3.55x_2$. As stated previously, the main limitation is the lower bound requirement of carbohydrates, fat, and proteins. The combination of units of type P and of type Q should have the minimum specified units of carbohydrates, fat, and proteins. The objective function of linear optimization model formulation of the given diet problem is:

Minimize: $C = 2.55x_1 + 3.55x_2$
Subject to the following constraints:

$$9.75x_1 + 22.95x_2 \geq 5045$$
$$18.15x_1 + 12.15x_2 \geq 450.75$$
$$13.95x_1 + 18.85x_2 \geq 325.15$$
$$x_1 \geq 0$$
$$x_2 \geq 0$$

20.6.4 Case Study 4

The owners of a farm acquired a loan of $16,850.00 to produce three types of crops—corn, barley, and wheat—on 140 acres of land. An acre of land can produce an average of 135 bushels of corn, 45 of barley, or 100 bushels of wheat. The net profit per bushel of barley is $3.05, for corn is $1.70, and for wheat is $2.25. After the harvest, these crops must be stored in relatively large containers. At present, the farm can store 3895 bushels. The total expenses to plant an acre of land are $95.00 for corn, $205.00 for barley, and $115.00 for wheat. What amount of land should the farm plan to dedicate to each crop in order to optimize profit?

20.6.4.1 Understanding the Problem

As in previous case studies, the data given are represented in a table for easy understanding and for deriving the mathematical formulation of the problem.

There are three types of resources that will impose constraints on the problem formulation: the total storage capacity of the farm for the crops, the total available funds, and the total amount of land. The data for this problem is represented as follows:

Resource	Total	Corn	Barley	Wheat
Storage (bushels)	3895	135	45	100
Funds ($)	16,850.00	95.00	205.00	115.00
Land (acres)	140	x_1	x_2	x_3

20.6.4.2 Mathematical Formulation

Let x_1 denote the amount of land in acres allotted to corn, x_2 the amount of land allotted to barley, and x_3 the amount of land dedicated to wheat. This problem is to optimize profit. The total net profit is denoted by P and for each crop it consists of net profit per bushel times the number of bushels per acre, times the number of acres to plant. The constraints are derived from the resource limitations of the problem. The arithmetic expression for P is given by:

$$P = 135 \times 1.70 \times x_1 + 45 \times 3.05 \times x_2 + 100 \times 2.25 \times x_3.$$

The objective function of the linear optimization formulation of the given problem is:

Maximize: $P = 229.5x_1 + 137.25x_2 + 225.00x_3$

Subject to the following constraints:

$$
\begin{array}{rrrcl}
135x_1 & +45x_2 & +100x_3 & \leq & 3895 \\
95x_1 & +205x_2 & +115x_3 & \leq & 16850.00 \\
x_1 & +x_2 & +x_3 & \leq & 140 \\
& & x_1 & \geq & 0 \\
& & x_2 & \geq & 0 \\
& & x_3 & \geq & 0 \\
\end{array}
$$

20.7 SUMMARY

Linear optimization modeling consists of formulating a mathematical linear model that includes the linear function to be optimized (maximized or minimized), a set of decision variables, and a set of constraints. The computational models of this type are also known as Linear Programming Models. The actual solution of linear optimization problems is performed by software tools known as LP solvers. The Simplex algorithm and its derivatives are discussed in some detail.

Key Terms

Simplex algorithm	two-phase method	standard form
feasible region	optimal solution	slack variables
excess variables	basic variables	convergence
feasible region	simplex tableau	degeneracy
linear optimization	decision variables	constraints
linear programming	linear problems	problem formulation
sign restriction	objective function	technological coefficient

20.8 EXERCISES

20.1 A factory produces three types of bed, A, B, and C. The company that owns the factory sells beds of type A for \$250.00 each, beds of type B for \$320.00 each, and beds of type C for \$625.00 each. Management estimates that all beds produced of types A and C will be sold. The number of beds of type B that can be sold is at most 45. The production of different types of bed requires different amounts of resources such as basic labor hours, specialized hours, and materials. The following table provides these data.

Type A	Type B	Type C	Resource
10 ft.	60ft.	80 ft.	Material
15 h	20 h	40 h	Basic labor
5 h	15 h	20 h	Specialized labor

The amounts of resources available are 450 ft of material, 210 hours of basic labor, and 95 hours of specialized labor. The problem is to optimize profit. Write the formulation of the mathematical optimization problem.

20.2 An automobile factory needs to have a different number of employees working for every day of the week. Each employee has to work 5 consecutive days a week and have two days of rest. The following are the requirements of the factory: Monday needs 330 employees, Tuesday needs 270 employees, Wednesday needs 300 employees, Thursday needs 390 employees, Friday needs 285 employees, Saturday needs 315 employees, and Sunday needs 225 employees. To improve profitability, the factory is required to optimize the number of employees. (Hint: the key decision is how many employees begin work on each day. Employees who begin work on Monday are those who do not begin on Tuesday or Wednesday). Write the formulation of the mathematical optimization problem.

20.3 Consider the problem in Exercise 20.2 with the following changes: the problem is to optimize the labor costs of the auto factory based on amount of labor time in hours worked by employees. Full-time employees work 8 hours per day and cost the factory $12.50 per hour. Part-time employees work 4 hours a day and cost $8.50 per hour. The total amount of part-time labor should be at most 30% of total labor time. Write the formulation of the mathematical optimization problem.

20.4 A factory manufactures 5 different products: P_1, P_2, P_3, P_4, and P_5. The factory needs to maximize profit. Each product requires machine time on three different devices, A, B, and C, each of which is available 135 hours per week. The following table provides the data (machine time in minutes):

Product	Device A	Device B	Device C
P_1	20	15	8
P_2	13	17	12
P_3	15	7	11
P_4	16	4	7
P_5	18	14	6

The unit sale price for products P_1, P_2, and P_3 is $7.50, $6.00, and $8.25, respectively. The first 26 units of P_4 and P_5 have a sale price of $5.85 each, and all excess units have a sale price of $4.50 each. The operational costs of devices A and B are $5.25 per hour, and $5.85 for device C. The cost of materials for products P_1 and P_4 are $2.75. For products P_2, P_3, and P_5 the materials cost is $2.25. Hint: assume that x_i, $i = 1 \ldots 5$ are the number of units produced of the various products and use two new variables y_4 and y_5 for the units produced of P_4 and P_5, respectively, in excess of 26. Write the formulation of the mathematical optimization problem.

20.5 A company produces a certain number of products and distributes these to various customer distribution centers that request specified numbers of units of the product. There are two production facilities P_1 and P_2. There are three storage facilities, S_1, S_2, and S_3. The company also has four distribution centers, D_1, D_2, D_3, and D_4. This problem can be represented as a network in which there is a cost per unit to ship products from a production facility to a storage facility, and from a storage facility to a distribution center. The problem must optimize the cost of distribution of the products. The following table provides the costs per unit shipped from a source node to a destination node.

Source node	S_1	S_2	S_3	D_1	D_2	D_3	D_4
P_1	4	6	0	0	0	0	0
P_2	10	4	7	0	0	0	0
S_1	0	0	0	17	20	0	0
S_2	0	0	0	20	15	25	0
S_3	0	0	0	0	30	18	11

## CHAPTER 21

Solving Linear Optimization Models

21.1   INTRODUCTION

This chapter presents the general principles and the basic concepts of numerical solution to linear optimization models that represent real problems. Several case studies are presented using the modeling capabilities of several software packages using Python, which use an underlying software solver such as GLPK.

21.2   LINEAR OPTIMIZATION MODELS WITH PYTHON

Python is a very good language used to model linear optimization problems. Two important Python features facilitate this modeling:

- The syntax of Python is very clean and it lends itself to naturally adapt to expressing (linear) mathematical programming models.

- Python has the built-in data structures necessary to build and manipulate models.

Python uses a linear optimization solver, such as GLPK, to compute the actual optimization. Therefore, with Python there will always be an underlying efficient linear optimization solver.

Several Python libraries or packages are available for modeling linear optimization problems. Some of the best known are:

- Pyomo - Coopr

- Pulp

- PyGLPK

- PyLPSolve

- PyMathProg

- PyCplex

21.3 MODELING WITH PYOMO

The Python Optimization Modeling Objects, also known as Pyomo, is a software package that supports the formulation and analysis of mathematical models for complex optimization applications. A linear optimization model in Pyomo is comprised of modeling components that define different aspects of the model. Pyomo uses index sets, symbolic parameters that are used to specify decision variables, objective functions, and constraints. Two types of models can be specified with Pyomo:

1. A *concrete* model, in which the problem data is embedded in the mathematical model itself.

2. An *abstract* model, in which the problem data is separated from the symbolic (mathematical) model.

A concrete model is generally more convenient for simple and relatively small problems. An abstract model is more appropriate for larger problems, which often have larger data sets.

21.3.1 Formulating Case Study 1

The following listing includes the model of Case Study 1 (previous chapter) using Pyomo. Note that this is a concrete model and is stored in file `casestud1.py`. The problem data is specified in lines 13–34 using Python lists and dictionaries. The model decision variables are declared in line 60 with Pyomo class *Var*. The variables are indexed by list *Products* and the values are limited to be non-negative reals.

The objective function is specified in lines 63–65 with Pyomo class *Objective*. The decision variables and the dictionary *MPrice* are indexed by list *Products*.

This model has two types of constraints. The first type is the material constraints, which are generated for every material type in the model. The second type of constraint is the production constraints, for which only one constraint is generated.

Constraints are generated with Pyomo class *Constraint* and indicated the appropriate expression. Lines 68–73 define a constraint function with parameter p that is used in line 76 to specify one constraint for each material type. This statement generates the constraints indexed by list *Materials*. The production constraint is specified in line 80.

```
 1 """
 2 Case Study 1 Python Formulation for the Pyomo Modeler
 3 An industrial chemical plant produces substances A and B
 4 The company needs to optimize the amount of A and B to
 5 maximize sales. Concrete model.
 6 J M Garrido, September  2014. Usage: pyomo casestud1.py
 7 """
 8 print
 9 print "Case Study 1: Chemical Plant Production"
10 # Import
11 from coopr.pyomo import *
12
13 # Data for Linear Optimization Problem
14 N = 2   # number of products
15 Products = range(1, N+1) # list of indices for decision var
16 IndxProd = 1        # index of product with limit
17 ProdLimit = 18.5   # limit of product 1 (pounds)
18 numprod = range(N)
19
20 Price = [12.75, 15.25]  # price per pound for each product
21 MPrice = {Products[i] : Price[i] for i in numprod}
22
23 M = 2  # M: number of types of material
24 Material = range(1, M+1) # list of indices for materials
25 nummat = range(M)
26
27 #Capacity of available material (pounds)
28 CapMat = [21.85, 29.5]
29 AvailMat = {Material[i] : CapMat[i] for i in nummat}
30
31 # requirement of materials for every pound of product
32 MatReq = [[0.25, 0.125],
33          [0.15, 0.350]]
34 RequireMat = {(Products[i], Material[j]) : MatReq[i][j] for j in
       nummat for i in numprod}
35
36 #Print Data
37 print
38 print "Price (per pound) of product: "
39 for i in numprod:
40    print "Product",Products[i], ":", MPrice[Products[i]]
41 print "Product 1 limit: ", ProdLimit
42 print
43 print "Available Material:"
44 for j in nummat:
```

```
45     print "Material",Material[j], ":", AvailMat[Material[j]]
46 print
47 print "Requirements of Material "
48 for i in numprod:
49     for j in nummat:
50         print "Product",Products[i], "-", "Material",Material[j],
                ":", MatReq[i][j]
51 print
52
53 #Concrete Model
54 model = ConcreteModel()
55
56 #Decision Variables
57 # The 2 variables x1, x2 are created with a lower limit of zero
58 # x1 is the amount of product 1 to produce
59 # x2 is the amount of product 2 to produce
60 model.Prod = Var(Products, within=NonNegativeReals)
61
62 # The objective function
63 model.obj = Objective(expr=
64             sum(MPrice[i] * model.Prod[i] for i in Products),
65             sense = maximize)
66
67 # Capacity Constraints
68 def CapacityRule(model, p):
69     """
70     This function has the Pyomo model as the first positional
            parameter,
71     and a material requirement index as a second positional
            parameter
72     """
73     return sum(RequireMat[i,p] * model.Prod[i] for i in Products)
            <= AvailMat[p]
74
75 # Generate one constraint for each material type
76 model.Capacity = Constraint(Material, rule = CapacityRule)
77
78 # Production Constraint
79 # Limit of production for Products[0]
80 model.ProdRestriction = Constraint(expr=model.Prod[IndxProd]
            <= ProdLimit)
```

The command line that runs this linear optimization model with Pyomo and the results are shown in the following listing. The first part displays the values of the input data for verification purposes and some messages about solving the optimization problem. In the second part of the output, the Solution Summary is displayed. The value computed of the decision variables (*Prod*) at the optimal point are 18.6 and 77.67867. The optimal value of the objective function is 1420.47321429.

```
$ pyomo casestud1.py --summary
[    0.00] Setting up Pyomo environment
[    0.00] Applying Pyomo preprocessing actions

Case Study 1: Chemical Plant Production

Price (per pound) of product:
Product 1 : 12.75
Product 2 : 15.25
Product 1 limit:  18.5

Available Material:
Material 1 : 21.85
Material 2 : 29.5

Requirements of Material
Product 1 - Material 1 : 0.25
Product 1 - Material 2 : 0.125
Product 2 - Material 1 : 0.15
Product 2 - Material 2 : 0.35

[    0.01] Creating model
[    0.01] Applying solver
[    0.05] Processing results
    Number of solutions: 1
    Solution Information
      Gap: 0.0
      Status: feasible
      Function Value: 1420.47321429
    Solver results file: results.json

=============================================================
Solution Summary
=============================================================

Model Chemical Plant Production

  Variables:
    Prod : Size=2, Index=Prod_index, Domain=NonNegativeReals
        Key : Lower : Value        : Upper : Initial : Fixed : Stale
```

```
          1 :      0 :           18.5 :  None :    None : False : False
          2 :      0 : 77.6785714286 :  None :    None : False : False

  Objectives:
     obj : Size=1, Index=None, Active=True
         Key  : Active : Value
         None :   True : 1420.47321429

  Constraints:
     Capacity : Size=2
         Key : Lower : Body          : Upper
           1 :  None : 16.2767857143 : 21.85
           2 :  None :          29.5 : 29.5
     ProdRestriction : Size=1
         Key  : Lower : Body : Upper
         None :  None : 18.5 : 18.5

[    0.06] Applying Pyomo postprocessing actions
[    0.06] Pyomo Finished
```

21.3.2 An Abstract Model Case Study 1

As mentioned previously, in an abstract model the data are separated from the mathematical model. The abstract data structures in the model are declared as parameters and sets and the actual data values are given in a data file. So essentially two files are used when invoking Pyomo, the file with the symbolic model and the file with the data values.

The following list shows the abstract model in file `casestud1abs.py`. In this model, the first important line is the declaration of the abstract model and this appears in line 13.

In line 16, a parameter is declared with name m that corresponds to the number of types of products and it is constrained to be a non-negative integer. Parameter n is similarly in line 21 and corresponds to the number of types of materials.

Parameter m is used in line 19 to declare a set with name *Products* that will be used to index parameters *MPrice* (line 31), *RequireMat* (line 37), and to index the decision variables *Prod* in line 44 .

Parameter n is similarly used to declare a set with name *Material* in line 23 that will be used to index parameters *AvailMat* in line 34, *RequireMat* in line 37, and the material capacity constraints in line 60.

```
1 """
2 Case Study 1 Python Formulation for the Pyomo Modeler
3 An industrial chemical plant produces substances A and B
```

```
 4 The company needs to optimize the amount of A and B to
 5 maximize sales. Abstract model.
 6 J M Garrido, September  2014
 7 usage: pyomo casestud1abs.py casestud1.dat --summary
 8 """
 9 # Linear Optimization Problem
10 print
11 print "Case Study 1: Chemical Plant Production"
12 from coopr.pyomo import *
13
14 model = AbstractModel()
15
16 model.m = Param(within=NonNegativeIntegers) # Number products
17
18 # Set of indices of Types of items produced
19 model.Products = RangeSet(1, model.m)
20
21 model.n = Param(within=NonNegativeIntegers) # Number materials
22 # Set of indices of Types of Substances required in production
23 model.Material = RangeSet(1, model.n)
24
25 # Limit of product 1
26 model.ProdLimit = Param(within=PositiveReals)
27 # Index for type of individual material with limit
28 model.K = Param(within=NonNegativeIntegers)
29 model.IndxProd = RangeSet(model.K, model.K)
30
31 model.MPrice = Param(model.Products, within=PositiveReals)
32
33 #Available material
34 model.AvailMat = Param(model.Material, within=PositiveReals)
35
36 # requirement of materials for every pound of product
37 model.RequireMat = Param(model.Products, model.Material)
38
39 #Decision Variables
40 # The 2 variables x1, x2 are created with a lower limit of zero
41 # x1 is the amount of product 1 to produce
42 # x2 is the amount of product 2 to produce
43
44 model.Prod = Var(model.Products, domain=NonNegativeReals)
45
46 # The objective function
47 def objective_expr(model):
48     return summation(model.MPrice, model.Prod)
```

```
49 model.obj = Objective(rule=objective_expr, sense = maximize)
50
51 # Material Capacity Constraints
52 def CapacityF(model,p):
53     """
54     This function for material constraint, the pyomo model as
55   the first positional parameter, and a material requirement index
56   as the second positional parameter.
57     """
58     return sum(
        model.RequireMat[i,p] * model.Prod[i] for i in model.Products)
            <= model.AvailMat[p]
59 # Generate one constraint for each material
60 model.Capacity = Constraint(model.Material, rule=CapacityF)
61
62 # Production Constraint - Limit of production for Product 1
63 def ProductLimit(model, i):
64     return (model.Prod[i] <= model.ProdLimit)
65 model.ProdRestriction = Constraint(model.IndxProd,
            rule=ProductLimit)
```

The data for the linear optimization model of Case Study 1 is stored in file casestud1.dat and is shown in the following listing.

```
# Data file: casestud1.dat. Case Study 1: Chemical Plant Production

param m := 2;    # Number of types of products
param n := 2;    # Number of types of materials

# Limit of production of (first) product
param ProdLimit   := 18.5 ;
# Index of (first) product with production limit
param K := 1 ;

# Market Price of every pound of product
param MPrice := 1 12.75
                2 15.25
  ;

# Available material, amount of every material
param AvailMat := 1 21.85
                  2 29.5
  ;
```

```
# Amount of material required for producing each pound of product
# One product per row (product, material, material_amount)
param RequireMat := 1 1 0.25 1 2 0.125
                    2 1 0.15 2 2 0.350

    ;
```

The other three case studies are formulated as concrete models with Pyomo and are stored in files casestud2.py, casestud3.py, and casestud4.py. In addition to these case studies, several sample models are included in the pyomo directory.

21.4 MODELING WITH PULP

Pulp is another good modeler for linear optimization models and is written in Python. The general setup of the problem model is similar to Pyomo. It is convenient to use Python lists and dictionaries for the problem data.

The following listing contains the model specification for Case Study 1 and is stored in file casestud1b.py. A simplified model of this problem is stored in file casestud1.py and several additional models are stored in the directory pulp_models.

```
 1 """
 2 Case Study 1 Python Formulation for the PuLP Modeller
 3 An industrial chemical plant produces substances A and B
 4 J M Garrido, September  2014
 5 Usage: python casestud1b.py
 6 """
 7
 8 # Import PuLP modeler functions
 9 from pulp import *
10
11 N = 2          # number of product types
12 Products = range(1, N+1)   # list of products
13 IndxProd = 1       # product with production limit
14 ProdLimit = 18.5
15 numprod = range(N) # index list of product types
16
17 Price = [12.75, 15.25] # price per pound for each product
18 MPrice = {Products[i] : Price[i] for i in numprod}
19
20 M = 2    # number of material types
21 Materials = range(1, M+1) # list material types
22 nummat = range(M)         # index list of materials
23
24 #Capacity of available material (pounds)
```

```
25 CapMat = [21.85, 29.5]
26 AvailMat = {Materials[i] : CapMat[i] for i in nummat}
27
28 # requirement of materials for every pound of product
29 MatReq = [[0.25, 0.125],
30           [0.15, 0.350]]
31 RequireMat = {(Products[i], Materials[j]) : MatReq[i][j] for j
       in nummat for i in numprod}
32
33 #Print Data
34 print
35 print "Price (per pound) of product: "
36 for i in numprod:
37     print "Product",Products[i], ":", MPrice[Products[i]]
38 print "Product 1 limit: ", ProdLimit
39 print
40 print "Available Material:"
41 for j in nummat:
42     print "Material",Materials[j], ":", AvailMat[Materials[j]]
43 print
44 print "Requirements of Material "
45 for i in numprod:
46     for j in nummat:
47         print "Product",Products[i], "-", "Material",Materials[j],
                ":", MatReq[i][j]
48 print
49
50 # Create the model to contain the problem data
51 model = LpProblem("Case study 1", LpMaximize)
52
53 # Decision variables
54 Prod = LpVariable.dicts("ProdVar", Products, 0, None)
55
56 # The objective function
57 model += lpSum([MPrice[i]*Prod[i] for i in Products]),"Total Sales"
58
59 # The constraints of available material
60 for p in Materials:
61     model += lpSum([RequireMat[i,p] * Prod[i] for i in Products])
            <= AvailMat[p], "Maximum of Material %d"%p
62
63 # Production Constraint
64 # Limit of production for Products[IndxProd]
65 model += Prod[IndxProd] <= ProdLimit, "Production limit"
66
```

```
67 # Write the problem data to an .lp file
68 model.writeLP("casestud1b.lp")
69
70 # Solve the optimization problem using the specified PuLP Solver
71 model.solve(GLPK())
72
73 # Print the status of the solution
74 print "Status:", LpStatus[model.status]
75
76 # Print each of the variables with its resolved optimum value
77 for v in model.variables():
78     print v.name, "=", v.varValue
79
80 # Print the optimised value of the objective function
81 print "Optimal sales", value(model.objective)
```

The data are set with Python lists and dictionaries on lines 11–31 and is exactly the same as in the model using Pyomo. Lines 33–48 displays the problem data for verification purposes.

The Pulp model for the problem is declared in line 51 using Pulp class *LPProblem* with argument *LPMaximize*. The decision variables are created in line 54 as a dictionary indexed by list *Products*. The lower bound for the value of the variables is zero and there is no upper bound (specified as *None*). By default, the variables created are of type real and type integer is specified with the argument *LpInteger*.

The objective function is specified in line 57 and the summation is indexed by list *Products*. The materials constraints are specified in lines 60–61 and are indexed by list *Materials*. The single production constraint is specified with variable *Prod* indexed by the scalar parameter *IndxProd*.

The model is solved with the GLPK specified solver in line 71. The status of the solution is displayed in line 74. The computed optimal values of the decision variables are displayed in lines 77–78 and the optimal value of the objective function is displayed in line 81.

The following listing shows the output produced after running the model. The command line used is the first line shown in the listing.

```
$python casestud1b.py
/usr/local/lib/python2.7/dist-packages/pulp_or-1.4.6-py2.7.
  egg/pulp

Price (per pound) of product:
Product 1 : 12.75
Product 2 : 15.25
Product 1 limit:  18.5
```

```
Available Material:
Material 1 : 21.85
Material 2 : 29.5

Requirements of Material
Product 1 - Material 1 : 0.25
Product 1 - Material 2 : 0.125
Product 2 - Material 1 : 0.15
Product 2 - Material 2 : 0.35


GLPSOL: GLPK LP/MIP Solver, v4.54
Parameter(s) specified in the command line:
 --cpxlp /tmp/3556-pulp.lp -o /tmp/3556-pulp.sol
Reading problem data from '/tmp/3556-pulp.lp'...
3 rows, 2 columns, 5 non-zeros
11 lines were read
GLPK Simplex Optimizer, v4.54
3 rows, 2 columns, 5 non-zeros
Preprocessing...
2 rows, 2 columns, 4 non-zeros
Scaling...
 A: min|aij| =  1.250e-01  max|aij| =  3.500e-01  ratio =
    2.800e+00
Problem data seem to be well scaled
Constructing initial basis...
Size of triangular part is 2
*    0: obj =   0.000000000e+00  infeas =  0.000e+00 (0)
*    2: obj =   1.420473214e+03  infeas =  0.000e+00 (0)
OPTIMAL LP SOLUTION FOUND
Time used:   0.0 secs
Memory used: 0.0 Mb (37657 bytes)
Writing basic solution to '/tmp/3556-pulp.sol'...
Status: Optimal
ProdVar_1 = 18.5
ProdVar_2 = 77.6786
Optimal sales 1420.47365
```

21.5 SOFTWARE LINEAR OPTIMIZATION SOLVERS

There are many software linear optimization solvers, some free open source and others commercial solvers. This section mentions two software tools that are cross-platform (Linux, MacOS, and MS Windows), and solve linear optimization problems using the Simplex algorithm and variations of it. These are LP_solve and GLPK, and problem formulation and executions are shown with several examples. Several Python linear optimization modelers use GLPK as the default solver.

21.6 SHORT LIST OF OPTIMIZATION SOLVERS

Commercial linear optimization solvers are available; some only for MS Windows and others for multiple platforms. These are generally faster than the free open-source solvers and some of the most widely used are:

- Gurobi
- Cplex
- Xpress
- Lindo

The most widely used of the open-source and cross-platform software linear optimization solvers are:

- The Gnu Linear Programming Kit also known as GLPK
- LPsolve
- SCIP
- CLP (Coin-or)

21.7 SUMMARY

Python is a very good language to model linear optimization problems. Two modeling packages used with Python are Pyomo and Pulp. The solution to linear optimization modeling is performed by software tools known as LP solvers. Some solvers require that the linear optimization problem formulation be in standard form, in which all constraints are equations and all variables non-negative. The solvers discussed and used in this chapter are open source programs. There are also several commercial proprietary solvers as well, such as Lindo.

Key Terms

modeling software	Pyomo	Pulp
lists	dictionaries	optimal solution
concrete model	abstract model	model data

21.8 EXERCISES

21.1 Model and solve the following linear optimization problem using Pyomo and Pulp.

Minimize: $2x_1 + 3x_2$

Subject to the following constraints:

$$
\begin{array}{rcl}
0.5x_1 + 0.25x_2 & \leq & 4.0 \\
x_1 + 3x_2 & \geq & 36 \\
x_1 + x_2 & = & 10 \\
x_1 & \geq & 0 \\
x_2 & \geq & 0
\end{array}
$$

21.2 Model and solve the following linear optimization problem using Pyomo and Pulp.

Maximize: $2x_1 + 3x_2 + x_3$

Subject to the following constraints:

$$
\begin{array}{rcl}
x_1 + x_2 + x_3 & \leq & 40 \\
2x_1 + x_2 - x_3 & \geq & 10 \\
- x_2 + x_3 & \geq & 10 \\
x_1 & \geq & 0 \\
x_2 & \geq & 0 \\
x_3 & \geq & 0
\end{array}
$$

21.3 A factory produces three types of beds, A, B, and C. The company that owns the factory sells beds of type A for $250.00 each, beds of type B for $320.00 each, and beds of type C for $625.00 each. Management estimates that all beds produced of types A and C will be sold. The number of beds of type B that can be sold is at the most 45. The production of different types of beds requires a different amount of resources such as basic labor hours, specialized hours, and materials. The following table provides these data.

Type A	Type B	Type C	Resource
10 ft.	60 ft.	80 ft.	Material
15 min	20 min	40 min	Basic labor
5 min	15 min	20 min	Specialized labor

The amounts of resources available are 450 ft of material, 210 hours of basic labor, and 95 hours of specialized labor. The problem is to optimize sales. Write the formulation of the mathematical optimization problem. Use Pyomo and Pulp to find a numerical solution to the linear optimization problem.

21.4 An automobile factory needs to have a different number of employees working for every day of the week. Each employee has to work 5 consecutive days a week and have two days of rest. The following are the requirements of the factory: Monday needs 330 employees, Tuesday needs 270 employees, Wednesday needs 300 employees, Thursday needs 390 employees, Friday needs 285 employees, Saturday needs 315 employees, and Sunday needs 225 employees. To improve profitability, the factory is required to optimize the number of employees. (Hint: the key decision is how many employees begin work on each day. Employees who begin work on Monday are those who do not begin on Tuesday or Wednesday). Write the formulation of the mathematical optimization problem. Use Pyomo and Pulp to find a numerical solution to the linear optimization problem.

21.5 Consider the problem in Exercise 21.4 with the following changes: the problem is to optimize the labor costs of the auto factory based on the amount of labor time in hours worked by employees. Full-time employees work 8 hours per day and cost the factory $12.50 per hour. Part-time employees work 4 hours a day and cost $8.50 per hour. The total amount of part-time labor should be at the most 30% of total labor time. Write the formulation of the mathematical optimization problem. Use Pyomo and Pulp to find a numerical solution to the linear optimization problem.

21.6 A factory manufactures 5 different products: P_1, P_2, P_3, P_4, and P_5. The factory needs to maximize profit. Each product requires machine time (in minutes) on three different devices: A, B, and C, each of which is available 135 hours per week. The following table provides the data on machine time needed (in minutes):

Product	Device A	Device B	Device C
P_1	20	15	8
P_2	13	17	12
P_3	15	7	11
P_4	16	4	7
P_5	18	14	6

The unit sale price for products P_1, P_2, and P_3 is $7.50, $7.00, and $8.25, respectively. The first 26 units of P_4 and P_5 have sale prices of $5.85 each; all excess units have a sale price of $4.50 each. The operational costs of devices A and B are $5.25 per hour, and $5.85 for device C. The cost of materials for products P_1 and P_4 are $2.75. For products P_2, P_3, and P_5 the materials cost is $2.25. Hint: assume that x_i, $i = 1 \ldots 5$ are the number of units produced of the various products and use two new variables y_4 and y_5 for the units produced of P_4 and P_5, respectively in excess of 26. Write the formulation of the mathematical optimization problem. Use Pyomo and Pulp to find a numerical solution to the linear optimization problem.

21.7 A company produces a certain number of products and distributes these to various customer distribution centers that request a specified number of units of the product. There are two production facilities P_1 and P_2. There are three storage facilities: S_1, S_2, and S_3. The company also has four distribution centers: D_1, D_2, D_3, and D_4. This problem can be represented as a network in which there is a cost per unit to ship products from a production facility to a storage facility, and from a storage facility to a distribution center. The problem must optimize the cost of distribution of the products. The following table provides the cost per unit shipped from a source node to a destination node. Use Pyomo and Pulp to find a numerical solution to the linear optimization problem.

Source node	S_1	S_2	S_3	D_1	D_2	D_3	D_4
P_1	4	6	0	0	0	0	0
P_2	10	4	7	0	0	0	0
S_1	0	0	0	17	20	0	0
S_2	0	0	0	20	15	25	0
S_3	0	0	0	0	30	18	11

## CHAPTER 22

Sensitivity Analysis and Duality

22.1 INTRODUCTION

This chapter presents the general concepts and techniques of sensitivity analysis and duality. With sensitivity analysis, we can find out how relatively small changes in the parameters of a linear optimization problem can cause changes in the optimal solution computed.

The concepts of *marginal values*, which are also known as *shadow prices*, and the *reduced costs* are extremely useful in sensitivity analysis. *Duality* helps to better understand sensitivity analysis as well as the nature of linear optimization.

22.2 SENSITIVITY ANALYSIS

Sensitivity analysis deals with the effect of independent, multiple changes in the values of:

- the coefficients of the objective function of linear program models having a unique solution and

- the right-hand side of the constraints.

Other useful points in the analysis are the study of the change in the solution to a problem that occurs when a new constraint is added, and the changes when a new variable is added to the problem.

22.2.1 Coefficients of the Objective Function

How much can the objective function coefficients change before the values of the variables change? Or when the objective function coefficient of a single variable is changed, for what range of values of this coefficient will the optimal values of the decision variables be retained?

The concept of *reduced cost* is associated with the coefficients of the variables in the objective function. It is a very useful concept and applies to a variable with value zero in the optimal value of the objective function. The reduced cost of a variable is the amount by which the objective function will decrease when the variable is forced to a value of 1.

The simplest and most direct way to find the reduced costs of variables and the allowable changes to the coefficients in the objective function is to observe the output results in the computer solution to a linear optimization problem. These are briefly discussed next using the Pulp and Pyomo modeling tools that apply the GLPK solver.

The two modeling tools are used to solve the following linear optimization model, Example 1:

Maximize: $60x_1 + 30x_2 + 20x_3$
Subject to the following constraints:

$$
\begin{array}{rrrrr}
8x_1 & + 6x_2 & + x_3 & \leq & 48 \\
4x_1 & + 2x_2 & + 1.5x_3 & \leq & 20 \\
2x_1 & + 1.5x_2 & + 0.5x_3 & \leq & 8 \\
& & x_1 & \geq & 0 \\
& & x_2 & \geq & 0 \\
& & x_3 & \geq & 0
\end{array}
$$

22.2.2 Using Pulp: Example 1

The formulation of the problem with the Pulp modeler is stored in file `sensit1.py` and is shown in the following listing.

```
1  """
2  Python Formulation for the PuLP Modeler
3  An example of sensitivity analysis
4  J M Garrido, September  2014
5  Usage: python sensit1.py
6  """
7
8  # Import PuLP modeler functions
9  from pulp import *
10
11 # Create the model for the problem
12 prob = LpProblem("Sensit 1",LpMaximize)
13
14 # The 3 variables x1, x2, and x3 have a lower limit of zero
15 x1=LpVariable("x1",0,None)
16 x2=LpVariable("x2",0)
```

```
17 x3=LpVariable("x3",0)
18
19 # The objective function
20 prob += 60.0*x1 + 30*x2 + 20*x3, "Objective"
21
22 # The three constraints are
23 prob += 8.0*x1 + 6.0*x2 + x3      <= 48.0, "Constraint 1"
24 prob += 4.0*x1 + 2.0*x2 + 1.5*x3 <= 20.0, "Constraint 2"
25 prob += 2.0*x1 + 1.5*x2 + 0.5*x3 <= 8.0,  "Constraint 3"
26
27 # Write the problem data to an .lp file
28 prob.writeLP("sensit1.lp")
29
30 # Solve the optimization problem using the specified Solver
31 prob.solve(GLPK(options=['--ranges sensit1.sen']))
32
33 # Print the status of the solution
34 print "Status:", LpStatus[prob.status]
35
36 # Print each of the variables with it's resolved optimum value
37 for v in prob.variables():
38     print v.name, "=", v.varValue
39
40 # Print the optimised value of the objective function
41 print "Objective", value(prob.objective)
```

In line 31, the program invokes the solver GLPK and specifies the option: --ranges sensit1.sen. This option indicates that GLPK will generate the sensitivity analysis report in file sensit1.sen. The results of running the program are stored in file sensit1.sen and only the second part is shown in the following listing.

```
GLPK 4.55 - SENSITIVITY ANALYSIS REPORT                Page   2


Problem:    sensit1
Objective:  z = 280 (MAXimum)
```

No.	Col	St	Acty	Obj coef Marginal	Lower Upper	Activity range
1	x1	BS	2.000	60.000 .	. +Inf	-Inf 4.00000
2	x2	NL	.	30.000	.	-4.00000

```
                                -5.00000        Inf         1.60000

3   x3   BS      8.0000         20.0000           .          -16.000
                                   .            +Inf         13.3333


Col   Obj coef    Obj value     Limiting
        range     break pt      variable
----  ----------  -----------  ----------
 X1    56.000      272.000         x2
       80.000      320.000         C2

 X2    -Inf        300.000         x3
       35.000      272.000         x1

 X3    15.000      240.000         C2
       22.500      300.000         C3

End of report
```

Column "Acty" (activity) shows the actual values of variables $x1$, $x2$, and $x3$. Note that variable $x2$ has a value of zero and column "Activity range" shows the value 1.6, which is the value that the variable will take when its coefficient reaches the value of 35.

The range of values of the coefficients in the objective function appear in column "Obj coef range," for the three decision variables $x1$, $x2$, and $x3$. These ranges of values correspond to the coefficients of variables that retain the optimal value of the objective function. For example, the value of the coefficient of $x1$ can range from 56.0 to 80.0, the coefficient of $x2$ can range from a very low value up to 35.0, and the coefficient of variable $x3$ can range from 15.0 to 22.5. Any values of the coefficient outside these ranges will change the conditions of the objective function to a suboptimal value.

Variable $x2$ has a *reduced cost* of −5 and is shown in column "Marginal." This is the amount the objective function would change if the value of $x2$ is changed to 1. Variables $x1$ and $x3$ are basic variables and have a zero reduced cost.

22.2.3 Using Pyomo: Example 1

The formulation of the problem with the Pyomo modeler is stored in file sensit1.py and is shown in the following listing.

```
1 """
2 Python Formulation for the Pyomo Modeler
3 Example for sensitivity analysis
4 J M Garrido, September  2014. File sensit1.py
```

```
 5 Usage: pyomo sensit1.py
 6 """
 7 from coopr.pyomo import *
 8
 9 # Data for Linear Optimization Problem
10 xlist = [1, 2, 3]
11
12 #Concrete Model
13 model = ConcreteModel(name="Sensitivity 1")
14
15 #Decision Variables
16 model.x = Var(xlist, within=NonNegativeReals)
17
18 # The objective function
19 model.obj = Objective(expr= 60.0*model.x[1] + 30.0*model.x[2] +
      20.0 * model.x[3],sense = maximize)
20
21 # Constraints
22 model.Constraint1 = Constraint(expr= 8.0 * model.x[1] +
      6.0 * model.x[2] + model.x[3] <= 48.0)
23 model.Constraint2 = Constraint(expr= 4.0 * model.x[1] +
      2.0 * model.x[2] + 1.5 * model.x[3] <= 20.0)
```

To run the model, the command line to use is indicated at the top of the following listing. It indicates to the Pyomo modeler to save the model in GLPK format in file sensit1.lp. The reason for this is that with Pyomo, it is sometimes difficult to pass options to the GLPK solver.

```
pyomo sensit1.py --summary --save-model sensit1.lp
[    0.00] Setting up Pyomo environment
[    0.00] Applying Pyomo preprocessing actions
Sensitivity 1
[    0.00] Creating model
[    0.00] Applying solver
[    0.05] Processing results
    Number of solutions: 1
    Solution Information
      Gap: 0.0
      Status: feasible
      Function Value: 280.0
    Solver results file: results.json


============================================================
Solution Summary
============================================================
```

```
Model Sensitivity 1

  Variables:
    x : Size=3, Index=x_index, Domain=NonNegativeReals
        Key : Lower : Value : Upper : Initial : Fixed : Stale
          1 :     0 :   2.0 :  None :    None : False : False
          2 :     0 :   0.0 :  None :    None : False : False
          3 :     0 :   8.0 :  None :    None : False : False

  Objectives:
    obj : Size=1, Index=None, Active=True
        Key  : Active : Value
        None :   True : 280.0

  Constraints:
    Constraint1 : Size=1
        Key  : Lower : Body : Upper
        None :  None : 24.0 :  48.0
    Constraint2 : Size=1
        Key  : Lower : Body : Upper
        None :  None : 20.0 :  20.0
    Constraint3 : Size=1
        Key  : Lower : Body : Upper
        None :  None :  8.0 :   8.0

[    0.08] Applying Pyomo postprocessing actions
[    0.08] Pyomo Finished
```

The next step is to run the model in file **sensit1.lp** directly using the GLPK command line executable, **glpsol**. The option to produce a report with the sensitivity analysis to the specified file is **--ranges sensit1.sen**. This generates the same output discussed previously and the command line is:

```
glpsol -m sensit1.lp --lp --ranges sensit1.sen
```

22.2.4   Right-Hand Side of Constraints

As mentioned previously, another part of sensitivity analysis involves finding out how changes in the right-hand side of a constraint would change the basis of the optimal solution to a problem.

The *dual value* (marginal value or shadow price) in a constraint is the amount by which the objective function will decrease when the right-hand side of the constraint in incremented by one. Another way to define a dual value is the rate of change of the objective function with respect to the right-hand side of the constraint.

The right-hand side of a single constraint can vary within a specified range. The dual value of constraint will only hold while the right-hand side of the constraint is in the range of values.

A constraint that is not active has a dual value of zero. This is a constraint that is not binding because its actual value is less (or greater) than the value of the right-hand side specified in the problem formulation.

This second set of data is provided by computer LP solvers such as GLPK *glpsol* and *LP_solve*. These solvers display output results that can be used for sensitivity analysis.

22.3 DUALITY

For every linear optimization problem, known as the *primal*, there is an associated problem known as its *dual*. The relationship between these two problems helps to understand the connection between the reduced cost and the shadow price.

22.3.1 Formulating the Dual Problem

Assume that the primal linear maximization problem has the generalized standard form:

Maximize $f = c_1 x_1 + c_2 x_2 + \ldots + c_n x_n$
Subject to the following constraints:

$$
\begin{array}{cccccc}
a_{1,1}x_1 & + a_{1,2}x_2 & + \ldots & + a_{1,n}x_n & \leq & b_1 \\
a_{2,1}x_1 & + a_{2,2}x_2 & + \ldots & + a_{2,n}x_n & \leq & b_2 \\
\vdots & \vdots & & \vdots & & \vdots \\
a_{m,1}x_1 & + a_{m,2}x_2 & + \ldots & + a_{m,n}x_n & \leq & b_m
\end{array}
\tag{22.1}
$$

$$x_i \geq 0, \quad i = 1, 2, \ldots, n.$$

The constraints in matrix form are as follows:

$$
\begin{bmatrix}
a_{1,1} & a_{1,2} & \cdots & a_{1,n} \\
a_{2,1} & a_{2,2} & \cdots & a_{2,n} \\
\vdots & \vdots & \ddots & \vdots \\
a_{m,1} & a_{2,m} & \cdots & a_{m,n}
\end{bmatrix}
\begin{bmatrix}
x_1 \\
x_2 \\
\vdots \\
x_n
\end{bmatrix}
\leq
\begin{bmatrix}
b_1 \\
b_2 \\
\vdots \\
b_m
\end{bmatrix}.
\tag{22.2}
$$

This linear problem can also be written as maximize $f = C'X$ such that $AX \leq B$, in which C is a column vector of size n, X is a column vector of size n, A is an $m \times n$ matrix, and B is a column vector of size m. The dual problem is a minimization problem formulated in standard form as follows:

Minimize $g = b_1 y_1 + b_2 y_2 + \ldots + b_m y_m$

Subject to the following constraints:

$$
\begin{array}{llll}
a_{1,1} y_1 & + a_{2,1} y_2 & + \ldots & + a_{m,1} y_m & \geq c_1 \\
a_{1,2} y_1 & + a_{2,2} y_2 & + \ldots & + a_{m,2} y_m & \geq c_2 \\
\vdots & \vdots & & \vdots & \vdots \\
a_{1,n} y_1 & + a_{2,n} y_2 & + \ldots & + a_{m,n} y_m & \geq c_n
\end{array}
\tag{22.3}
$$

$$
y_i \geq 0, \quad i = 1, 2, \ldots, m.
$$

The constraints in matrix form are as follows:

$$
\begin{bmatrix}
a_{1,1} & a_{2,1} & \cdots & a_{m,1} \\
a_{1,2} & a_{2,2} & \cdots & a_{m,2} \\
\vdots & \vdots & \ddots & \vdots \\
a_{1,n} & a_{2,n} & \cdots & a_{m,n}
\end{bmatrix}
\begin{bmatrix}
y_1 \\
y_2 \\
\vdots \\
y_m
\end{bmatrix}
\geq
\begin{bmatrix}
c_1 \\
c_2 \\
\vdots \\
c_n
\end{bmatrix}.
\tag{22.4}
$$

The matrix form of this dual linear problem can also be written as minimize $g = B'Y$ such that $A'Y \geq C$, in which C is a column vector of size n, X is a column vector of size n, A is an $m \times n$ matrix, and B is a column vector of size m. In a similar manner, if the primal problem is a minimization problem, its dual is a maximization problem. For example, the following primal maximization problem is in standard form.

Maximize $f = 144x_1 + 60x_2$

Subject to the following constraints:

$$
\begin{array}{lll}
120x_1 & + 210x_2 & \leq 15000 \\
110x_1 & + 30x_2 & \leq 4000 \\
x_1 & + x_2 & \leq 75
\end{array}
$$

$$
x_1 \geq 0, \quad x_2 \geq 0.
$$

The dual problem of the given primal is formulated as follows:

Minimize $g = 15000y_1 + 4000y_2 + 75y_3$

Subject to the following constraints:

$$
\begin{array}{llll}
120y_1 & + 110y_2 & + y_3 & \geq 144 \\
210y_1 & + 30y_2 & + y_3 & \geq 60
\end{array}
$$

$$
y_1 \geq 0, \quad y_2 \geq 0.
$$

22.3.2 Transforming a Problem to Standard Form

A primal linear problem has to be formulated in standard form before its dual can be formulated. A standard maximization form is also known as the **normal maximization** form that has been explained previously.

When a linear problem is not in standard (or normal) form, then a few transformations are necessary. For example, the following linear maximization problem formulation is not in standard form.

Maximize $f = 144x_1 + 60x_2$
Subject to the following constraints:

$$
\begin{array}{rcl}
120x_1 + 210x_2 & \leq & 15000 \\
110x_1 + 30x_2 & = & 4000 \\
x_1 + x_2 & \geq & 75
\end{array}
$$

$$x_1 \geq 0, \quad x_2 \ urs.$$

This problem has a constraint with an equal sign (=), a constraint with a $\geq$ sign, and a variable with unrestricted sign. Therefore, the problem is not formulated in standard (normal) form. The following transformation steps are necessary:

- To transform a constraint with a $\geq$ to a constraint with a $\leq$ sign, the constraint must be multiplied by -1. For example, the constraint $x_1 + x_2 \geq 75$, is transformed to $-x_1 - x_2 \leq -75$.

- To transform a constraint with an equal sign, it must be replaced by two inequality constraints, one with a $\leq$ sign and another with a $\geq$ sign. For example, the second constraint in the problem ($110x_1 + 30x_2 = 4000$) is replaced by the two constraints: $110x_1 + 30x_2 \geq 4000$ and $110x_1 + 30x_2 \leq 4000$. The first of these constraints is transformed to a constraint with a $\leq$ sign by multiplying it by -1.

- When a decision variable x_i is unrestricted in sign (urs), it means that it can take positive, negative, and zero values. The equivalence $x_i = x'_i - x''_i$ is applied; therefore, variable x_i is replaced by $x'_i - x''_i$. In the problem previously discussed, variable x_2 is unrestricted in sign, so it is replaced by $x'_2 - x''_2$.

The transformed linear maximization (primal) problem formulation is now expressed as follows:

Maximize $f = 144x_1 + 60x'_2 - 60x''_2$
Subject to the following constraints:

$$
\begin{array}{rcl}
120x_1 + 210x'_2 - 210x''_2 & \leq & 15000 \\
110x_1 + 30x'_2 - 30x''_2 & \leq & 4000 \\
-110x_1 - 30x'_2 + 30x''_2 & \leq & -4000 \\
-x_1 - x'_2 + x''_2 & \leq & -75
\end{array}
$$

$$x_1 \geq 0, \quad x_2' \geq 0, \quad x_2'' \geq 0.$$

The final step is to find the dual problem of the primal problem discussed. Because the primal problem is a maximization problem, its dual is a minimization problem. The formulation of the dual problem is as expressed as follows:

Minimize $g = 15000y_1 + 4000y_2 - 4000y_3 - 75y_4$
Subject to the following constraints:

$$
\begin{array}{rrrrl}
120y_1 & + \, 110y_2 & - \, 110y_3 & - \, y_4 & \geq \ 144 \\
210y_1 & + \, 30y_2 & - \, 30y_3 & - \, y_4 & \geq \ 60 \\
-210y_1 & - \, 30y_2 & + \, 30y_2 & + \, y_4 & \leq \ -60
\end{array}
$$

$$y_1 \geq 0, \quad y_2 \geq 0, \quad y_3 \geq 0, \quad y_4 \geq 0.$$

In a similar manner, a linear minimization (primal) problem that is not in standard form can be transformed to a standard form. To transform a constraint with a $\leq$ to a constraint with a $\geq$ sign, the constraint must be multiplied by -1. The other steps are the same as outlined previously.

22.3.3 Duality Discussion

The constraint values of the primal problem are related to the variables of the dual problem. These variables are known as *shadow prices*.

The *weak duality* theorem states that the objective function of value g of the dual problem at any feasible solution $y_1, y_2, \ldots, y_m$, is always greater than or equal to the objective value z of the primal problem at any feasible solution $x_1, x_2, \ldots x_n$. This can be expressed in matrix form as follows:

$$g = B'Y \quad \geq \quad C'X = z.$$

The *strong duality* theorem specifies that the primal and dual problems have equal optimal values of the objective function. This theorem can be used to solve the primal linear optimization problem.

Recall that the *shadow price* of constraint i of a linear maximization problem is the amount by which the optimal value of the objective function increases when the right-hand value of constraint i is increased by 1. The strong dual theorem can be used to calculate the shadow price of constraint i.

22.4 SUMMARY

Sensitivity analysis and duality are important concepts and the information is valuable in addition to the data on optimality. Sensitivity analysis helps to find the effect of small changes in the parameters of the formulation of a linear optimization problem. Duality helps to improve the understanding of a linear optimization problem.

Key Terms

sensitivity	duality	objective coefficients
right-hand constraints	marginal values	shadow prices
reduced costs	dual value	primal problem
dual problem	normal maximization	

22.5   EXERCISES

22.1   A factory manufactures 5 different products: P_1, P_2, P_3, P_4, and P_5. The factory needs to maximize profit. Each product requires machine time on three different devices: A, B, and C, each of which is available 135 hours per week. The following table provides the data:

Product	Device A	Device B	Device C
P_1	20	15	8
P_2	13	17	12
P_3	15	7	11
P_4	16	4	7
P_5	18	14	6

The unit sale price for products P_1, P_2, and P_3 is $7.50, $6.00, and $8.25, respectively. The first 26 units of P_4 and P_5 have a sale price of $5.85 each; all excess units have a sale price of $4.50 each. The operational costs of devices A and B are $5.25 per hour, and $5.85 for device C. The cost of materials for products P_1 and P_4 is $2.75. For products P_2, P_3, and P_5 the materials cost is $2.25. Hint: assume that x_i, $i = 1 \ldots 5$ are the number of units produced of the various products and use two new variables y_4 and y_5 for the units produced of P_4 and P_5, respectively, in excess of 26. Write the formulation of the mathematical optimization problem with Pyomo and Pulp. If the price of P_1 changes by 35%, how does the optimal solution change?

22.2   Find the dual of Exercise 22.1. Formulate with Pyomo and Pulp then solve the problem.

## CHAPTER 23

Transportation Models

23.1 INTRODUCTION

This chapter presents the general concepts and formulation of transportation and transshipment problems. These are special cases of linear optimization problems.

The main goal is to formulate these problems as linear optimization problems and compute minimum cost to transport a product as the optimal solution computed. For formulating these problems, the two modelers Pyomo and Pulp are used.

Transportation problems deal with finding the optimal manner by which a product or commodity produced or available at different supply points can be transported to a number of destinations or demand points. Typically, the objective function is the cost of transportation subject to capacity constraints at the supply points and demand constraints at the demand points.

23.2 MODEL OF A TRANSPORTATION PROBLEM

A transportation problem is formulated as a standard linear optimization problem. The objective function is defined to minimize the cost of transportation, subject to demand and supply constraints.

Assume there are m *supply points* and n *demand points* in a problem. Let $c_{i,j}$ denote the given unit cost of transportation from supply point i to demand (destination) point j. Let $x_{i,j}$ denote the amount of product to be transported from supply point i to demand (destination) point j. The objective function can then be expressed as follows:

Minimize z,

$$z = \sum_{i=1}^{i=m} \sum_{j=1}^{j=n} x_{i,j} c_{i,j}.$$

This equation can be expanded and written with equations in which each row of the right-hand side of the equation represents the cost of transportation from a supply point. For example, row 1 of the right-hand side of the equation represents the cost of transportation from supply point 1; row 2 represents the cost of transportation from supply point 2, the last row represents the cost of transportation from supply point m.

$$z = \begin{matrix} c_{1,1}x_{1,1} & + c_{1,2}x_{1,2} & + \cdots & + c_{1,n}x_{1,n} + \\ c_{2,1}x_{2,1} & + c_{2,2}x_{2,2} & + \cdots & + c_{2,n}x_{2,n} + \\ \vdots & \vdots & \ddots & \vdots \quad + \\ c_{m,1}x_{m,1} & + c_{m,2}x_{m,2} & + \cdots & + c_{m,n}x_{m,n} \end{matrix}$$

In transportation problems there are two types of constraints: *supply constraints* and *demand constraints*. Let s_i denote the amount of product at the supply point i. Let d_j denote the amount of product at the demand point j.

The supply constraints have the right-hand side as an upper bound. There are m supply constraints, the constraint of supply point i is expressed as follows:

$$\sum_{j=1}^{j=n} x_{i,j} \leq s_i.$$

This equation can be expanded to show all the quantities of the product to be shipped from supply point i. There are m supply constraints, with each row representing the total quantity of product transported from an indicated supply point. For example, the first row (1) represents the quantities of product shipped from supply point 1. The second row (2) represents the quantities of product shipped from supply point 2. The last row (m) represents the quantities of product shipped from supply point m.

$$\begin{matrix} x_{1,1} & + x_{1,2} & + \cdots & + x_{1,n} & \leq & s_1 \\ x_{2,1} & + x_{2,2} & + \cdots & + x_{2,n} & \leq & s_2 \\ \vdots & \vdots & \ddots & \vdots & \vdots \\ x_{m,1} & + x_{m,2} & + \cdots & + x_{m,n} & \leq & s_m \end{matrix}$$

The demand constraints have the right-hand side as a lower bound. There are n demand constraints, and the constraint of supply point j is expressed as follows:

$$\sum_{i=1}^{i=m} x_{i,j} \geq d_j.$$

This equation can be expanded to show all the quantities of the product to be received at demand point j. There are n demand constraints, with each row representing the total quantity of product to be transported and received by the indicated demand point. For example, the first row (1) represents the quantities of product at demand point 1. The second row (2) represents the quantities of product at demand point 2. The last row (n) represents the quantities of product at demand point n.

$$
\begin{array}{ccccccc}
x_{1,1} & + x_{2,1} & + \cdots & + x_{m,1} & \geq d_1 \\
x_{1,2} & + x_{2,2} & + \cdots & + x_{m,2} & \geq d_2 \\
\vdots & \vdots & \ddots & \vdots & \vdots \\
x_{1,n} & + x_{2,n} & + \cdots & + x_{m,n} & \geq d_m
\end{array}
$$

The decision variables $x_{i,j}$ have a sign constraint: $x_{i,j} \geq 0$, for $i = 1, 2, \ldots m$ and $j = 1, 2, \ldots n$.

The following three case studies help to illustrate the modeling of transportation problems.

23.3   TRANSPORTATION CASE STUDY 1

The distribution manager of a company needs to minimize global transport costs between a set of three factories (supply points) S1, S2, and S3, and a set of four distributors (demand points) D1, D2, D3, and D4. The following table shows the transportation cost from each supply point to every demand point, the supply of the product at the supply points, and the demand of the product at the demand points.

	D1	D2	D3	D4	Supply
S1	20	40	70	50	400
S2	100	60	90	80	1500
S3	10	110	30	200	900
Demand	700	600	1000	500	

The transportation unit costs for every supply point are shown from columns 2 to 5. The transportation unit cost from supply point S1 to demand point D1 is 20. The transportation unit cost from supply point S1 to demand point D2 is 40. The transportation unit cost from supply point S2 to demand point D3 is 90, and so on.

The last column in the table shows the supply capacity of the supply point, in quantity of the product. The capacity of supply point S1 is 400. The summation of the values in the last column is the total supply in the system; this value is 2800. The last row of the table shows the demand of each demand point, in quantity of the product. The demand of demand point D1 is 700, of demand point D2 is 600, and so on. The summation of the values in the last row is the total demand in the system; this value is 2800.

Note that the total supply is 2800, and the total demand is also 2800. This is calculated by summing the values of the last column and the last row. Because the value for the amount of product of total supply and the amount of total demand is the same, this transportation problem is said to be *balanced*. The objective function can be expressed as follows:

Minimize z,

$$z = \sum_{i=1}^{1=m} \sum_{j=1}^{j=n} x_{i,j} c_{i,j}.$$

This problem has $m = 3$ supply points and $n = 4$ demand points. The objective function can be completely written with the unit cost values of the product to be transported, given in the table shown previously. The objective function is expressed as follows:

$$z = \begin{array}{llll} 20x_{1,1} & + 40x_{1,2} & + 70x_{1,3} & + 50x_{1,4} + \\ 100x_{2,1} & + 60x_{2,2} & + 90x_{2,3} & + 80x_{2,4} + \\ 10x_{3,1} & + 110x_{3,2} & + 30x_{3,3} & + 200x_{3,4}. \end{array}$$

The supply constraints are:

$$\begin{array}{lllll} x_{1,1} & + x_{1,2} & + x_{1,3} & + x_{1,4} & \leq 400 \\ x_{2,1} & + x_{2,2} & + x_{2,3} & + x_{2,4} & \leq 1500 \\ x_{3,1} & + x_{3,2} & + x_{3,3} & + x_{3,4} & \leq 900. \end{array}$$

The demand constraints are:

$$\begin{array}{llll} x_{1,1} & + x_{2,1} & + x_{3,1} & \geq 700 \\ x_{1,2} & + x_{2,2} & + x_{3,2} & \geq 600 \\ x_{1,3} & + x_{2,3} & + x_{3,3} & \geq 1000 \\ x_{1,4} & + x_{2,4} & + x_{3,4} & \geq 500. \end{array}$$

23.3.1 Formulation Using the Pyomo Modeler

The formulation of this problem using the Pyomo modeler is shown in the following listing and stored in file **transport1.py**. It defines a two-dimensional index list, *xindx*, using list comprehension and is shown in line 19. This allows creating two-dimensional decision variables in line 26 and using these variables for specifying the objective function (lines 28–32) and the constraints (lines 35–40 and lines 43–50.

```
1  """
2  Python Formulation for the Pyomo Modeler
3  Example transportation problem. File: transport1.py
4  J M Garrido, September  2014
5  usage: pyomo transport1.py --summary
6  """
7  print "Transportation Problem 1"
```

```
 8 # Import
 9 from coopr.pyomo import *
10
11 # Data for Linear Optimization Problem
12 M = 3   # Supply points
13 N = 4   # Demand points
14 a = range(1, M+1)
15 al = range(M)
16 b = range(1,N+1)
17 bl = range(N)
18 # Index list for decision variables x
19 xindx = [(a[i],b[j]) for j in bl for i in al]
20
21 #Concrete Model
22 model = ConcreteModel(name="Transportation Problem 1")
23
24 #Decision Variables
25 model.x = Var(xindx, within=NonNegativeReals)
26
27 # The objective function
28 model.obj = Objective(expr=
29 20.0 * model.x[1,1] + 40.0 * model.x[1,2] + 70.0 * model.x[1,3] +
      50.0*model.x[1,4]
30  + 100*model.x[2,1] + 60.0*model.x[2,2] + 90.0*model.x[2,3] +
      80.0*model.x[2,4]
31  +10.0*model.x[3,1] + 110.0*model.x[3,2] + 30.0*model.x[3,3] +
      200*model.x[3,4],
32  sense = minimize)
33
34 # Supply Constraints
35 model.SConstraint1 = Constraint(expr=
36   model.x[1,1] + model.x[1,2] +  model.x[1,3] + model.x[1,4]
      <= 400.0)
37 model.SConstraint2 = Constraint(expr=
38   model.x[2,1] + model.x[2,2] + model.x[2,3] + model.x[2,4]
      <= 1500.0)
39 model.SConstraint3 = Constraint(expr=
40   model.x[3,1] + model.x[3,2] + model.x[3,3] + model.x[3,4]
      <=  900.0)
41
42 # Demand Constraints
43 model.DConst1 = Constraint(expr=
44   model.x[1,1]  +  model.x[2,1] + model.x[3,1]  >= 700.0)
45 model.DConst2 = Constraint(expr=
46   model.x[1,2]  +  model.x[2,2] + model.x[3,2]  >= 600.0)
```

```
47 model.DConst3 = Constraint(expr=
48    model.x[1,3]  +  model.x[2,3] + model.x[3,3]  >= 1000)
49 model.DConst4 = Constraint(expr=
50    model.x[1,4] + model.x[2,4] + model.x[3,4]   >= 500.0)
```

Running the model produces the following output listing. Note that the optimal total transportation cost is $141,000$ and some of the values of x are zero. If $x_{i,j} = 0$, then the amount of the product to be transported from supply point i to demand point j is zero. In this problem, the total demand of demand point D1 is satisfied by the amount 400 from supply point S1, and the amount 300 from supply point S3. There was no supply from supply point D2 to demand point D1, therefore $x_{2,1} = 0$.

```
$ pyomo transport1.py --summary
[     0.00] Setting up Pyomo environment
[     0.00] Applying Pyomo preprocessing actions
Transportation Problem 1
[     0.00] Creating model
[     0.01] Applying solver
[     0.05] Processing results
    Number of solutions: 1
    Solution Information
      Gap: 0.0
      Status: feasible
      Function Value: 141000.0
    Solver results file: results.json


===========================================================
Solution Summary
===========================================================


Model Transportation Problem 1

    Variables:
      x : Size=12, Index=x_index, Domain=NonNegativeReals
          Key     : Lower : Value : Upper : Initial : Fixed : Stale
          (1, 1) :     0 : 400.0 :  None :    None : False : False
          (1, 2) :     0 :   0.0 :  None :    None : False : False
          (1, 3) :     0 :   0.0 :  None :    None : False : False
          (1, 4) :     0 :   0.0 :  None :    None : False : False
          (2, 1) :     0 :   0.0 :  None :    None : False : False
          (2, 2) :     0 : 600.0 :  None :    None : False : False
          (2, 3) :     0 : 400.0 :  None :    None : False : False
          (2, 4) :     0 : 500.0 :  None :    None : False : False
```

```
      (3, 1) :     0 : 300.0 :  None :    None : False : False
      (3, 2) :     0 :   0.0 :  None :    None : False : False
      (3, 3) :     0 : 600.0 :  None :    None : False : False
      (3, 4) :     0 :   0.0 :  None :    None : False : False

Objectives:
  obj : Size=1, Index=None, Active=True
      Key  : Active : Value
      None :   True : 141000.0

Constraints:
  SConstraint1 : Size=1
      Key  : Lower : Body   : Upper
      None :  None : 400.0  : 400.0
  SConstraint2 : Size=1
      Key  : Lower : Body    : Upper
      None :  None : 1500.0  : 1500.0
  SConstraint3 : Size=1
      Key  : Lower : Body   : Upper
      None :  None : 900.0  : 900.0
  DConst1 : Size=1
      Key  : Lower : Body   : Upper
      None : 700.0 : 700.0  :  None
  DConst2 : Size=1
      Key  : Lower : Body   : Upper
      None : 600.0 : 600.0  :  None
  DConst3 : Size=1
      Key  : Lower  : Body    : Upper
      None : 1000.0 : 1000.0  :  None
  DConst4 : Size=1
      Key  : Lower : Body   : Upper
      None : 500.0 : 500.0  :  None

[    0.14] Applying Pyomo postprocessing actions
[    0.14] Pyomo Finished
```

23.3.2 Formulation Using the Pulp Modeler

The following listing shows the formulation of the transportation problem with Pulp and is stored in file transort1.py in the directory pulp_models. Note that setting the data is similar to the model that was formulated with Pyomo.

```
"""
```

```
Python Formulation for the Pulp Modeler
Example transportation problem. File: transport1.py
J M Garrido, September  2014
usage: python transport1.py
"""
print "Transportation Problem 1"
# Import PuLP modeler functions
from pulp import *

# Data for Linear Optimization Problem
M = 3  # Supply points
N = 4  # Demand points
a = range(1, M+1)
al = range(M)
b = range(1,N+1)
bl = range(N)
# Index list for decision variables x
xindx = [(a[i],b[j]) for j in bl for i in al]

# Create the model to contain the problem data
model = LpProblem("Transportation Problem 1",LpMinimize)

# Decision variables
x = LpVariable.dicts("X", xindx,0,None)

# The Pulp objective function
model += 20.0 * x[1,1] + 40.0 * x[1,2] + 70.0 * x[1,3] + 50.0*x[1,4] \
   + 100*x[2,1] + 60.0*x[2,2] + 90.0*x[2,3] + 80.0*x[2,4] \
   + 10.0*x[3,1] + 110.0*x[3,2] + 30.0*x[3,3] + 200*x[3,4], \
     "Transportation cost"

# Supply Constraints
model += x[1,1] + x[1,2] + x[1,3] + x[1,4] <= 400.0, "Supply Pt 1"
model += x[2,1] + x[2,2] + x[2,3] + x[2,4] <= 1500.0, "Supply Pt 2"
model += x[3,1] + x[3,2] + x[3,3] + x[3,4] <=  900.0, "Supply Pt 3"

# Demand Constraints
model += x[1,1]  +  x[2,1] + x[3,1]  >= 700.0, "Demand Pt 1"
model += x[1,2]  +  x[2,2] + x[3,2]  >= 600.0, "Demand Pt 2"
model += x[1,3]  +  x[2,3] + x[3,3]  >= 1000, "Demand Pt 3"
model += x[1,4] + x[2,4] + x[3,4]    >= 500.0, "Demand Pt 4"

# Solve the optimization problem using the specified PuLP Solver
model.solve(GLPK())
```

```
# Print the status of the solution
print "Status:", LpStatus[model.status]

# Print each of the variables with it's resolved optimum value
for v in model.variables():
    print v.name, "=", v.varValue

# Print the optimised value of the objective function
print "Objective Function", value(model.objective)
```

23.4   UNBALANCED PROBLEM: CASE STUDY 2

The transportation problem discussed in the previous section is an example of a balanced problem. In this case, the total supply is equal to the total demand and is expressed mathematically as:

$$\sum_{i=1}^{i=m} s_i = \sum_{j=1}^{j=n} d_j.$$

If the transportation problem is not balanced, the total supply may be less than the total demand, or the total supply may be greater than the total demand. When total supply is greater than the total demand, the problem is *unbalanced* and its formulation must include a *dummy demand* point to balance the problem, and the transportation costs to this demand point are zero.

The following problem has a small variation to the one discussed in the previous section. The distribution manager of a company needs to minimize global transport costs between a set of three factories (supply points) S1, S2, and S3, and a set of four distributors (demand points) D1, D2, D3, and D4.

The following table shows the transportation cost from each supply point to every demand point, the supply of the product at the supply points, and the demand of the product at the demand points.

	D1	D2	D3	D4	D5	Supply
S1	20	40	70	50	0	600
S2	100	60	90	80	0	1500
S3	10	110	30	200	0	900
Demand	700	600	1000	500	200	

The last column in the table shows the supply capacity of the supply point, in quantity of the product. The summation of the values in the last column is the total supply in the system; this value is 3000. The last row of the table shows the demand of each demand point, in quantity of the product. The demand of demand point D1 is 700, of demand point D2 is 600, and so on. The summation of the values in the last row is the total demand in the system, and this value is 2800.

Note that the total supply is 3000 and the total demand is 2800. This is calculated by summing the values of the last column and the last row. Because the value for the amount of product of total supply is greater than the amount of total demand, this transportation problem is said to be *unbalanced*.

The transportation unit costs for every supply point are shown from columns 2 to 5. The transportation unit cost from supply point S1 to demand point D1 is 20. The transportation unit cost from supply point S1 to demand point D2 is 40. The transportation unit cost from supply point S2 to demand point D3 is 90, and so on.

Because there is an *excess supply* of 200, the formulation of the problem must include a *dummy demand* point, D5, with a demand of 200. The transportation costs to demand point D5 are zero, and this is expressed as follows:

$$c_{i,5} = 0, \quad i = 1, \dots, 3.$$

This problem now has $m = 3$ supply points and $n = 5$ demand points. The objective function can be completely written with the unit cost values of the product to be transported, given in the table shown previously. The objective function is expressed as follows:

$$z = \begin{array}{llll} 20x_{1,1} & + 40x_{1,2} & + 70x_{1,3} & + 50x_{1,4} + \\ 100x_{2,1} & + 60x_{2,2} & + 90x_{2,3} & + 80x_{2,4} + \\ 10x_{3,1} & + 110x_{3,2} & + 30x_{3,3} & + 200x_{3,4}. \end{array}$$

The *supply constraints* have the right-hand side as an upper bound. There are 3 supply constraints, and the constraint of supply point i is expressed as follows:

$$\sum_{j=1}^{j=n} x_{i,j} \le s_i.$$

There are 3 supply constraints, with each row representing the total quantity of product transported from an indicated supply point. For example, the first row (1) represents the quantities of product shipped from supply point 1. The second row (2) represents the quantities of product shipped from supply point 2. The last row (3) represents the quantities of product shipped from supply point 3.

$$\begin{array}{llllll} x_{1,1} & + x_{1,2} & + x_{1,3} & + x_{1,4} & + x_{1,5} & \le 600 \\ x_{2,1} & + x_{2,2} & + x_{2,3} & + x_{2,4} & + x_{2,5} & \le 1500 \\ x_{3,1} & + x_{3,2} & + x_{3,3} & + x_{3,4} & + x_{3,5} & \le 900 \end{array}$$

The *demand constraints* have the right-hand side as a lower bound. There are 5 demand constraints, and the constraint of supply point j is expressed as follows:

$$\sum_{i=1}^{i=m} x_{i,j} \geq d_j.$$

There are 5 demand constraints, with each row representing the total quantity of product to be transported and received by the indicated demand point.

$$
\begin{array}{rrrr}
x_{1,1} & + x_{2,1} & + x_{3,1} & \geq 700 \\
x_{1,2} & + x_{2,2} & + x_{3,2} & \geq 600 \\
x_{1,3} & + x_{2,3} & + x_{3,3} & \geq 1000 \\
x_{1,4} & + x_{2,4} & + x_{3,4} & \geq 500 \\
x_{1,5} & + x_{2,5} & + x_{3,5} & \geq 200
\end{array}
$$

The decision variables $x_{i,j}$ have the sign constraint $x_{i,j} \geq 0$, for $i = 1, 2, 3$ and $j = 1, 2, \ldots, 5$.

23.4.1 Formulation with the Pyomo Modeler

The following listing shows the Python script that contains the formulation of the model using Pyomo. The main difference with the previous model is that now there are five demand points (line 13), three additional decision variables, and the index list is now a (3×5) list (line 19). There is also an additional demand constraint (lines 51–52).

```
1  """
2  Python Formulation for the Pyomo Modeler
3  Example unbalanced transportation problem.
      File: transport1u.py
4  J M Garrido, September  2014
5  usage: pyomo transport1u.py --summary
6  """
7  print "Unbalanced Transportation Problem"
8  # Import
9  from coopr.pyomo import *
10
11  # Data for Linear Optimization Problem
12  M = 3   # Supply points
13  N = 5   # Demand points
14  a = range(1, M+1)
15  al = range(M)
16  b = range(1,N+1)
17  bl = range(N)
18  # Index list for decision variables x
```

```
19 xindx = [(a[i],b[j]) for j in bl for i in al]
20
21 #Concrete Model
22 model = ConcreteModel(name="Unbalanced Transportation Problem 1")
23
24 #Decision Variables
25 model.x = Var(xindx, within=NonNegativeReals)
26
27 # The objective function
28 model.obj = Objective(expr=
29 20.0 * model.x[1,1] + 40.0 * model.x[1,2] + 70.0 * model.x[1,3] +
      50.0*model.x[1,4]
30  + 100*model.x[2,1] + 60.0*model.x[2,2] + 90.0*model.x[2,3] +
      80.0*model.x[2,4]
31  +10.0*model.x[3,1] + 110.0*model.x[3,2] + 30.0*model.x[3,3] +
      200*model.x[3,4],
32  sense = minimize)
33
34 # Supply Constraints
35 model.SConstraint1 = Constraint(expr=
36   model.x[1,1] + model.x[1,2] +  model.x[1,3] + model.x[1,4] +
      model.x[1,5] <= 600.0)
37 model.SConstraint2 = Constraint(expr=
38   model.x[2,1] + model.x[2,2] + model.x[2,3] + model.x[2,4] +
      model.x[2,5] <= 1500.0)
39 model.SConstraint3 = Constraint(expr=
40   model.x[3,1] + model.x[3,2] + model.x[3,3] + model.x[3,4] +
      model.x[3,5] <=  900.0)
41
42 # Demand Constraints
43 model.DConst1 = Constraint(expr=
44   model.x[1,1]  +  model.x[2,1] + model.x[3,1]  >= 700.0)
45 model.DConst2 = Constraint(expr=
46   model.x[1,2]  +  model.x[2,2] + model.x[3,2]  >= 600.0)
47 model.DConst3 = Constraint(expr=
48   model.x[1,3]  +  model.x[2,3] + model.x[3,3]  >= 1000)
49 model.DConst4 = Constraint(expr=
50   model.x[1,4] + model.x[2,4] + model.x[3,4]   >= 500.0)
51 model.DConst5 = Constraint(expr=
52   model.x[1,5] + model.x[2,5] + model.x[3,5]   >= 200.0)
```

Running the model with Pyomo produces the following output listing.

```
$ pyomo transport1u.py --summary
```

```
[    0.00] Setting up Pyomo environment
[    0.00] Applying Pyomo preprocessing actions
Unbalanced Transportation Problem
[    0.02] Creating model
[    0.02] Applying solver
[    0.05] Processing results
    Number of solutions: 1
    Solution Information
      Gap: 0.0
      Status: feasible
      Function Value: 131000.0
    Solver results file: results.json


===========================================================
Solution Summary
===========================================================


Model Unbalanced Transportation Problem 1

  Variables:
    x : Size=15, Index=x_index, Domain=NonNegativeReals
        Key     : Lower : Value : Upper : Initial : Fixed : Stale
        (1, 1) :     0 : 600.0 :  None :    None : False : False
        (1, 2) :     0 :   0.0 :  None :    None : False : False
        (1, 3) :     0 :   0.0 :  None :    None : False : False
        (1, 4) :     0 :   0.0 :  None :    None : False : False
        (1, 5) :     0 :   0.0 :  None :    None : False : False
        (2, 1) :     0 :   0.0 :  None :    None : False : False
        (2, 2) :     0 : 600.0 :  None :    None : False : False
        (2, 3) :     0 : 200.0 :  None :    None : False : False
        (2, 4) :     0 : 500.0 :  None :    None : False : False
        (2, 5) :     0 : 200.0 :  None :    None : False : False
        (3, 1) :     0 : 100.0 :  None :    None : False : False
        (3, 2) :     0 :   0.0 :  None :    None : False : False
        (3, 3) :     0 : 800.0 :  None :    None : False : False
        (3, 4) :     0 :   0.0 :  None :    None : False : False
        (3, 5) :     0 :   0.0 :  None :    None : False : False

  Objectives:
    obj : Size=1, Index=None, Active=True
        Key  : Active : Value
        None :   True : 131000.0

  Constraints:
    SConstraint1 : Size=1
```

```
     Key   : Lower : Body   : Upper
     None  :  None : 600.0  : 600.0
 SConstraint2 : Size=1
     Key   : Lower : Body    : Upper
     None  :  None : 1500.0  : 1500.0
 SConstraint3 : Size=1
     Key   : Lower : Body   : Upper
     None  :  None : 900.0  : 900.0
 DConst1 : Size=1
     Key   : Lower : Body   : Upper
     None  : 700.0 : 700.0  :  None
 DConst2 : Size=1
     Key   : Lower : Body   : Upper
     None  : 600.0 : 600.0  :  None
 DConst3 : Size=1
     Key   : Lower : Body    : Upper
     None  : 1000.0 : 1000.0 :  None
 DConst4 : Size=1
     Key   : Lower : Body   : Upper
     None  : 500.0 : 500.0  :  None
 DConst5 : Size=1
     Key   : Lower : Body   : Upper
     None  : 200.0 : 200.0  :  None


[    0.41] Applying Pyomo postprocessing actions
[    0.41] Pyomo Finished
```

23.4.2   Formulation with the Pulp Modeler

The Python script with the model formulated for Pulp is shown in the following listing. The output listing produced when running the model shows the same results as with the model using the Pyomo modeler.

```
"""
Python Formulation for the Pulp Modeler
Example transport problem. File: transport1u.py
J M Garrido, September  2014
usage: python transport1u.py
"""
print "Unbalanced Transportation Problem 1"
# Import PuLP modeler functions
from pulp import *

# Data for Linear Optimization Problem
M = 3  # Supply points
```

```
N = 5  # Demand points
a = range(1, M+1)
al = range(M)
b = range(1,N+1)
bl = range(N)
# Index list for decision variables x
xindx = [(a[i],b[j]) for j in bl for i in al]

# Create the model to contain the problem data
model = LpProblem("Unbalanced Transportation Problem 1",LpMinimize)

# Decision variables
x = LpVariable.dicts("X", xindx,0,None)

# The Pulp objective function
model += 20.0 * x[1,1] + 40.0 * x[1,2] + 70.0 * x[1,3] + 50.0*x[1,4] \
+ 100*x[2,1] + 60.0*x[2,2] + 90.0*x[2,3] + 80.0*x[2,4] \
+10.0*x[3,1] + 110.0*x[3,2] + 30.0*x[3,3] + 200*x[3,4], \
"Transportation cost"

# Supply Constraints
model += x[1,1] + x[1,2] + x[1,3] + x[1,4] + x[1,5] <= 600.0,
  "Supply Pt 1"
model += x[2,1] + x[2,2] + x[2,3] + x[2,4] + x[2,5] <= 1500.0,
  "Supply Pt 2"
model += x[3,1] + x[3,2] + x[3,3] + x[3,4] + x[3,5] <=  900.0,
  "Supply Pt 3"

# Demand Constraints
model += x[1,1]  +  x[2,1] + x[3,1]  >= 700.0, "Demand Pt 1"
model += x[1,2]  +  x[2,2] + x[3,2]  >= 600.0, "Demand Pt 2"
model += x[1,3]  +  x[2,3] + x[3,3]  >= 1000.0, "Demand Pt 3"
model += x[1,4] + x[2,4] + x[3,4]    >= 500.0, "Demand Pt 4"
model += x[1,5] + x[2,5] + x[3,5]    >= 200.0, "Dummy Demand Pt"

# Solve the optimization problem using the specified PuLP Solver
model.solve(GLPK())

# Print the status of the solution
print "Status:", LpStatus[model.status]

# Print each of the variables with it's resolved optimum value
for v in model.variables():
    print v.name, "=", v.varValue
```

```
# Print the optimised value of the objective function
print "Objective Function", value(model.objective)
```

23.5   UNBALANCED PROBLEM: CASE STUDY 3

When the demand exceeds the supply, the problem formulation includes a *penalty* associated with the *unmet demand*. Suppose that in the original problem discussed previously, supply point S1 produces 300 units of the product (instead of 400). This problem is now unbalanced, with an unmet demand of 100.

A dummy supply point, S4, with a supply of 100, is added to the problem formulation. The penalty for unmet demand at demand point D1 is 125; at demand point D2 is 147; at demand point D3 is 95; and at demand point D4 is 255.

The following table shows the transportation cost from each supply point to every demand point, the supply of the product at the supply points, and the demand of the product at the demand points.

	D1	D2	D3	D4	Supply
S1	20	40	70	50	300
S2	100	60	90	80	1500
S3	10	110	30	200	900
S4	125	147	95	255	100
Demand	700	600	1000	500	

Note that the dummy supply point S4 has been included in the table. The penalty amounts have also been included for this supply point. The total supply and demand is now 2800.

This problem has $m = 4$ supply points and $n = 4$ demand points. The objective function can be completely written with the unit cost values of the product to be transported, given in the table shown previously. The objective function is expressed as follows:

$$z = \begin{aligned} & 20x_{1,1} & + 40x_{1,2} & + 70x_{1,3} & + 50x_{1,4} + \\ & 100x_{2,1} & + 60x_{2,2} & + 90x_{2,3} & + 80x_{2,4} + \\ & 10x_{3,1} & + 110x_{3,2} & + 30x_{3,3} & + 200x_{3,4} + \\ & 125x_{4,1} & + 147x_{4,2} & + 95x_{4,3} & + 255x_{4,4}. \end{aligned}$$

There are 4 supply constraints, with each row representing the total quantity of product transported from an indicated supply point. For example, the first row (1) represents the quantities of product shipped from supply point 1. The second row (2) represents the quantities of product shipped from supply point 2. The last row (4) represents the quantities of product shipped from supply point 4.

$$
\begin{array}{rrrrr}
x_{1,1} & + x_{1,2} & + x_{1,3} & + x_{1,4} & \leq 300 \\
x_{2,1} & + x_{2,2} & + x_{2,3} & + x_{2,4} & \leq 1500 \\
x_{3,1} & + x_{3,2} & + x_{3,3} & + x_{3,4} & \leq 900 \\
x_{4,1} & + x_{4,2} & + x_{4,3} & + x_{4,4} & \leq 100
\end{array}
$$

The demand constraints have the right-hand side as a lower bound. There are 4 demand constraints, with each row representing the total quantity of product to be transported and received by the indicated demand point.

$$
\begin{array}{rrrrr}
x_{1,1} & + x_{2,1} & + x_{3,1} & + x_{4,1} & \geq 700 \\
x_{1,2} & + x_{2,2} & + x_{3,2} & + x_{4,2} & \geq 600 \\
x_{1,3} & + x_{2,3} & + x_{3,3} & + x_{4,3} & \geq 1000 \\
x_{1,4} & + x_{2,4} & + x_{3,4} & + x_{4,4} & \geq 500
\end{array}
$$

The decision variables $x_{i,j}$ have the sign constraint: $x_{i,j} \geq 0$, for $i = 1, \ldots, 4$ and $j = 1, \ldots, 4$.

23.5.1 Formulation with the Pyomo Modeler

The following listing shows the Python script that contains the formulation of the model using Pyomo and is stored in file **transport1ub.py** (in folder pyomo.

```
"""
Python Formulation for the Pyomo Modeler
Example unbalanced transportation problem with excess supply.
File: transport1u.py J M Garrido, September  2014
usage: pyomo transport1ub.py --summary
"""
print "Unbalanced Transportation Problem 2"
# Import
from coopr.pyomo import *

# Data for Linear Optimization Problem
M = 4  # Supply points
N = 4  # Demand points
a = range(1, M+1)
al = range(M)
b = range(1,N+1)
bl = range(N)
# Index list for decision variables x
xindx = [(a[i],b[j]) for j in bl for i in al]

#Concrete Model
model = ConcreteModel(name="Unbalanced Transportation Problem 2")
```

```
#Decision Variables
model.x = Var(xindx, within=NonNegativeReals)

# The objective function
model.obj = Objective(expr=
20.0 * model.x[1,1] + 40.0 * model.x[1,2] + 70.0 * model.x[1,3]
    + 50.0*model.x[1,4]
 + 100*model.x[2,1] + 60.0*model.x[2,2] + 90.0*model.x[2,3]
    + 80.0*model.x[2,4]
+10.0*model.x[3,1] + 110.0*model.x[3,2] + 30.0*model.x[3,3]
    + 200*model.x[3,4]
+125.0*model.x[4,1] + 147.0*model.x[4,2] + 95.0*model.x[4,3]
    + 255.0*model.x[4,4],
 sense = minimize)

# Supply Constraints
model.SConstraint1 = Constraint(expr=
  model.x[1,1] + model.x[1,2] +  model.x[1,3] + model.x[1,4] <= 300.0)
model.SConstraint2 = Constraint(expr=
  model.x[2,1] + model.x[2,2] + model.x[2,3] + model.x[2,4] <= 1500.0)
model.SConstraint3 = Constraint(expr=
  model.x[3,1] + model.x[3,2] + model.x[3,3] + model.x[3,4] <=  900.0)
model.SConstraint4 = Constraint(expr=
  model.x[4,1] + model.x[4,2] + model.x[4,3] + model.x[4,4] <=  100.0)

# Demand Constraints
model.DConst1 = Constraint(expr=
  model.x[1,1]   +   model.x[2,1] + model.x[3,1] + model.x[4,1] >= 700.0)
model.DConst2 = Constraint(expr=
  model.x[1,2]   +   model.x[2,2] + model.x[3,2] + model.x[4,2] >= 600.0)
model.DConst3 = Constraint(expr=
  model.x[1,3]   +   model.x[2,3] + model.x[3,3] + model.x[4,3] >= 1000)
model.DConst4 = Constraint(expr=
  model.x[1,4] + model.x[2,4] + model.x[3,4] +model.x[4,4]   >= 500.0)
```

The following listing is produced after running the model with the Pyomo modeler.

```
$ pyomo transport1ub.py --summary
[    0.00] Setting up Pyomo environment
[    0.00] Applying Pyomo preprocessing actions
Unbalanced Transportation Problem 2
[    0.02] Creating model
```

```
[    0.02] Applying solver
[    0.06] Processing results
     Number of solutions: 1
     Solution Information
       Gap: 0.0
       Status: feasible
       Function Value: 146500.0
     Solver results file: results.json


============================================================
Solution Summary
============================================================


Model Unbalanced Transportation Problem 2

  Variables:
    x : Size=16, Index=x_index, Domain=NonNegativeReals
        Key    : Lower : Value : Upper : Initial : Fixed : Stale
        (1, 1) :     0 : 300.0 :  None :    None : False : False
        (1, 2) :     0 :   0.0 :  None :    None : False : False
        (1, 3) :     0 :   0.0 :  None :    None : False : False
        (1, 4) :     0 :   0.0 :  None :    None : False : False
        (2, 1) :     0 :   0.0 :  None :    None : False : False
        (2, 2) :     0 : 600.0 :  None :    None : False : False
        (2, 3) :     0 : 400.0 :  None :    None : False : False
        (2, 4) :     0 : 500.0 :  None :    None : False : False
        (3, 1) :     0 : 400.0 :  None :    None : False : False
        (3, 2) :     0 :   0.0 :  None :    None : False : False
        (3, 3) :     0 : 500.0 :  None :    None : False : False
        (3, 4) :     0 :   0.0 :  None :    None : False : False
        (4, 1) :     0 :   0.0 :  None :    None : False : False
        (4, 2) :     0 :   0.0 :  None :    None : False : False
        (4, 3) :     0 : 100.0 :  None :    None : False : False
        (4, 4) :     0 :   0.0 :  None :    None : False : False

  Objectives:
    obj : Size=1, Index=None, Active=True
        Key  : Active : Value
        None :   True : 146500.0

  Constraints:
    SConstraint1 : Size=1
        Key  : Lower : Body  : Upper
        None :  None : 300.0 : 300.0
    SConstraint2 : Size=1
```

```
        Key   : Lower : Body    : Upper
        None :   None : 1500.0 : 1500.0
  SConstraint3 : Size=1
        Key   : Lower : Body   : Upper
        None :   None : 900.0 : 900.0
  SConstraint4 : Size=1
        Key   : Lower : Body   : Upper
        None :   None : 100.0 : 100.0
  DConst1 : Size=1
        Key   : Lower : Body   : Upper
        None : 700.0 : 700.0 :   None
  DConst2 : Size=1
        Key   : Lower : Body   : Upper
        None : 600.0 : 600.0 :   None
  DConst3 : Size=1
        Key   : Lower  : Body    : Upper
        None : 1000.0 : 1000.0 :   None
  DConst4 : Size=1
        Key   : Lower : Body   : Upper
        None : 500.0 : 500.0 :   None


[    0.08] Applying Pyomo postprocessing actions
[    0.08] Pyomo Finished
```

23.5.2 Formulation with the Pulp Modeler

The following listing shows the Python script with the formulation of the model using Pulp modeler and is stored in file **transport1ub.py** in directory pulp_models.

```
"""
Python Formulation for the Pulp Modeler
Example transport problem with excess demand.
File: transport1ub.py
J M Garrido, September  2014
usage: python transport1ub.py
"""
print "Unbalanced Transportation Problem 2"
# Import PuLP modeler functions
from pulp import *

# Data for Linear Optimization Problem
M = 4  # Supply points
N = 4  # Demand points
```

```
a = range(1, M+1)
al = range(M)
b = range(1,N+1)
bl = range(N)
# Index list for decision variables x
xindx = [(a[i],b[j]) for j in bl for i in al]

# Create the model to contain the problem data
model = LpProblem("Unbalanced Transportation Problem 2",LpMinimize)

# Decision variables
x = LpVariable.dicts("X", xindx,0,None)

# The Pulp objective function
model += 20.0 * x[1,1] + 40.0 * x[1,2] + 70.0 * x[1,3] + 50.0*x[1,4] \
 + 100*x[2,1] + 60.0*x[2,2] + 90.0*x[2,3] + 80.0*x[2,4] \
 +10.0*x[3,1] + 110.0*x[3,2] + 30.0*x[3,3] + 200*x[3,4] \
 +125.0*x[4,1] + 147.0*x[4,2] + 95.0*x[4,3] + 255.0*x[4,4], \
 "Transportation cost"

# Supply Constraints
model += x[1,1] + x[1,2] + x[1,3] + x[1,4] <= 300.0,"Supply Pt 1"
model += x[2,1] + x[2,2] + x[2,3] + x[2,4] <= 1500.0, "Supply Pt 2"
model += x[3,1] + x[3,2] + x[3,3] + x[3,4] <=  900.0, "Supply Pt 3"
model += x[4,1] + x[4,2] + x[4,3] + x[4,4] <=  100.0, "Dummy Supply Pt"

# Demand Constraints
model += x[1,1] + x[2,1] + x[3,1] + x[4,1] >= 700.0, "Demand Pt 1"
model += x[1,2] + x[2,2] + x[3,2] + x[4,2] >= 600.0, "Demand Pt 2"
model += x[1,3] + x[2,3] + x[3,3] + x[4,3] >= 1000.0, "Demand Pt 3"
model += x[1,4] + x[2,4] + x[3,4] + x[4,4] >= 500.0, "Demand Pt 4"

# Solve the optimization problem using the specified PuLP Solver
model.solve(GLPK())

# Print the status of the solution
print "Status:", LpStatus[model.status]

# Print each of the variables with it's resolved optimum value
for v in model.variables():
    print v.name, "=", v.varValue

# Print the optimised value of the objective function
print "Objective Function", value(model.objective)
```

23.6   TRANSSHIPMENT MODELS

A transshipment model includes *intermediate* or *transshipment points* in a transportation model. A transshipment point is an intermediate point between one or more supply points and one or more demand points. Quantities of a product can be sent from a supply point directly to a demand point or via a transshipment point.

A transshipment point can be considered both a supply point and a demand point. At a transshipment point, k, the total quantity of product shipped to this point must equal the total quantity of the product shipped from this intermediate point and can be expressed as follows:

$$\sum_{i=1}^{i=m} x_{i,k} = \sum_{j=1}^{j=n} x_{k,j},$$

where m is the number of supply points, n is the number of demand points, and $x_{i,j}$ is the amount of the product shipped from supply point i to demand point j. From the previous equation, the *transshipment constraint* of a point, k, is expressed as follows:

$$\sum_{i=1}^{i=m} x_{i,k} - \sum_{j=1}^{j=n} x_{k,j} = 0. \tag{23.1}$$

23.7   TRANSSHIPMENT PROBLEM: CASE STUDY 4

Various quantities of a product are shipped from two cities (supply points), S1 and S2, to three destinations (demand points), D1, D2, and D3. The products are first shipped to three warehouses (transshipment points), T1, T2, and T3, then shipped to their final destinations.

The following table shows the transportation cost from each supply point to the intermediate points and to every demand point. The table also includes the supply of the product at the supply points, and the demand of the product at the demand points.

	T1	T2	T3	D1	D2	D3	Supply
S1	16	10	12	0	0	0	300
S2	15	14	17	0	0	0	300
T1	0	0	0	6	8	10	0
T2	0	0	0	7	11	11	0
T3	0	0	0	4	5	12	0
Demand	0	0	0	200	100	300	

Note that the total supply of the product is 600 and the total demand is also 600. Therefore, this is a balanced problem.

The problem can be formulated directly as a standard transportation problem, using the data in the previous table. The conventional notation for the quantity of product is used and $x_{i,j}$ denotes the quantity of product shipped from point i to point j. For example: $x_{2,3}$ denotes the quantity of product shipped from supply point S2 to transshipment point T3.

This problem has $m = 2$ supply points, $n = 3$ demand points, and 3 transshipment points. Because a transshipment point can be a supply point and a demand point, the problem can be formulated with a total of 5 supply points and 6 demand points.

The objective function can be completely written with the unit cost values of the product to be transported given in the table. The objective function is expressed as follows:

$$z = \begin{array}{l} 16x_{1,1} + 10x_{1,2} + 12x_{1,3} + \\ 15x_{2,1} + 14x_{2,2} + 17x_{2,3} + \\ 6x_{3,4} + 8x_{3,5} + 10x_{3,6} + \\ 7x_{4,4} + 11x_{4,5} + 11x_{4,6} + \\ 4x_{5,4} + 5x_{5,5} + 12x_{5,6}. \end{array}$$

There are three transshipment constraints, one for each transshipment point. These constraints are:

$$
\begin{array}{llllll}
x_{1,1} & + x_{2,1} & - x_{3,4} & - x_{3,5} & - x_{3,6} & = 0 \quad \text{(T1)} \\
x_{1,2} & + x_{2,2} & - x_{4,4} & - x_{4,5} & - x_{4,6} & = 0 \quad \text{(T2)} \\
x_{1,3} & + x_{2,3} & - x_{5,4} & - x_{5,5} & - x_{5,6} & = 0 \quad \text{(T3).}
\end{array}
$$

There are two supply constraints, each representing the total quantity of product transported from an indicated supply point.

$$
\begin{array}{llll}
x_{1,1} & + x_{1,2} & + x_{1,3} & \leq 300 \\
x_{2,1} & + x_{2,2} & + x_{2,3} & \leq 300
\end{array}
$$

There are three demand constraints, each representing the total quantity of product transported to the indicated demand point.

$$
\begin{array}{llll}
x_{3,6} & + x_{4,6} & + x_{5,6} & \geq 300 \\
x_{3,5} & + x_{4,5} & + x_{5,5} & \geq 100 \\
x_{3,4} & + x_{4,4} & + x_{5,4} & \geq 200
\end{array}
$$

There are three transshipment constraints applying the balance equations (Equation 23.1) at each one.

$$
\begin{array}{llllll}
x_{1,1} & + x_{2,1} & - x_{3,4} & - x_{3,5} & - x_{3,6} & = 0 \\
x_{1,2} & + x_{2,2} & - x_{4,4} & - x_{4,5} & - x_{4,6} & = 0 \\
x_{1,3} & + x_{2,3} & - x_{5,4} & - x_{5,5} & - x_{5,6} & = 0.
\end{array}
$$

The decision variables $x_{i,j}$ have the sign constraint $x_{i,j} \geq 0$, for $i = 1, \dots, 5$ and $j = 1, \dots, 6$.

23.7.1   Formulation with the Pyomo Modeler

The following listing shows the Python script with the model using Pyomo and is stored in file `transship1.py`. Note that a list of artificial variables were included in the formulation because Pyomo does not accept constraints with an equal sign. The size of this list, y, is the number of transshipment points. With this list, the right-hand side of the transshipment constraints were written with ≥ 0.0 in lines 55–60.

```
1  """
2  Python Formulation for the Pyomo Modeler
3  Example a transshipment problem.
4  File: transship1.py - J M Garrido, September  2014
5  usage: pyomo transship1.py --summary
6  """
7  print "Transshipment Problem 1"
8  # Import
9  from coopr.pyomo import *
10
11 # Data for Linear Optimization Problem
12 M = 5  # Supply points
13 N = 6  # Demand points
14 NT = 3 # Number of transshipment points
15 a = range(1, M+1)
16 al = range(M)
17 b = range(1,N+1)
18 bl = range(N)
19 # Index list for decision variables x
20 xindx = [(a[i],b[j]) for j in bl for i in al]
21 tindx = range(1, NT+1) # index list for y variables
22
23 #Concrete Model
24 model = ConcreteModel(name="Transshipment Problem 1")
25
26 # Decision Variables
27 model.x = Var(xindx, within=NonNegativeReals)
28 # Artificial variables
29 model.y = Var(tindx, within=NonNegativeReals)
30
31 # The objective function
32 model.obj = Objective(expr=
33 16.0 * model.x[1,1] + 10.0 * model.x[1,2] + 12.0 * model.x[1,3]
34  + 15*model.x[2,1] + 14.0*model.x[2,2] + 17.0*model.x[2,3]
35  +6.0*model.x[3,4] + 8.0*model.x[3,5] + 10.0*model.x[3,6]
36  +7.0*model.x[4,4] + 11.0*model.x[4,5] + 11.0*model.x[4,6]
37  +4.0*model.x[5,4] + 5.0*model.x[5,5] + 12.0*model.x[5,6],
```

```
38  sense = minimize)
39
40 # Supply Constraints
41 model.SConstraint1 = Constraint(expr=
42    model.x[1,1] + model.x[1,2] +  model.x[1,3] <= 300.0)
43 model.SConstraint2 = Constraint(expr=
44    model.x[2,1] + model.x[2,2] + model.x[2,3]  <= 300.0)
45
46 # Demand Constraints
47 model.DConst1 = Constraint(expr=
48    model.x[3,4] + model.x[4,4] + model.x[5,4] >= 200.0)
49 model.DConst2 = Constraint(expr=
50    model.x[3,5] + model.x[4,5] + model.x[5,5] >= 100.0)
51 model.DConst3 = Constraint(expr=
52    model.x[3,6] + model.x[4,6] + model.x[5,6] >= 300.0)
53
54 # Transshipment Constraints
55 model.TConst1 = Constraint(expr=
56    model.x[1,1] + model.x[2,1] - model.x[3,4] - model.x[3,5]
      - model.x[3,6] - model.y[1] >= 0.0)
57 model.TConstraint2 = Constraint(expr=
58    model.x[1,2] + model.x[2,2] - model.x[4,4] - model.x[4,5]
      - model.x[4,6] - model.y[2] >= 0.0)
59 model.TConstraint3 = Constraint(expr=
60    model.x[1,3] + model.x[2,3] - model.x[5,4] - model.x[5,5]
      - model.x[5,6] - model.y[3] >= 0.0)
```

After running the model with Pyomo, the following listing is produced. Note that the value of the objective function is 12400.0, which is the optimal value.

```
$ pyomo transship1.py --summary
[    0.00] Setting up Pyomo environment
[    0.00] Applying Pyomo preprocessing actions
Transshipment Problem 1
[    0.02] Creating model
[    0.02] Applying solver
[    0.05] Processing results
    Number of solutions: 1
    Solution Information
      Gap: 0.0
      Status: feasible
      Function Value: 12400.0
    Solver results file: results.json
```

```
==============================================================
Solution Summary
==============================================================

Model Transshipment Problem 1

  Variables:
    x : Size=30, Index=x_index, Domain=NonNegativeReals
        Key    : Lower : Value : Upper : Initial : Fixed : Stale
        (1, 1) :    0 :   0.0 : None :    None : False : False
        (1, 2) :    0 :   0.0 : None :    None : False : False
        (1, 3) :    0 : 300.0 : None :    None : False : False
        (1, 4) :    0 :  None : None :    None : False :  True
        (1, 5) :    0 :  None : None :    None : False :  True
        (1, 6) :    0 :  None : None :    None : False :  True
        (2, 1) :    0 : 300.0 : None :    None : False : False
        (2, 2) :    0 :  -0.0 : None :    None : False : False
        (2, 3) :    0 :   0.0 : None :    None : False : False
        (2, 4) :    0 :  None : None :    None : False :  True
        (2, 5) :    0 :  None : None :    None : False :  True
        (2, 6) :    0 :  None : None :    None : False :  True
        (3, 1) :    0 :  None : None :    None : False :  True
        (3, 2) :    0 :  None : None :    None : False :  True
        (3, 3) :    0 :  None : None :    None : False :  True
        (3, 4) :    0 :   0.0 : None :    None : False : False
        (3, 5) :    0 :   0.0 : None :    None : False : False
        (3, 6) :    0 : 300.0 : None :    None : False : False
        (4, 1) :    0 :  None : None :    None : False :  True
        (4, 2) :    0 :  None : None :    None : False :  True
        (4, 3) :    0 :  None : None :    None : False :  True
        (4, 4) :    0 :   0.0 : None :    None : False : False
        (4, 5) :    0 :   0.0 : None :    None : False : False
        (4, 6) :    0 :   0.0 : None :    None : False : False
        (5, 1) :    0 :  None : None :    None : False :  True
        (5, 2) :    0 :  None : None :    None : False :  True
        (5, 3) :    0 :  None : None :    None : False :  True
        (5, 4) :    0 : 200.0 : None :    None : False : False
        (5, 5) :    0 : 100.0 : None :    None : False : False
        (5, 6) :    0 :   0.0 : None :    None : False : False
    y : Size=3, Index=y_index, Domain=NonNegativeReals
        Key : Lower : Value : Upper : Initial : Fixed : Stale
          1 :    0 :   0.0 : None :    None : False : False
          2 :    0 :   0.0 : None :    None : False : False
          3 :    0 :   0.0 : None :    None : False : False
```

```
Objectives:
  obj : Size=1, Index=None, Active=True
      Key  : Active : Value
      None :   True : 12400.0

Constraints:
  SConstraint1 : Size=1
      Key  : Lower : Body  : Upper
      None :  None : 300.0 : 300.0
  SConstraint2 : Size=1
      Key  : Lower : Body  : Upper
      None :  None : 300.0 : 300.0
  DConst1 : Size=1
      Key  : Lower : Body  : Upper
      None : 200.0 : 200.0 :  None
  DConst2 : Size=1
      Key  : Lower : Body  : Upper
      None : 100.0 : 100.0 :  None
  DConst3 : Size=1
      Key  : Lower : Body  : Upper
      None : 300.0 : 300.0 :  None
  TConst1 : Size=1
      Key  : Lower : Body : Upper
      None :   0.0 :  0.0 : None
  TConstraint2 : Size=1
      Key  : Lower : Body : Upper
      None :   0.0 :  0.0 : None
  TConstraint3 : Size=1
      Key  : Lower : Body : Upper
      None :   0.0 :  0.0 : None

[    0.08] Applying Pyomo postprocessing actions
[    0.08] Pyomo Finished
```

23.7.2 Formulation with the Pulp Modeler

The following listing shows the Python script with the model using Pulp and is stored in file **transship1.py**. Note that a list of artificial variables is included in the formulation because Pulp does not accept constraints with an equal sign. The size of this list, y, is the number of transshipment points. With this list, the right-hand side of the transshipment constraints were written with ≥ 0.0.

```
"""
Python Formulation for the Pulp Modeler
Transshipment problem. File: transship1.py
J M Garrido, September  2014
usage: python transship1.py
"""
print "Transshipment Problem"
# Import PuLP modeler functions
from pulp import *

# Data for Linear Optimization Problem
M = 5   # Supply points
N = 6   # Demand points
NT = 3 # Number of transshipment points
a = range(1, M+1)
al = range(M)
b = range(1,N+1)
bl = range(N)
# Index list for decision variables x
xindx = [(a[i],b[j]) for j in bl for i in al]
tindx = range(1, NT+1) # index list for y variables

# Create the model to contain the problem data
model = LpProblem("Transshipment Problem",LpMinimize)

# Decision variables
x = LpVariable.dicts("X", xindx,0,None)
y = LpVariable.dicts("Y", tindx,0, None)

# The Pulp objective function
model += 16.0 * x[1,1] + 10.0 * x[1,2] + 12.0 * x[1,3] \
+ 15.0*x[2,1] + 14.0*x[2,2] + 17.0*x[2,3]   \
+6.0*x[3,4] + 8.0*x[3,5] + 10.0*x[3,6] \
+7.0*x[4,4] + 11.0*x[4,5] + 11.0*x[4,6] \
+4.0*x[5,4] + 5.0*x[5,5] + 12.0*x[5,6], \
"Transportation cost"

# Supply Constraints
model += x[1,1] + x[1,2] + x[1,3] <= 300.0,"Supply Pt 1"
model += x[2,1] + x[2,2] + x[2,3] <= 300.0, "Supply Pt 2"

# Demand Constraints
model += x[3,4] + x[4,4] + x[5,4] >= 200.0, "Demand Pt 1"
model += x[3,5] + x[4,5] + x[5,5] >= 100.0, "Demand Pt 2"
model += x[3,6] + x[4,6] + x[5,6] >= 300.0, "Demand Pt 3"
```

```
# Transshipment Constraints
model += x[1,1] + x[2,1] - x[3,4] - x[3,5] -x[3,6] - y[1] >= 0.0,
    "Transshipment Pt 1"
model += x[1,2] + x[2,2] - x[4,4] - x[4,5] -x[4,6] - y[2] >= 0.0,
    "Transshipment Pt 2"
model += x[1,3] + x[2,3] - x[5,4] - x[5,5] -x[5,6] - y[3] >= 0.0,
    "Transshipment Pt 3"

# Solve the optimization problem using the specified PuLP Solver
model.solve(GLPK())

# Print the status of the solution
print "Status:", LpStatus[model.status]

# Print each of the variables with it's resolved optimum value
for v in model.variables():
    print v.name, "=", v.varValue

# Print the optimized value of the objective function
print "Objective Function", value(model.objective)
```

23.8 ASSIGNMENT PROBLEMS

An *assignment problem* is a special case of a transportation problem that is formulated as a linear optimization model. The basic goal is to find an optimal assignment of *resources* to tasks. The typical objective is to minimize the total time to complete a task or to minimize the cost of the assignment.

A simple description of the general assignment problem is: minimize the total cost of a set of workers assigned to a set of tasks. The constraints are: each worker is assigned no more that a specified number of jobs, and each job requires no more than a specified number of workers.

An assignment of a resource i to a job j is denoted by $x_{i,j}$. Let m be the number of resources and n the number of jobs. The decision variables $x_{i,j}$ have the constraint $x_{i,j} = 1$ or $x_{i,j} = 0$, for $i = 1 \ldots m$ and $j = 1 \ldots n$. The cost of resource i to complete the job j is denoted by $c_{i,j}$. The total cost of the resource allocation is:

$$\sum_{i=1}^{m} \sum_{j=1}^{n} c_{i,j} x_{i,j}.$$

Let P_j denote the maximum number of resources to be assigned to job j. Similarly, let Q_i denote the maximum number of jobs that resource i can be assigned to. In simple problems, all these parameters are equal to 1. The two types of constraints are m *resource constraints* and n *job constraints*. The job constraints are expressed as:

$$\sum_{i=1}^{m} x_{i,j} \le P_j, \quad j = 1 \dots n.$$

The resource constraints are expressed as:

$$\sum_{j=1}^{n} x_{i,j} \le Q_i, \quad i = 1 \dots m.$$

23.9 ASSIGNMENT PROBLEM: CASE STUDY 5

A factory has 3 machines: M1, M2, and M3. These are to be assigned to four jobs: T1, T2, T3, and T4. The following table is the cost matrix that gives the expected costs when a specific machine is assigned a specific job. The goal of the problem is to optimize the assignment of machines to jobs. In this problem, each machine can only be assigned to one job, and each job can only receive one machine.

	T1	T2	T3	T4
M1	13	16	12	11
M2	15	2	13	20
M3	5	7	10	6

Note that the total number of machines is 3 and the total number of jobs is 4, so this is an unbalanced problem. A dummy machine, M4, is included with zero cost when assigned to a job.

The problem can be formulated directly using the data in the previous table. The conventional notation for an assignment of resource i to job j is used and denoted by $x_{i,j}$. For example, $x_{2,3}$ denotes the assignment of machine M2 to job T3.

The goal of the problem is to minimize the objective function, which is the cost of the assignment of the various machines to the jobs and expressed as follows:

$$
\begin{aligned}
z = \quad & 13x_{1,1} & + \ 16x_{1,2} & + \ 12x_{1,3} & + \ 11x_{1,4} & + \\
& 15x_{2,1} & + \ 2x_{2,2} & + \ 13x_{2,3} & + \ 20x_{2,4} & + \\
& 5x_{3,1} & + \ 7x_{3,2} & + \ 10x_{3,3} & + \ 6x_{3,4} & +.
\end{aligned}
$$

The dummy machine assignments are not included in the objective function because their costs are zero. There are four resource constraints, each representing the possible assignments of a machine.

$$x_{1,1} + x_{1,2} + x_{1,3} + x_{1,4} = 1 \quad \text{(M1)}$$
$$x_{2,1} + x_{2,2} + x_{2,3} + x_{2,4} = 1 \quad \text{(M2)}$$
$$x_{3,1} + x_{3,2} + x_{3,3} + x_{3,4} = 1 \quad \text{(M3)}$$
$$x_{4,1} + x_{4,2} + x_{4,3} + x_{4,4} = 1 \quad \text{(M4)}$$

There are four job constraints, each representing the possible assignments of the machines to it.

$$x_{1,1} + x_{2,1} + x_{3,1} + x_{4,1} = 1 \quad \text{(T1)}$$
$$x_{1,2} + x_{2,2} + x_{3,2} + x_{4,2} = 1 \quad \text{(T2)}$$
$$x_{1,3} + x_{2,3} + x_{3,3} + x_{4,3} = 1 \quad \text{(T3)}$$
$$x_{1,4} + x_{2,4} + x_{3,4} + x_{4,4} = 1 \quad \text{(T4)}$$

The decision variables $x_{i,j}$ have the constraint $x_{i,j} = 0$ or $x_{i,j} = 1$, for $i = 1, \ldots, 4$ and $j = 1, \ldots, 4$.

23.9.1 Formulation with the Pyomo Modeler

The following listing shows the Python script with the model using Pyomo and stored in file assign1.py. As with the previous models, it was necessary to introduce artificial variables.

```
"""
Python Formulation for the Pyomo Modeler
Assignment problem. File: assign1.py
J M Garrido, September  2014
usage: pyomo assign1.py --summary
"""
print "Assignment Problem 1"
# Import
from coopr.pyomo import *

# Data for Linear Optimization Problem
M = 4   # Number of Jobs
N = 6   # Number of Machines
NAV = M+N # Number of artificial variables
a = range(1, M+1)
al = range(M)
b = range(1,N+1)
bl = range(N)
# Index list for decision variables x
xindx = [(a[i],b[j]) for j in bl for i in al]
tindx = range(1, NAV+1) # index list for y variables

#Concrete Model
```

```python
model = ConcreteModel(name="Assignment Problem 1")

# Decision Variables
model.x = Var(xindx, within=NonNegativeReals)
# Artificial variables
model.y = Var(tindx, within=NonNegativeReals)

# The objective function
model.obj = Objective(expr=
13.0 * model.x[1,1] + 16.0 * model.x[1,2] + 12.0 * model.x[1,3]
    + 11.0*model.x[1,4]
 + 15*model.x[2,1] + 2.0*model.x[2,2] + 13.0*model.x[2,3]
    + 20.0*model.x[2,4]
+5.0*model.x[3,1] + 7.0*model.x[3,2] + 10.0*model.x[3,3]
    + 6.0*model.x[3,4],
 sense = minimize)

# Job Constraints
model.JConstraint1 = Constraint(expr=
  model.x[1,1] + model.x[2,1] +  model.x[3,1] + model.x[4,1]
    - model.y[1] >= 1)
model.JConstraint2 = Constraint(expr=
  model.x[1,2] + model.x[2,2] + model.x[3,2] + model.x[4,2]
    - model.y[2] >= 1)
model.JConstraint3 = Constraint(expr=
  model.x[1,3] + model.x[2,3] + model.x[3,3] + model.x[4,3]
    - model.y[3] >= 1)
model.JConstraint4 = Constraint(expr=
  model.x[1,4] + model.x[2,4] + model.x[3,4] + model.x[4,4]
    - model.y[4] >= 1)

# Machine Constraints
model.MConst1 = Constraint(expr=
  model.x[1,1] + model.x[1,2] + model.x[1,3] + model.x[1,4]
    - model.y[5] >= 1)
model.MConst2 = Constraint(expr=
  model.x[2,1] + model.x[2,2] + model.x[2,3] + model.x[2,4]
    - model.y[6] >= 1)
model.MConst3 = Constraint(expr=
  model.x[3,1] + model.x[3,2] + model.x[3,3] + model.x[3,4]
    - model.y[7] >= 1)
model.MConst4 = Constraint(expr=
  model.x[4,1] + model.x[4,2] + model.x[4,3] + model.x[4,4]
    - model.y[8] >= 1)
```

After running the model with Pyomo, the following listing is produced. Note that the optimum value of the objective function is 18.

```
$ pyomo assign1.py --summary
[    0.00] Setting up Pyomo environment
[    0.00] Applying Pyomo preprocessing actions
Assignment Problem 1
[    0.00] Creating model
[    0.00] Applying solver
[    0.05] Processing results
    Number of solutions: 1
    Solution Information
      Gap: 0.0
      Status: feasible
      Function Value: 18.0
    Solver results file: results.json


==============================================================
Solution Summary
==============================================================


Model Assignment Problem 1

  Variables:
    x : Size=24, Index=x_index, Domain=NonNegativeReals
        Key     : Lower : Value : Upper : Initial : Fixed : Stale
        (1, 1) :    0 :   0.0 :  None :    None : False : False
        (1, 2) :    0 :   0.0 :  None :    None : False : False
        (1, 3) :    0 :   0.0 :  None :    None : False : False
        (1, 4) :    0 :   1.0 :  None :    None : False : False
        (1, 5) :    0 :  None :  None :    None : False :  True
        (1, 6) :    0 :  None :  None :    None : False :  True
        (2, 1) :    0 :   0.0 :  None :    None : False : False
        (2, 2) :    0 :   1.0 :  None :    None : False : False
        (2, 3) :    0 :   0.0 :  None :    None : False : False
        (2, 4) :    0 :   0.0 :  None :    None : False : False
        (2, 5) :    0 :  None :  None :    None : False :  True
        (2, 6) :    0 :  None :  None :    None : False :  True
        (3, 1) :    0 :   1.0 :  None :    None : False : False
        (3, 2) :    0 :   0.0 :  None :    None : False : False
        (3, 3) :    0 :   0.0 :  None :    None : False : False
        (3, 4) :    0 :   0.0 :  None :    None : False : False
        (3, 5) :    0 :  None :  None :    None : False :  True
        (3, 6) :    0 :  None :  None :    None : False :  True
        (4, 1) :    0 :   0.0 :  None :    None : False : False
```

```
    (4, 2) :      0 :    0.0 :  None :    None : False : False
    (4, 3) :      0 :    1.0 :  None :    None : False : False
    (4, 4) :      0 :    0.0 :  None :    None : False : False
    (4, 5) :      0 :   None :  None :    None : False :  True
    (4, 6) :      0 :   None :  None :    None : False :  True
y : Size=10, Index=y_index, Domain=NonNegativeReals
    Key : Lower : Value : Upper : Initial : Fixed : Stale
      1 :     0 :   0.0 :  None :    None : False : False
      2 :     0 :   0.0 :  None :    None : False : False
      3 :     0 :   0.0 :  None :    None : False : False
      4 :     0 :   0.0 :  None :    None : False : False
      5 :     0 :   0.0 :  None :    None : False : False
      6 :     0 :   0.0 :  None :    None : False : False
      7 :     0 :   0.0 :  None :    None : False : False
      8 :     0 :   0.0 :  None :    None : False : False
      9 :     0 :  None :  None :    None : False :  True
     10 :     0 :  None :  None :    None : False :  True

Objectives:
  obj : Size=1, Index=None, Active=True
      Key   : Active : Value
      None :   True :  18.0

Constraints:
  JConstraint1 : Size=1
      Key  : Lower : Body : Upper
      None :   1.0 :  1.0 :  None
  JConstraint2 : Size=1
      Key  : Lower : Body : Upper
      None :   1.0 :  1.0 :  None
  JConstraint3 : Size=1
      Key  : Lower : Body : Upper
      None :   1.0 :  1.0 :  None
  JConstraint4 : Size=1
      Key  : Lower : Body : Upper
      None :   1.0 :  1.0 :  None
  MConst1 : Size=1
      Key  : Lower : Body : Upper
      None :   1.0 :  1.0 :  None
  MConst2 : Size=1
      Key  : Lower : Body : Upper
      None :   1.0 :  1.0 :  None
  MConst3 : Size=1
      Key  : Lower : Body : Upper
      None :   1.0 :  1.0 :  None
```

```
        MConst4 : Size=1
            Key  : Lower : Body : Upper
            None :   1.0 :  1.0 :  None

[     0.09] Applying Pyomo postprocessing actions
[     0.09] Pyomo Finished
```

23.9.2   Formulation with the Pulp Modeler

The following listing shows the Python script with the model using Pulp and
stored in file assign1.py. As with the previous models, it was necessary to
introduce artificial variables.

```
"""
Python Formulation for the Pulp Modeler
Assignment problem. File: assign1.py
J M Garrido, September  2014
usage: python assign1.py
"""
print "Transshipment Problem"
# Import PuLP modeler functions
from pulp import *

# Data for Linear Optimization Problem
M = 4  # Number of jobs
N = 4  # Number of resources
NAV = M+N # Number of artificial variables
a = range(1, M+1)
al = range(M)
b = range(1,N+1)
bl = range(N)
# Index list for decision variables x
xindx = [(a[i],b[j]) for j in bl for i in al]
tindx = range(1, NAV+1)

# Create the model to contain the problem data
model = LpProblem("Assignment Problem",LpMinimize)

# Decision variables
x = LpVariable.dicts("X", xindx,0,None)
y = LpVariable.dicts("Y", tindx,0,None)

# The Pulp objective function
model += 13.0 * x[1,1] + 16.0 * x[1,2] + 12.0 * x[1,3] + 11.0*x[1,4] \
 + 15.0*x[2,1] + 2.0*x[2,2] + 13.0*x[2,3] + 20.0*x[2,4]   \
```

```
+5.0*x[3,1] + 7.0*x[3,2] + 10.0*x[3,3] + 6.0*x[3,4], \
"Assignment cost"

# Job Constraints
model += x[1,1] + x[2,1] + x[3,1] + x[4,1] -y[1] >= 1,"Job 1"
model += x[1,2] + x[2,2] + x[3,2] + x[4,2] -y[2] >= 1,"Job 2"
model += x[1,3] + x[2,3] + x[3,3] + x[4,3] -y[3] >= 1,"Job 3"
model += x[1,4] + x[2,4] + x[3,4] + x[4,4] -y[4] >= 1,"Job 4"

# Machine Constraints
model += x[1,1] + x[1,2] + x[1,3] + x[1,4] -y[5] >= 1, "Machine 1"
model += x[2,1] + x[2,2] + x[2,3] + x[2,4] -y[6] >= 1, "Machine 2"
model += x[3,1] + x[3,2] + x[3,3] + x[3,4] -y[7] >= 1, "Machine 3"
model += x[4,1] + x[4,2] + x[4,3] + x[4,4] -y[8] >= 1, "Machine 4"

# Solve the optimization problem using the specified PuLP Solver
model.solve(GLPK())

# Print the status of the solution
print "Status:", LpStatus[model.status]

# Print each of the variables with it's resolved optimum value
for v in model.variables():
    print v.name, "=", v.varValue

# Print the optimized value of the objective function
print "Objective Function", value(model.objective)
```

23.10   SUMMARY

Two important application areas of linear optimization are transportation and transshipment problems. The goal of the first type of problem is finding the minimum cost to transport a product. The goal of the second type of problem is finding the optimal manner to transport products to destination or demand points.

Key Terms

supply points	demand points	destination points
transportation cost	demand constraints	supply constraints
balanced problem	unbalanced problem	transshipment points
transshipment constraints	job constraints	resource constraints
dummy assignments	artificial variables	

23.11 EXERCISES

23.1 A factory of automobile spare parts has two supply locations: S_1 and S_2. There are three demand points that require 35, 42, and 50 parts, respectively. Supply point S_1 has a capacity of 85 parts and S_2 a capacity of 65 parts. The following table shows transportation costs and the costs (penalty) for excess demand of each supply point. Formulate and solve the optimization problem using Pyomo and Pulp.

	D1	D2	D3	Supply
S1	25	45	42	85
S2	20	60	50	65
Demand	35	42	50	
Penalty	100	85	150	

23.2 In Exercise 23.1, management of the factory decided to meet excess demand by purchasing additional parts and sending them to the demand points. The cost of purchasing and sending each of these parts is $140.00. Formulate and solve the optimization problem using Pyomo and Pulp.

23.3 A computer manufacturer produces three types of computer laptops: T1, T2, and T3. These are built in three facilities: F1, F2, and F3. In a week of 44 hours, the demand is for a total of 10 computers that must be produced. Formulate and solve the minimum cost of producing the computers. The following table includes the time (in hours) and the cost of producing every type of computer.

Facility	Time	T1	T2	T3
F1	5	$180	$120	$84
F2	4	$150	$90	$90
F3	3.5	$129	$60	$60

23.4 In Exercise 23.3, each computer type has a different time (in hours) that it takes to manufacture it depending on the facility where it is built. The following table includes this data. Formulate and solve the minimum cost of producing the computers.

Facility	T1	T2	T3
F1	4.5	4	4.5
F2	4.5	4.3	5
F3	3.5	3.5	4

## CHAPTER 24

Network Models

24.1 INTRODUCTION

This chapter presents the general concepts, formulation, and solution of problems that can be described with network models. These are special cases of linear optimization problems. The main goal is to formulate these problems as linear optimization models and compute the minimum cost or the maximum flow from a source point to a destination point in the network.

Examples of these types of problems are shortest path problems, maximum flow problems, and minimum spanning tree problems.

24.2 GRAPHS

A *graph* is used as a visual representation of a network. A graph consists of a finite set of *nodes* (also known as vertices) and a finite set of *arcs* that connect pairs of nodes. A directed arc connects an ordered pair of vertices; if an arc starts at node P (head) and ends at node Q (tail), it is denoted as (P, Q). An arc will typically have an associated weight or length.

A *path* is a sequence of arcs with the property that for every arc its tail vertex is the head vertex of the next arc. The length (or cost) of the path is the summation of the lengths of the individual arcs in it.

24.3 SHORTEST PATH PROBLEM

A shortest path problem consists of finding the path from an initial node in the graph to a final node with the shortest length. Several algorithms have been developed to solve this general problem, the most important of which is Dijkstra's algorithm.

This section discusses the formulation of shortest path problems as transshipment problems. For this, the techniques of the previous chapters are applied. As discussed previously, a transshipment model includes *intermediate* or transshipment points in a transportation model. A transshipment point is an intermediate point between one or more supply points and one or more demand points.

The general problem consists of transporting one unit of a product from a source point P to a destination point Q. The intermediate nodes are the transshipment points. Units of the product can be sent from the source point to the destination point using one of several possible paths.

When in a graph there is no arc between two points, an *artificial arc* is included and its length is given a relatively large value, H. An arc from a node to the same node will have zero length.

The *cost* of sending 1 unit of product from node i to node j is the length of the arc and denoted by $c_{i,j}$. For example, $c_{2,3}$ denotes the cost of sending 1 unit of product shipped from node 2 to node 3. The shipment of 1 unit of the product from node i to node j is denoted by $x_{i,j}$. Therefore, the value of $x_{i,j}$ is 1 or zero.

The objective function indicates the total cost of transporting 1 unit of the product from the source node to the destination node and can be expressed as follows:

Minimize: z,

$$z = \sum_{i=1}^{1=n} \sum_{j=1}^{j=n} c_{i,j} x_{i,j}.$$

The number of nodes is n and there is one supply point and one demand point. A transshipment point can be considered both a supply point and a demand point. At a transshipment point, k, the total inputs to this point must equal the total outputs from this intermediate point and can be expressed as follows:

$$\sum_{i=1}^{i=p} x_{i,k} = \sum_{j=1}^{j=q} x_{k,j}.$$

where p is the number of supply points, q is the number of demand points, and $x_{i,j}$ denotes a unit of (1) product or no product (0) shipped from point i to point j. From the previous equation, the transshipment constraint of a point, k, is expressed as follows:

$$\sum_{i=1}^{i=p} x_{i,k} - \sum_{j=1}^{j=q} x_{k,j} = 0.$$

24.4 SHORTEST PATH: CASE STUDY 1

A product is shipped from a supply city (supply point), represented by node 1, to a destination city (demand point), represented by node 6. The product is first shipped to one or more cities that are represented by intermediate nodes

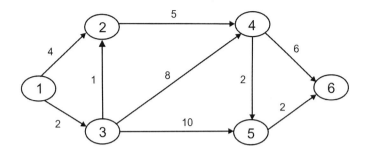

Figure 24.1 Graph of shortest path problem.

2–5. The goal of the problem is to find the shortest path from the supply city to the destination city. The distance between the cities is shown by the arcs between nodes in the graph of Figure 24.1.

The problem can be formulated directly as a standard transportation problem, using the data in the previous figure. This problem has one supply point and one demand point. The objective function can be completely written with the unit cost values of the product to be transported. The objective function is to minimize the following expression:

$$
z = \begin{aligned}
& 0x_{1,1} && + 4x_{1,2} && + 2x_{1,3} && + Hx_{1,4} && + Hx_{1,5} && + Hx_{1,6} && + \\
& Hx_{2,1} && + 0x_{2,2} && + Hx_{2,3} && + 5x_{2,4} && + Hx_{2,5} && + Hx_{2,6} && + \\
& Hx_{3,1} && + 1x_{3,2} && + 0x_{3,3} && + 8x_{3,4} && + 10x_{3,5} && + Hx_{3,6} && + \\
& Hx_{4,1} && + Hx_{4,2} && + Hx_{4,3} && + 0x_{4,4} && + 2x_{4,5} && + 6x_{4,6} && + \\
& Hx_{5,1} && + Hx_{5,2} && + Hx_{5,3} && + Hx_{5,4} && + 0x_{5,5} && + 2x_{5,6} && + \\
& Hx_{6,1} && + Hx_{6,2} && + Hx_{6,3} && + Hx_{6,4} && + Hx_{6,5} && + 0x_{6,6}.
\end{aligned}
$$

There are two source–destination constraints, each representing 1 unit of product transported from the indicated source node to the indicated destination node.

$$
\begin{aligned}
x_{1,2} + x_{1,3} &= 1 \quad \text{(Source node)} \\
x_{4,6} + x_{5,6} &= 1 \quad \text{(Destination node)}
\end{aligned}
$$

There are four intermediate nodes, so there are four transshipment constraints, one for each intermediate node. These constraints are:

$$
\begin{aligned}
x_{1,2} + x_{3,2} - x_{4,2} &= 0 \quad \text{(Node 2)} \\
x_{1,3} - x_{3,2} - x_{3,4} - x_{3,5} &= 0 \quad \text{(Node 3)} \\
x_{2,4} + x_{3,4} - x_{4,5} - x_{4,6} &= 0 \quad \text{(Node 4)} \\
x_{3,5} + x_{4,5} - x_{5,6} &= 0 \quad \text{(Node 5)}.
\end{aligned}
$$

The decision variables $x_{i,j}$ have the sign constraint $x_{i,j} = 0$ or $x_{i,j} = 0$, for $i = 1, \ldots, 6$ and $j = 1, \ldots, 6$.

24.4.1 Formulation Using the Pyomo Modeler

The following listing shows the formulation of the model using the Pyomo modeler. The model is stored in file `network1.py`.

```
"""
Python Formulation using Pyomo Modeler
Network problem, shortest path. File: network1.py
J M Garrido, September  2014
usage: pyomo network1.py --summary
"""
print "Assignment Problem 1"
# Import
from coopr.pyomo import *


# Data for Linear Optimization Problem
N = 6   # Number of nodes in network
M = 2   # number of end nodes (source and destination)
INT = 4   # Number of intermediate nodes
H = 10000.0 # A very high cost constant
a = range(1, N+1)
al = range(N)
b = range(1,N+1)
bl = range(N)
# Index list for decision variables x
xindx = [(a[i],b[j]) for j in bl for i in al]
T = INT + M # number of artificial variables (y)
tindx = range(1, T+1) # index list for y variables

#Concrete Model
model = ConcreteModel(name="Shortest Path Problem 1")

# Decision Variables
model.x = Var(xindx, within=NonNegativeReals)
# Artificial variables
model.y = Var(tindx, within=NonNegativeReals)

# The objective function
model.obj = Objective(expr=
0.0 * model.x[1,1] + 4.0 * model.x[1,2] + 2.0 * model.x[1,3]
      + H*model.x[1,4] +H*model.x[1,5] + H*model.x[1,6]
 + H*model.x[2,1] + 0.0*model.x[2,2] + H*model.x[2,3]
      + 5.0*model.x[2,4] + H*model.x[2,5] + H*model.x[2,6]
 + H*model.x[3,1] + 1.0*model.x[3,2] + 0.0*model.x[3,3]
      + 8.0*model.x[3,4] + 10.0*model.x[3,5] + H*model.x[3,6]
 + H*model.x[4,1] + H*model.x[4,2] + H*model.x[4,3]
```

```
            + 0.0*model.x[4,4] + 2.0*model.x[4,5] + 6.0*model.x[4,6]
  + H*model.x[5,1] + H*model.x[5,2] + H*model.x[5,3]
            + H*model.x[5,4] + 0.0*model.x[5,5] + 2.0*model.x[5,6]
  + H*model.x[6,1] + H*model.x[6,2] + H*model.x[6,3]
            + H*model.x[6,4] + H*model.x[6,5] + 0.0*model.x[6,6]
  , sense = minimize)


# Source and Destination Constraints
model.SConstraint1 = Constraint(expr=
  model.x[1,2] + model.x[1,3] -model.y[1] >= 1)
model.DConstraint1 = Constraint(expr=
  model.x[4,6] + model.x[5,6] -model.y[2] >= 1)


# Intermediate Node Constraints
model.IntConst1 = Constraint(expr=
  model.x[1,2] + model.x[3,2] - model.x[2,4] - model.y[3] >= 0)
model.IntConst2 = Constraint(expr=
  model.x[1,3] - model.x[3,2] - model.x[3,4] - model.x[3,5]
        - model.y[4] >= 0)
model.IntConst3 = Constraint(expr=
  model.x[2,4] + model.x[3,4] - model.x[4,5] - model.x[4,6]
        - model.y[5] >= 0)
model.IntConst4 = Constraint(expr=
  model.x[3,5] + model.x[4,5] - model.x[5,6] - model.y[6] >= 0)
```

After running the model with Pyomo, the following listing is produced. Note that the optimal value of the objective function is 12.0. The results show that the minimum path has length = 12. The path selected consists of the following arcs: $x_{1,3}$, $x_{3,2}$, $x_{2,4}$, $x_{4,5}$, and $x_{5,6}$.

```
$ pyomo network1.py --summary
[    0.00] Setting up Pyomo environment
[    0.00] Applying Pyomo preprocessing actions
Assignment Problem 1
[    0.01] Creating model
[    0.01] Applying solver
[    0.05] Processing results
    Number of solutions: 1
    Solution Information
      Gap: 0.0
      Status: feasible
      Function Value: 12.0
    Solver results file: results.json
```

```
============================================================
Solution Summary
============================================================
```

Model Shortest Path Problem 1

Variables:
  x : Size=36, Index=x_index, Domain=NonNegativeReals

Key	: Lower	: Value	: Upper	: Initial	: Fixed	: Stale
(1, 1) :	0 :	None :	None :	None :	False :	True
(1, 2) :	0 :	0.0 :	None :	None :	False :	False
(1, 3) :	0 :	1.0 :	None :	None :	False :	False
(1, 4) :	0 :	0.0 :	None :	None :	False :	False
(1, 5) :	0 :	0.0 :	None :	None :	False :	False
(1, 6) :	0 :	0.0 :	None :	None :	False :	False
(2, 1) :	0 :	0.0 :	None :	None :	False :	False
(2, 2) :	0 :	None :	None :	None :	False :	True
(2, 3) :	0 :	0.0 :	None :	None :	False :	False
(2, 4) :	0 :	1.0 :	None :	None :	False :	False
(2, 5) :	0 :	0.0 :	None :	None :	False :	False
(2, 6) :	0 :	0.0 :	None :	None :	False :	False
(3, 1) :	0 :	0.0 :	None :	None :	False :	False
(3, 2) :	0 :	1.0 :	None :	None :	False :	False
(3, 3) :	0 :	None :	None :	None :	False :	True
(3, 4) :	0 :	0.0 :	None :	None :	False :	False
(3, 5) :	0 :	0.0 :	None :	None :	False :	False
(3, 6) :	0 :	0.0 :	None :	None :	False :	False
(4, 1) :	0 :	0.0 :	None :	None :	False :	False
(4, 2) :	0 :	0.0 :	None :	None :	False :	False
(4, 3) :	0 :	0.0 :	None :	None :	False :	False
(4, 4) :	0 :	None :	None :	None :	False :	True
(4, 5) :	0 :	1.0 :	None :	None :	False :	False
(4, 6) :	0 :	0.0 :	None :	None :	False :	False
(5, 1) :	0 :	0.0 :	None :	None :	False :	False
(5, 2) :	0 :	0.0 :	None :	None :	False :	False
(5, 3) :	0 :	0.0 :	None :	None :	False :	False
(5, 4) :	0 :	0.0 :	None :	None :	False :	False
(5, 5) :	0 :	None :	None :	None :	False :	True
(5, 6) :	0 :	1.0 :	None :	None :	False :	False
(6, 1) :	0 :	0.0 :	None :	None :	False :	False
(6, 2) :	0 :	0.0 :	None :	None :	False :	False
(6, 3) :	0 :	0.0 :	None :	None :	False :	False
(6, 4) :	0 :	0.0 :	None :	None :	False :	False
(6, 5) :	0 :	0.0 :	None :	None :	False :	False
(6, 6) :	0 :	None :	None :	None :	False :	True

```
 y : Size=6, Index=y_index, Domain=NonNegativeReals
     Key : Lower : Value : Upper : Initial : Fixed : Stale
       1 :     0 :   0.0 :  None :    None : False : False
       2 :     0 :   0.0 :  None :    None : False : False
       3 :     0 :   0.0 :  None :    None : False : False
       4 :     0 :   0.0 :  None :    None : False : False
       5 :     0 :   0.0 :  None :    None : False : False
       6 :     0 :   0.0 :  None :    None : False : False

Objectives:
  obj : Size=1, Index=None, Active=True
      Key  : Active : Value
      None :   True :  12.0

Constraints:
  SConstraint1 : Size=1
      Key  : Lower : Body : Upper
      None :   1.0 :  1.0 :  None
  DConstraint1 : Size=1
      Key  : Lower : Body : Upper
      None :   1.0 :  1.0 :  None
  IntConst1 : Size=1
      Key  : Lower : Body : Upper
      None :   0.0 :  0.0 :  None
  IntConst2 : Size=1
      Key  : Lower : Body : Upper
      None :   0.0 :  0.0 :  None
  IntConst3 : Size=1
      Key  : Lower : Body : Upper
      None :   0.0 :  0.0 :  None
  IntConst4 : Size=1
      Key  : Lower : Body : Upper
      None :   0.0 :  0.0 :  None

[   0.11] Applying Pyomo postprocessing actions
[   0.11] Pyomo Finished
```

24.4.2   Formulation Using the Pulp Modeler

The following listing shows the formulation of the model using the Pulp modeler. The model is stored in file network1.py.

```
" " "
```

```
Python Formulation using the Pulp Modeler
Network problem, shortest path. File: network1.py
J M Garrido, September  2014
usage: python network1.py
"""
print "Network Problem, shortest path"
# Import PuLP modeler functions
from pulp import *

# Data for Linear Optimization Problem
N = 6   # Number of nodes in network
M = 2   # number of end nodes (source and destination)
INT = 4 # Number of intermediate nodes
H = 10000.0 # A very high cost constant
a = range(1, N+1)
al = range(N)
b = range(1,N+1)
bl = range(N)
# Index list for decision variables x
xindx = [(a[i],b[j]) for j in bl for i in al]
T = INT + M # number of artificial variables (y)
tindx = range(1, T+1)

# Create the model to contain the problem data
model = LpProblem("Shortest Path Problem",LpMinimize)

# Decision variables
x = LpVariable.dicts("X", xindx,0,None)
y = LpVariable.dicts("Y", tindx,0,None)

# The Pulp objective function
model += \
0.0*x[1,1] + 4.0*x[1,2] + 2.0*x[1,3] + H*x[1,4] +H*x[1,5] + H*x[1,6] \
 + H*x[2,1] + 0.0*x[2,2] + H*x[2,3] + 5.0*x[2,4] + H*x[2,5] \
       + H*x[2,6]  \
+ H*x[3,1] + 1.0*x[3,2] + 0.0*x[3,3] + 8.0*x[3,4] + 10.0*x[3,5] \
       + H*x[3,6] \
+ H*x[4,1] + H*x[4,2] + H*x[4,3] + 0.0*x[4,4] + 2.0*x[4,5] \
       + 6.0*x[4,6] \
+ H*x[5,1] + H*x[5,2] + H*x[5,3] + H*x[5,4] + 0.0*x[5,5] + 2.0*x[5,6] \
+ H*x[6,1] + H*x[6,2] + H*x[6,3] + H*x[6,4] + H*x[6,5] + 0.0*x[6,6] \
, "Transportation cost"

# Source and Constraints
model += x[1,2] + x[1,3] -y[1] >= 1,"Source node"
```

```
model += x[4,6] + x[5,6] -y[2] >= 1,"Destination node"

# Intermediate Node Constraints
model += x[1,2] + x[3,2] - x[2,4] - y[3] >= 0,"Node 2"
model += x[1,3] - x[3,2] - x[3,4] - x[3,5] - y[4] >= 0,"Node 3"
model += x[2,4] + x[3,4] - x[4,5] - x[4,6] - y[5] >= 0, "Node 4"
model += x[3,5] + x[4,5] - x[5,6] - y[6] >= 0, "Node 5"

# Solve the optimization problem using the PuLP Solver
model.solve(GLPK())

# Print the status of the solution
print "Status:", LpStatus[model.status]

# Print each of the variables with it's resolved optimum value
for v in model.variables():
    print v.name, "=", v.varValue

# Print the optimized value of the objective function
print "Objective Function", value(model.objective)
```

24.5   MAXIMUM FLOW PROBLEMS

The are many problems in which the goal is to send the *maximum quantity* of a product from a source node to a destination node of a network. The main limitation is the *capacity* of each segment of the network represented by the arcs. For these problems, the Ford–Fulkerson algorithm was developed. In this section, the linear optimization formulation is discussed.

The following maximum flow problem illustrates the basic approach to formulate a linear optimization problem. An airline company needs to plan and set up an optimal number of flights from Chicago to Rio de Janeiro. The intermediate stops that need to be included are: first Atlanta, then Bogota and/or Caracas. Table 24.1 shows the routes, the corresponding arcs in Figure 24.2, and the capacity (maximum number of flights allowed) of each route.

The problem can be formulated directly as a standard transportation problem, using the data in Table 24.1 and Figure 24.2.

Let $x_{i,j}$ denote the number of flights from node i to node j, and $c_{i,j}$ the capacity between nodes i and j. This problem has one supply point (source) and one demand point (destination). The objective function can be written by observing that the maximum flow from node 1 to node 5 is actually the flow from node 1 to node 2, which is represented by the variable $x_{1,2}$. The objective function is then to maximize the expression $z = x_{1,2}$.

Table 24.1 Airline routes from Chicago to Rio.

Route	Arc	Max number of flights
Chicago to Atlanta	$x_{1,2}$	7
Atlanta to Bogota	$x_{2,3}$	4
Atlanta to Caracas	$x_{2,4}$	3
Bogota to Caracas	$x_{3,4}$	6
Bogota to Rio	$x_{3,5}$	3
Caracas to Rio	$x_{4,5}$	4

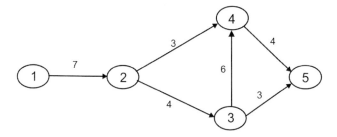

Figure 24.2 Graph of airline routes to Rio.

To simplify the formulation of the problem, three types of constraints are considered:

- Source–destination constraints

- Transshipment constraints

- Capacity constraints

There is one source–destination constraint, which represents the balance in the flow from the source node and the flow into the destination node in Figure 24.2. This constraint is expressed as follows:

$$x_{1,2} = x_{3,5} + x_{4,5}.$$

There are three intermediate nodes (2, 3, and 4), so there are three transshipment constraints, one for each intermediate node. These constraints are:

$$
\begin{array}{llll}
x_{1,2} & - \; x_{2,3} & - \; x_{2,4} & = 0 \quad \text{(Node 2)} \\
x_{2,3} & - \; x_{3,4} & - \; x_{3,5} & = 0 \quad \text{(Node 3)} \\
x_{2,4} & + \; x_{3,4} & - \; x_{4,5} & = 0 \quad \text{(Node 4)}.
\end{array}
$$

The capacity constraints represent the maximum number of flights possible on the indicated route:

$$x_{i,j} \leq c_{i,j}.$$

These are shown as the value of the arcs in Figure 24.2. There are six arcs in the graph, therefore there are six capacity constraints.

$$
\begin{aligned}
x_{1,2} &\leq 7 \\
x_{2,3} &\leq 4 \\
x_{2,4} &\leq 3 \\
x_{3,4} &\leq 6 \\
x_{3,5} &\leq 3 \\
x_{4,5} &\leq 4
\end{aligned}
$$

The decision variables $x_{i,j}$ have the sign constraint $x_{i,j} \geq 0$, for $i = 1, \ldots, 5$ and $j = 1, \ldots, 5$.

24.5.1 Formulation Using the Pyomo Modeler

The following listing shows the formulation of the model using the Pyomo modeler and is stored in file `maxflow1.py`. Note that this is a maximization model and the artificial variables are used with a + sign and the right-hand side of the respective constraints are `<= 0`. This is shown in line 36 and in lines 54–58.

```
1 """
2 Python Formulation using Pyomo Modeler
3 Network problem, maximum flow. File: network1.py
4 J M Garrido, September  2014
5 usage: pyomo maxflow1.py --summary
6 """
7 print "Maximum Flow Problem"
8 # Import
9 from coopr.pyomo import *
10
11 # Data for Linear Optimization Problem
12 N = 5  # Number of nodes in network
13 INT = 3 # Number of intermediate nodes
14 a = range(1, N+1)
15 al = range(N)
16 b = range(1,N+1)
17 bl = range(N)
18 # Index list for decision variables x
19 xindx = [(a[i],b[j]) for j in bl for i in al]
20 T = INT + 1 # number of artificial variables (y)
```

```
21 tindx = range(1, T+1) # index list for y variables
22
23 #Concrete Model
24 model = ConcreteModel(name="Maximum Flow Problem")
25
26 # Decision Variables
27 model.x = Var(xindx, within=NonNegativeReals)
28 # Artificial variables
29 model.y = Var(tindx, within=NonNegativeReals)
30
31 # The objective function
32 model.obj = Objective(expr= model.x[1,2], sense = maximize)
33
34 # Source and Destination Constraint
35 model.SDConstraint1 = Constraint(expr=
36     model.x[1,2] - model.x[3,5] - model.x[4,5] + model.y[1] <= 0)
37
38 # Arc Capacity Constraints
39 model.ArcConstraint1 = Constraint(expr=
40     model.x[1,2] <= 7)
41 model.ArcConstraint2 = Constraint(expr=
42     model.x[2,3] <= 4)
43 model.ArcConstraint3 = Constraint(expr=
44     model.x[2,4] <= 3)
45 model.ArcConstraint4 = Constraint(expr=
46     model.x[3,4] <= 6)
47 model.ArcConstraint5 = Constraint(expr=
48     model.x[3,5] <= 3)
49 model.ArcConstraint6 = Constraint(expr=
50     model.x[4,5] <= 4)
51
52 # Intermediate Node Constraints
53 model.IntConst1 = Constraint(expr=
54     model.x[1,2] - model.x[2,3] - model.x[2,4] + model.y[2] <= 0)
55 model.IntConst2 = Constraint(expr=
56     model.x[2,3] - model.x[3,4] - model.x[3,5] + model.y[3] <= 0)
57 model.IntConst3 = Constraint(expr=
58     model.x[2,4] + model.x[3,4] - model.x[4,5] + model.y[4] <= 0)
```

After running the model with the Pyomo modeler, the following listing is produced. Note that optimum number of flights is 7 and the value of variable $x_{3,4}$ is 1, which means that the maximum number of flights from node 3 to node 4 is 1.

```
$ pyomo maxflow1.py --summary
[    0.00] Setting up Pyomo environment
[    0.00] Applying Pyomo preprocessing actions
Maximum Flow
[    0.02] Creating model
[    0.02] Applying solver
[    0.06] Processing results
    Number of solutions: 1
    Solution Information
      Gap: 0.0
      Status: feasible
      Function Value: 7.0
    Solver results file: results.json


============================================================
Solution Summary
============================================================


Model Maximum Flow Problem

  Variables:
    x : Size=25, Index=x_index, Domain=NonNegativeReals
        Key     : Lower : Value : Upper : Initial : Fixed : Stale
        (1, 1) :     0 :  None :  None :    None : False :  True
        (1, 2) :     0 :   7.0 :  None :    None : False : False
        (1, 3) :     0 :  None :  None :    None : False :  True
        (1, 4) :     0 :  None :  None :    None : False :  True
        (1, 5) :     0 :  None :  None :    None : False :  True
        (2, 1) :     0 :  None :  None :    None : False :  True
        (2, 2) :     0 :  None :  None :    None : False :  True
        (2, 3) :     0 :   4.0 :  None :    None : False : False
        (2, 4) :     0 :   3.0 :  None :    None : False : False
        (2, 5) :     0 :  None :  None :    None : False :  True
        (3, 1) :     0 :  None :  None :    None : False :  True
        (3, 2) :     0 :  None :  None :    None : False :  True
        (3, 3) :     0 :  None :  None :    None : False :  True
        (3, 4) :     0 :   1.0 :  None :    None : False : False
        (3, 5) :     0 :   3.0 :  None :    None : False : False
        (4, 1) :     0 :  None :  None :    None : False :  True
        (4, 2) :     0 :  None :  None :    None : False :  True
        (4, 3) :     0 :  None :  None :    None : False :  True
        (4, 4) :     0 :  None :  None :    None : False :  True
        (4, 5) :     0 :   4.0 :  None :    None : False : False
        (5, 1) :     0 :  None :  None :    None : False :  True
        (5, 2) :     0 :  None :  None :    None : False :  True
```

```
       (5, 3) :      0 :  None :  None :    None : False :  True
       (5, 4) :      0 :  None :  None :    None : False :  True
       (5, 5) :      0 :  None :  None :    None : False :  True
  y : Size=4, Index=y_index, Domain=NonNegativeReals
       Key : Lower : Value : Upper : Initial : Fixed : Stale
         1 :     0 :   0.0 :  None :    None : False : False
         2 :     0 :   0.0 :  None :    None : False : False
         3 :     0 :   0.0 :  None :    None : False : False
         4 :     0 :   0.0 :  None :    None : False : False

Objectives:
  obj : Size=1, Index=None, Active=True
       Key  : Active : Value
       None :   True :   7.0

Constraints:
  SDConstraint1 : Size=1
       Key  : Lower : Body : Upper
       None :  None :  0.0 :   0.0
  ArcConstraint1 : Size=1
       Key  : Lower : Body : Upper
       None :  None :  7.0 :   7.0
  ArcConstraint2 : Size=1
       Key  : Lower : Body : Upper
       None :  None :  4.0 :   4.0
  ArcConstraint3 : Size=1
       Key  : Lower : Body : Upper
       None :  None :  3.0 :   3.0
  ArcConstraint4 : Size=1
       Key  : Lower : Body : Upper
       None :  None :  1.0 :   6.0
  ArcConstraint5 : Size=1
       Key  : Lower : Body : Upper
       None :  None :  3.0 :   3.0
  ArcConstraint6 : Size=1
       Key  : Lower : Body : Upper
       None :  None :  4.0 :   4.0
  IntConst1 : Size=1
       Key  : Lower : Body : Upper
       None :  None :  0.0 :   0.0
  IntConst2 : Size=1
       Key  : Lower : Body : Upper
       None :  None :  0.0 :   0.0
  IntConst3 : Size=1
       Key  : Lower : Body : Upper
```

```
        None :   None :   0.0 :    0.0

[    0.08] Applying Pyomo postprocessing actions
[    0.08] Pyomo Finished
```

24.5.2   Formulation Using the Pulp Modeler

The following listing shows the formulation of the model using the Pulp modeler and is stored in file maxflow1.py.

```
"""
Python Formulation using the Pulp Modeler
Network problem, maximum flow. File: maxflow1.py
J M Garrido, September  2014
usage: python maxflow1.py
"""
print "Network Problem, shortest path"
# Import PuLP modeler functions
from pulp import *

# Data for Linear Optimization Problem
N = 5  # Number of nodes in network
INT = 3 # Number of intermediate nodes
a = range(1, N+1)
al = range(N)
b = range(1,N+1)
bl = range(N)
# Index list for decision variables x
xindx = [(a[i],b[j]) for j in bl for i in al]
T = INT + 1 # number of artificial variables (y)
tindx = range(1, T+1)

# Create the model to contain the problem data
model = LpProblem("Maximum Flow Problem",LpMaximize)

# Decision variables
x = LpVariable.dicts("X", xindx,0,None)
y = LpVariable.dicts("Y", tindx,0,None)

# The Pulp objective function
model += x[1,2], "Maximum Flow"

# Source and Destination Constraints
model += x[1,2] - x[4,5] - x[3,5] + y[1] <= 0,"Source to destination"
```

```
# Arc Capacity Constraints
model += x[1,2] <= 7,"Arc 1-2"
model += x[2,3] <= 4,"Arc 2-3"
model += x[2,4] <= 3,"Arc 2-4"
model += x[3,4] <= 6,"Arc 3-4"
model += x[3,5] <= 3,"Arc 3-5"
model += x[4,5] <= 4,"Arc 4-5"

# Intermediate Node Constraints
model += x[1,2] - x[2,3] - x[2,4] + y[2] <= 0,"Node 2"
model += x[2,3] - x[3,4] - x[3,5] + y[3] <= 0,"Node 3"
model += x[2,4] + x[3,4] - x[4,5] + y[4] <= 0,"Node 4"

# Solve the optimization problem using the PuLP Solver
model.solve(GLPK())

# Print the status of the solution
print "Status:", LpStatus[model.status]

# Print each of the variables with it's resolved optimum value
for v in model.variables():
    print v.name, "=", v.varValue

# Print the optimized value of the objective function
print "Objective Function", value(model.objective)
```

24.6   CRITICAL PATH METHOD

The critical path method (CPM) is a network model that can help in the *scheduling* of large projects. The important computations are the total time to complete the project and the interval that represents how long an activity of the project can be delayed without causing delays to the project. This method calculates the minimum completion time for a project and the possible start and finish times for the project activities.

A network model of a project typically consists of a sequence of the various activities that need to be performed, and the duration of each activity. A directed arc represents an *activity*, a node represents a *start* or *finish event* of an activity. A special initial node represents the start of the project, and a special end node represents the completion of the project.

In any project there are cases in which more than one activity needs to be completed before the next activity can start. Figure 24.3 illustrates this situation. Activity Aj and activity Ak have to be completed before activity Al can start. Another situation is an activity that needs to be completed before two or more activities can start. Figure 24.4 illustrates this by showing activity Al, which needs to be completed before activities Am, Ak, and Ap can start.

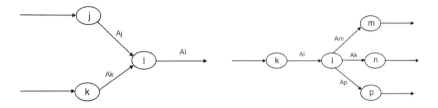

Figure 24.3 Several activities completed before next activity.

Figure 24.4 Sequencing of activities.

The following terms are used in project scheduling using the duration of the activities in the project. The *early event* time of node i, denoted by e_i, is the earliest time at which the event can occur. The *late event* time of node i, denoted by l_i, is the latest time at which the event can occur.

The *total float*, denoted by $f_{i,j}$ of an activity $A_{i,j}$ is the time interval by which the starting time of the activity can be delayed and not cause delay in the completion time of the project.

An activity that has a total float equal to zero is known as a *critical activity*. A critical path consists of a sequence of critical activities. Delays in the activities in a critical path will delay the completion of the project.

Recall that for any activity $A_{i,j}$, the start time of the activity is the event represented by node i and the completion time of the activity is the event represented by node j. Let x_k denote the time occurrence of event k and the duration of activity $A_{i,j}$ is denoted by $\Delta_{i,j}$. For every activity $A_{i,j}$, the completion time of the activity is given by the expression $x_j \geq x_i + \Delta_{i,j}$.

Let f denote the finish node of the project; the event time of the completion of the project is denoted by x_f. Similarly, node 1 is the start node of the project and the event time of the start of the project is denoted by x_1. The total time interval or duration of the entire project is given by the expression $x_f - x_1$. Let n denote the total number of nodes in the project network. This implies that $x_f = x_n$. The formulation of the linear optimization problem that finds the critical path of a project is given by the following expressions:

Minimize: $z = x_f - x_1$
Subject to:

$$x_j \geq x_i + \Delta_{i,j}, \quad i = 1, \ldots, n-1, \quad j = 2, \ldots, n$$

The variables $x_i, i = 1, \ldots, n$ are unrestricted in sign.

24.6.1 Critical Path Method: Case Study

A project has been defined with the activities and their duration given in Table 24.2. The goal of the problem is to find the critical path of the project. The various activities and their predecessors are shown by the arcs between nodes in the graph of Figure 24.5.

Table 24.2 Project data.

Activity	Predecessor	Duration
$A_{1,2}$	-	6
$A_{2,3}$	$A_{1,2}$	9
$A_{3,5}$	$A_{2,3}$	13
$A_{3,6}$	$A_{2,3}$	6
$A_{3,4}$	$A_{2,3}$	5
$A_{4,5}$	$A_{3,4}$	7
$A_{5,6}$	$A_{4,5}, A_{3,5}$	4

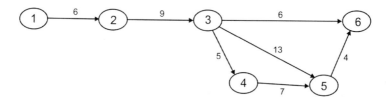

Figure 24.5 Graph of project activities.

The problem can be formulated directly as a standard transportation problem, using the data in Table 24.2 and in Figure 24.5.

The final node of the project in this problem is node 6. The objective function is then to minimize the expression $z = x_6 - x_1$. The constraints are expressed as follows:

$$
\begin{array}{lll}
x_2 & \geq x_1 + 6 & \text{Activity } A_{1,2} \\
x_3 & \geq x_2 + 9 & \text{Activity } A_{2,3} \\
x_4 & \geq x_3 + 5 & \text{Activity } A_{3,4} \\
x_5 & \geq x_4 + 7 & \text{Activity } A_{4,5} \\
x_5 & \geq x_3 + 13 & \text{Activity } A_{3,5} \\
x_6 & \geq x_3 + 6 & \text{Activity } A_{3,6} \\
x_6 & \geq x_5 + 4 & \text{Activity } A_{5,6}.
\end{array}
$$

The decision variables x_i have unrestricted sign constraint for $i = 1, \ldots, 6$.

24.6.2 Formulation Using the Pyomo Modeler

The following listing shows the Python script of the model written for the Pyomo modeler, and is stored in file cpm1.py.

```
"""
Python Formulation using Pyomo Modeler
Network problem, critical path method. File: cpm1.py
J M Garrido, September  2014
usage: pyomo cpm1.py --summary
"""
print "Maximum Flow"
# Import
from coopr.pyomo import *

# Data for Linear Optimization Problem
N = 6  # Number of nodes in network
# Index list for decision variables x
xindx =  range(1, N+1)

#Concrete Model
model = ConcreteModel(name="Critical Path Problem")

# Decision Variables
model.x = Var(xindx, within=NonNegativeReals)

# The objective function
model.obj = Objective(expr= model.x[6] - model.x[1], sense = minimize)

# Activity Constraint
model.ActConstraint1 = Constraint(expr=
   model.x[2] - model.x[1] >= 6)
model.ActConstraint2 = Constraint(expr=
   model.x[3] - model.x[2] >= 9)
model.ActConstraint3 = Constraint(expr=
   model.x[4] - model.x[3] >= 5)
model.ActConstraint4 = Constraint(expr=
   model.x[5] - model.x[4] >= 7)
model.ActConstraint5 = Constraint(expr=
   model.x[5] - model.x[3] >= 13)
model.ActConstraint6 = Constraint(expr=
   model.x[6] - model.x[3] >= 6)
model.ActConstraint7 = Constraint(expr=
   model.x[6] - model.x[5] >= 4)
```

The following listing shows the results after running the model with the Pyomo modeler. Note that the total time interval for completion of the project is 32 days and the start time of every activity is shown in the listing.

```
$ pyomo cpm1.py --summary
[    0.00] Setting up Pyomo environment
[    0.00] Applying Pyomo preprocessing actions
Maximum Flow
[    0.02] Creating model
[    0.02] Applying solver
[    0.06] Processing results
    Number of solutions: 1
    Solution Information
      Gap: 0.0
      Status: feasible
      Function Value: 32.0
    Solver results file: results.json


============================================================
Solution Summary
============================================================

Model Critical Path Problem

  Variables:
    x : Size=6, Index=x_index, Domain=NonNegativeReals
        Key : Lower : Value : Upper : Initial : Fixed : Stale
          1 :     0 :   0.0 :  None :    None : False : False
          2 :     0 :   6.0 :  None :    None : False : False
          3 :     0 :  15.0 :  None :    None : False : False
          4 :     0 :  20.0 :  None :    None : False : False
          5 :     0 :  28.0 :  None :    None : False : False
          6 :     0 :  32.0 :  None :    None : False : False

  Objectives:
    obj : Size=1, Index=None, Active=True
        Key  : Active : Value
        None :   True :  32.0

  Constraints:
    ActConstraint1 : Size=1
        Key  : Lower : Body : Upper
        None :   6.0 :  6.0 :  None
    ActConstraint2 : Size=1
        Key  : Lower : Body : Upper
```

```
      None :    9.0 :   9.0 :    None
   ActConstraint3 : Size=1
      Key   : Lower : Body : Upper
      None :    5.0 :   5.0 :    None
   ActConstraint4 : Size=1
      Key   : Lower : Body : Upper
      None :    7.0 :   8.0 :    None
   ActConstraint5 : Size=1
      Key   : Lower : Body : Upper
      None :   13.0 :  13.0 :    None
   ActConstraint6 : Size=1
      Key   : Lower : Body : Upper
      None :    6.0 :  17.0 :    None
   ActConstraint7 : Size=1
      Key   : Lower : Body : Upper
      None :    4.0 :   4.0 :    None


[    0.08] Applying Pyomo postprocessing actions
[    0.08] Pyomo Finished
```

24.6.3 Formulation Using the Pulp Modeler

The following listing shows the Python script of the model written for the Pyomo modeler, and is stored in file cpm1.py.

```
"""
Python Formulation using the Pulp Modeler
Network problem, critical path. File: cpm1.py
J M Garrido, September  2014
usage: python cpm1.py
"""
print "Network Problem, critical path method"
# Import PuLP modeler functions
from pulp import *

# Data for Linear Optimization Problem
N = 6  # Number of nodes in network
# Index list for decision variables x
xindx = range(1, N+1)

# Create the model to contain the problem data
model = LpProblem("Critical path Method",LpMinimize)

# Decision variables
```

```
x = LpVariable.dicts("X", xindx,0,None)

# The Pulp objective function
model += x[6] - x[1], "Duration of project"

# Activity Constraints
model += x[2] - x[1] >= 6,"Activity 1-2"
model += x[3] - x[2] >= 9,"Acivity  2-3"
model += x[4] - x[3] >= 5,"Activity 3-4"
model += x[5] - x[4] >= 7,"Activity 4-5"
model += x[5] - x[3] >= 13,"Activity 3-5"
model += x[6] - x[3] >= 6,"Activity 3-6"
model += x[6] - x[5] >= 4,"Activity 5-6"

# Solve the optimization problem using the PuLP Solver
model.solve(GLPK())

# Print the status of the solution
print "Status:", LpStatus[model.status]

# Print each of the variables with it's resolved optimum value
for v in model.variables():
    print v.name, "=", v.varValue

# Print the optimized value of the objective function
print "Objective Function", value(model.objective)
```

24.7 REDUCING THE TIME TO COMPLETE A PROJECT

When a decision is taken to *reduce* the total time to complete a project, additional resources must be allocated to the various activities. Linear optimization is used to minimize the total cost of allocating the additional resources to the project activities.

Let $r_{i,j}$ denote the number of days that the duration of activity $A_{i,j}$ is reduced, and $c_{i,j}$ denote the cost per day of allocating additional resources to activity $A_{i,j}$. Let R denote the time (in days) that an activity can be reduced, and let T denote the new total time (in days) of the project completion. The objective function to minimize is:

$$z = \sum_{i=1}^{n-1} \sum_{j=2}^{n} c_{i,j} r_{i,j}, \quad \text{for all activities } A_{i,j}.$$

The time-reduction constraints are expressed as:

$$r_{i,j} \leq R, \quad \text{for all activities } A_{i,j}.$$

The activity constraints are expressed as:

$$x_j = x_i + \Delta_{i,j} - r_{i,j}, \quad \text{for all activities } A_{i,j}.$$

The total time constraint is:

$$x_n - x_1 \leq T.$$

24.7.1 Reducing Time Case Study

Consider a reduction of 4 days in the total time to complete the project described in the previous problem. The completion time is now 28 days. The activity completion time can be reduced up to 2 days. The following table shows the cost per day of reducing each activity of the project.

Table 24.3 Project additional cost.

Activity	Cost
$A_{1,2}$	22.50
$A_{2,3}$	15.75
$A_{3,5}$	13.25
$A_{3,6}$	16.50
$A_{3,4}$	25.30
$A_{4,5}$	17.50
$A_{5,6}$	14.75

The various activities and their predecessors are shown by the arcs between nodes in the graph of Figure 24.5. This problem has node 6 as the final node of the project. The goal of the problem is to find the minimum cost of reducing the total completion time of the project. The objective function is then to minimize the expression:

$$z = 22.50\, r_{1,2} + 15.75\, r_{2,3} + 13.25\, r_{3,5} + 16.50\, r_{3,6} + 25.30\, r_{3,4} + 17.50\, r_{4,5} + 14.75\, r_{5,6}.$$

.

The time-reduction constraints are expressed as follows:

$$
\begin{aligned}
r_{1,2} &\leq 2, & \text{Activity } A_{1,2} \\
r_{2,3} &\leq 2, & \text{Activity } A_{2,3} \\
r_{3,4} &\leq 2, & \text{Activity } A_{3,4} \\
r_{4,5} &\leq 2, & \text{Activity } A_{4,5} \\
r_{3,5} &\leq 2, & \text{Activity } A_{3,5} \\
r_{3,6} &\leq 2, & \text{Activity } A_{3,6} \\
r_{5,6} &\leq 2, & \text{Activity } A_{5,6}.
\end{aligned}
$$

The activity constraints are expressed as follows:

$$
\begin{aligned}
x_2 &\geq x_1 + 6 - r_{1,2}, & \text{Activity } A_{1,2} \\
x_3 &\geq x_2 + 9 - r_{2,3}, & \text{Activity } A_{2,3} \\
x_4 &\geq x_3 + 5 - r_{3,4}, & \text{Activity } A_{3,4} \\
x_5 &\geq x_4 + 7 - r_{4,5}, & \text{Activity } A_{4,5} \\
x_5 &\geq x_3 + 13 - r_{3,5}, & \text{Activity } A_{3,5} \\
x_6 &\geq x_3 + 6 - r_{3,6}, & \text{Activity } A_{3,6} \\
x_6 &\geq x_5 + 4 - r_{5,6}, & \text{Activity } A_{5,6}.
\end{aligned}
$$

The decision variables x_i have unrestricted sign constraint for $i = 1, \ldots, 6$.

24.7.2 Formulation Using the Pyomo Modeler

The following listing shows the formulation of the model using the Pyomo modeler and is stored in file `cpm1b.py`.

```
"""
Python Formulation using Pyomo Modeler
Network problem, reducing time of project. File: cpm1b.py
J M Garrido, September 2014
usage: pyomo cpm1b.py --summary
"""
print "Reducing Time of Project"
# Import
from coopr.pyomo import *

# Data for Linear Optimization Problem
N = 6  # Number of nodes in project network
# Index list for decision variables x
xindx = range(1, N+1)
a = range(1, N+1)
al = range(N)
b = range(1,N+1)
bl = range(N)
# Index list for decision variables x
rindx = [(a[i],b[j]) for j in bl for i in al]
```

```
#Concrete Model
model = ConcreteModel(name="Project Time Reduction")

# Decision Variables
model.x = Var(xindx, within=NonNegativeReals)
model.r = Var(rindx, within=NonNegativeReals)

# The objective function
model.obj = Objective(expr= 22.5*model.r[1,2] + 15.75*model.r[2,3]
+ 13.25*model.r[3,5] + 16.5*model.r[3,6] + 25.3*model.r[3,4]
+ 17.5*model.r[4,5] + 14.75*model.r[5,6]
, sense = minimize)

# Time Reduction Constraints
model.TRConstraint1 = Constraint(expr=
  model.r[1,2] <= 2)
model.TRConstraint2 = Constraint(expr=
  model.r[2,3] <= 2)
model.TRConstraint3 = Constraint(expr=
  model.r[3,4] <= 2)
model.TRConstraint4 = Constraint(expr=
  model.r[4,5] <= 2)
model.TRConstraint5 = Constraint(expr=
  model.r[3,5] <= 2)
model.TRConstraint6 = Constraint(expr=
  model.r[3,6] <= 2)
model.TRConstraint7 = Constraint(expr=
  model.r[5,6] <= 2)

# Activity Constraint
model.ActConstraint1 = Constraint(expr=
  model.x[2] - model.x[1] + model.r[1,2] >= 6)
model.ActConstraint2 = Constraint(expr=
  model.x[3] - model.x[2] + model.r[2,3] >= 9)
model.ActConstraint3 = Constraint(expr=
  model.x[4] - model.x[3] + model.r[3,4] >= 5)
model.ActConstraint4 = Constraint(expr=
  model.x[5] - model.x[4] + model.r[4,5] >= 7)
model.ActConstraint5 = Constraint(expr=
  model.x[5] - model.x[3] + model.r[3,5] >= 13)
model.ActConstraint6 = Constraint(expr=
  model.x[6] - model.x[3] + model.r[3,6] >= 6)
model.ActConstraint7 = Constraint(expr=
  model.x[6] - model.x[5] + model.r[5,6] >= 4)
```

```
# Total Time of Project
model.TConstraint = Constraint(expr=
    model.x[6] - model.x[1] <= 28)
```

The following listing shows the results of running the model with the Pyomo modeler. The total cost of reducing the time interval for completion of the project is \$58.50. Note that the completion time of activities $A_{1,2}$, $A_{3,6}$, $A_{3,4}$, and $A_{4,5}$, were not reduced. However, the completion time of activity $A_{5,6}$ was reduced by 2 days.

```
[    0.00] Setting up Pyomo environment
[    0.00] Applying Pyomo preprocessing actions
Reducing Time of Project
[    0.00] Creating model
[    0.01] Applying solver
[    0.05] Processing results
    Number of solutions: 1
    Solution Information
      Gap: 0.0
      Status: feasible
      Function Value: 58.5
    Solver results file: results.json


================================================================
Solution Summary
================================================================

Model Project Time Reduction

  Variables:
    x : Size=6, Index=x_index, Domain=NonNegativeReals
        Key : Lower : Value : Upper : Initial : Fixed : Stale
          1 :    0 :   0.0 :  None :    None : False : False
          2 :    0 :   6.0 :  None :    None : False : False
          3 :    0 :  14.0 :  None :    None : False : False
          4 :    0 :  19.0 :  None :    None : False : False
          5 :    0 :  26.0 :  None :    None : False : False
          6 :    0 :  28.0 :  None :    None : False : False
    r : Size=36, Index=r_index, Domain=NonNegativeReals
        Key     : Lower : Value : Upper : Initial : Fixed : Stale
        (1, 1) :     0 :  None :  None :    None : False : True
        (1, 2) :     0 :   0.0 :  None :    None : False : False
        (1, 3) :     0 :  None :  None :    None : False : True
        (1, 4) :     0 :  None :  None :    None : False : True
```

```
         (1, 5) :       0 :  None :  None :       None : False :  True
         (1, 6) :       0 :  None :  None :       None : False :  True
         (2, 1) :       0 :  None :  None :       None : False :  True
         (2, 2) :       0 :  None :  None :       None : False :  True
         (2, 3) :       0 :   1.0 :  None :       None : False : False
         (2, 4) :       0 :  None :  None :       None : False :  True
         (2, 5) :       0 :  None :  None :       None : False :  True
         (2, 6) :       0 :  None :  None :       None : False :  True
         (3, 1) :       0 :  None :  None :       None : False :  True
         (3, 2) :       0 :  None :  None :       None : False :  True
         (3, 3) :       0 :  None :  None :       None : False :  True
         (3, 4) :       0 :   0.0 :  None :       None : False : False
         (3, 5) :       0 :   1.0 :  None :       None : False : False
         (3, 6) :       0 :   0.0 :  None :       None : False : False
         (4, 1) :       0 :  None :  None :       None : False :  True
         (4, 2) :       0 :  None :  None :       None : False :  True
         (4, 3) :       0 :  None :  None :       None : False :  True
         (4, 4) :       0 :  None :  None :       None : False :  True
         (4, 5) :       0 :   0.0 :  None :       None : False : False
         (4, 6) :       0 :  None :  None :       None : False :  True
         (5, 1) :       0 :  None :  None :       None : False :  True
         (5, 2) :       0 :  None :  None :       None : False :  True
         (5, 3) :       0 :  None :  None :       None : False :  True
         (5, 4) :       0 :  None :  None :       None : False :  True
         (5, 5) :       0 :  None :  None :       None : False :  True
         (5, 6) :       0 :   2.0 :  None :       None : False : False
         (6, 1) :       0 :  None :  None :       None : False :  True
         (6, 2) :       0 :  None :  None :       None : False :  True
         (6, 3) :       0 :  None :  None :       None : False :  True
         (6, 4) :       0 :  None :  None :       None : False :  True
         (6, 5) :       0 :  None :  None :       None : False :  True
         (6, 6) :       0 :  None :  None :       None : False :  True

Objectives:
  obj : Size=1, Index=None, Active=True
      Key  : Active : Value
      None :   True :  58.5

Constraints:
  TRConstraint1 : Size=1
      Key  : Lower : Body : Upper
      None :  None :  0.0 :   2.0
  TRConstraint2 : Size=1
      Key  : Lower : Body : Upper
      None :  None :  1.0 :   2.0
```

```
TRConstraint3 : Size=1
    Key   : Lower : Body : Upper
    None :   None :  0.0 :   2.0
TRConstraint4 : Size=1
    Key   : Lower : Body : Upper
    None :   None :  0.0 :   2.0
TRConstraint5 : Size=1
    Key   : Lower : Body : Upper
    None :   None :  1.0 :   2.0
TRConstraint6 : Size=1
    Key   : Lower : Body : Upper
    None :   None :  0.0 :   2.0
TRConstraint7 : Size=1
    Key   : Lower : Body : Upper
    None :   None :  2.0 :   2.0
ActConstraint1 : Size=1
    Key   : Lower : Body : Upper
    None :   6.0 :  6.0 :  None
ActConstraint2 : Size=1
    Key   : Lower : Body : Upper
    None :   9.0 :  9.0 :  None
ActConstraint3 : Size=1
    Key   : Lower : Body : Upper
    None :   5.0 :  5.0 :  None
ActConstraint4 : Size=1
    Key   : Lower : Body : Upper
    None :   7.0 :  7.0 :  None
ActConstraint5 : Size=1
    Key   : Lower : Body : Upper
    None :  13.0 : 13.0 :  None
ActConstraint6 : Size=1
    Key   : Lower : Body : Upper
    None :   6.0 : 14.0 :  None
ActConstraint7 : Size=1
    Key   : Lower : Body : Upper
    None :   4.0 :  4.0 :  None
TConstraint : Size=1
    Key   : Lower : Body : Upper
    None :   None : 28.0 :  28.0


[   0.05] Applying Pyomo postprocessing actions
[   0.05] Pyomo Finished
```

24.7.3   Formulation Using the Pulp Modeler

The following listing shows the formulation of the model using the Pulp modeler and is stored in file `cpm1b.py`.

```
"""
Python Formulation using the Pulp Modeler
Network problem, project time reduction. File: cpm1.py
J M Garrido, September  2014
usage: python cpm1b.py
"""
print "Network Problem, project time reduction"
# Import PuLP modeler functions
from pulp import *

# Data for Linear Optimization Problem
N = 6  # Number of nodes in network
# Index list for decision variables x
xindx = range(1, N+1)
a = range(1, N+1)
al = range(N)
b = range(1,N+1)
bl = range(N)
# Index list for decision variables x
rindx = [(a[i],b[j]) for j in bl for i in al]

# Create the model to contain the problem data
model = LpProblem("Reducing Time of Project",LpMinimize)

# Decision variables
x = LpVariable.dicts("X", xindx,0,None)
r = LpVariable.dicts("R", rindx,0,None)

# The Pulp objective function
model += 22.5*r[1,2] + 15.75*r[2,3] + 13.25*r[3,5] \
+ 16.5*r[3,6] + 25.3*r[3,4] + 17.5*r[4,5] + 14.75*r[5,6], \
"Time duration of project"

# Time Reduction Constraints
model += r[1,2] <= 2, "Reduc time Act 1-2"
model += r[2,3] <= 2, "Reduc time Act 2-3"
model += r[3,4] <= 2, "Reduc time Act 3-4"
model += r[4,5] <= 2, "Reduc time Act 4-5"
model += r[3,5] <= 2, "Reduc time Act 3-5"
model += r[3,6] <= 2, "Reduc time Act 3-6"
model += r[5,6] <= 2, "Reduc time Act 5-6"
```

```
# Activity Constraints
model += x[2] - x[1] + r[1,2] >= 6,"Activity 1-2"
model += x[3] - x[2] + r[2,3] >= 9,"Activity 2-3"
model += x[4] - x[3] + r[3,4] >= 5,"Activity 3-4"
model += x[5] - x[4] + r[4,5] >= 7,"Activity 4-5"
model += x[5] - x[3] + r[3,5] >= 13,"Activity 3-5"
model += x[6] - x[3] + r[3,6] >= 6,"Activity 3-6"
model += x[6] - x[5] + r[5,6] >= 4,"Activity 5-6"

# Total Time of Project
model += x[6] - x[1] <= 28, "Project Total Time"

# Solve the optimization problem using the PuLP Solver
model.solve(GLPK())

# Print the status of the solution
print "Status:", LpStatus[model.status]

# Print each of the variables with it's resolved optimum value
for v in model.variables():
    print v.name, "=", v.varValue

# Print the optimized value of the objective function
print "Objective Function", value(model.objective)
```

24.8 SUMMARY

Linear optimization modeling can be used to study and calculate various problems that are represented by networks. The typical problems are shortest path problems, maximum flow problems, and the critical path method for project management. A dummy activity has zero duration and is necessary to prevent two activities with the same start node and the same end node.

Key Terms

shortest path	routes	traffic
maximum flow	critical path	minimum spanning tree
graph	nodes	arcs
path	vertex	activity
event	early event time	late event time
total float	critical activity	transshipment point
transporting cost	artificial node	arc capacity

24.9 EXERCISES

24.1 A company manufactures bicycles in two facilities: F1 and F2. Facility F1 can build a maximum of 400 bicycles per year at a cost of $850 per unit. Facility F2 can build a maximum of 300 bicycles per year at a cost of $950. There are two main destinations D1 and D2. Destination D1 demands 400 bicycles per year and D2 demands 300 bicycles per year. The following table includes the transportation costs from a facility F1, F2, or the intermediate point D3 to destination points D1, D2, and D3. The bicycles may be sent to an intermediate location D3. Compute the minimum total cost that meets the demand.

Transportation Costs

From	D1	D2	D3
F1	$250	$200	$75
F2	$155	$130	$95
D3	$45	$35	0

24.2 Repeat the previous problem with the additional condition that the maximum number of bicycles transported to the intermediate point (D3) is 185.

24.3 A low-end computer laptop can be purchased for about $400 and can be used for five years with no salvage value. The maintenance of the computer is estimated to cost $85 for year 1, $140 for year 2, $210 for year 3, $250 for year 4, and $270 for year 5. Compute the minimum total cost of purchasing and using the computer for six years.

24.4 Traffic engineers are studying traffic patterns in part of a city. The immediate problem is to find the maximum flow of vehicles from a source point, S, to a destination point, D. Figure 25.1 and the following table show the flow capacity of the various roads (between nodes). Note that the direction of the traffic is important. Formulate and solve a linear optimization problem that computes the maximum traffic flow from point S to point D.

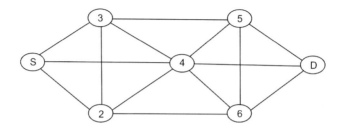

Figure 24.6 Graph of road capacity in a city.

Arc	Capacity	Arc	Capacity
S-3	6	4-5	5
3-S	1	5-4	1
S-4	7	4-6	7
4-S	1	6-4	1
S-2	5	4-D	2
2-S	0	D-4	0
2-4	1	5-D	8
4-2	4	D-5	0
2-3	2	5-6	2
3-2	1	6-5	2
3-4	2	6-D	8
4-3	1	D-6	2
3-5	4	5-3	3

## CHAPTER 25

Integer Linear Optimization Models

25.1 INTRODUCTION

This chapter presents the general concepts and formulation of problems that can be solved with modes of integer linear optimization. These are models with more constraint than the standard linear optimization problems.

An integer linear optimization problem in which all variables are required to be integers is called a *pure integer linear* problem. If some variables are restricted to be integers and others are not, the problem is a *mixed integer linear* problem. The special case of integer variables that are restricted to be 0 or 1 is very useful and are known as pure (mixed) 0–1 linear problems or pure (mixed) *binary* integer linear problems.

25.2 MODELING WITH INTEGER VARIABLES

An integer linear optimization problem is a conventional linear optimization problem with the additional constraints that the decision variables be integer variables. This implies that for a maximization integer linear problem, the optimal value of the objective function is less or equal to the optimal value of the linear optimization problem.

Removing the constraints that the variables must be integer variables results in a linear problem similar to the ones discussed in previous chapters. This linear problem is known as the linear problem *relaxation* of the integer linear problem.

In most situations, computing the optimal value of the two linear problems will produce very different results. The following pure integer linear problem illustrates this. The objective function is to maximize the following expression:

$$z = 20x_1 + 20x_2 + 10x_3,$$

subject to the following restrictions:

$$2x_1 + 20x_2 + 4x_3 \leq 15$$
$$6x_1 + 20x_2 + 4x_3 = 20$$

$$x_1, x_2, x_3 \geq 0, \quad \text{integer.}$$

The following listing shows the formulation of the problem with the Pyomo modeler and is stored in file `intprob.py`. Note that in line 20, the decision variables are declared as non-negative integers.

```
1  """
2  Python Formulation using Pyomo Modeler
3  Relaxed Optimization problem. File: intprobr.py
4  J M Garrido, September  2014
5  usage: pyomo intprobr.py --summary
6  """
7  print "Relaxed Optimization Model"
8  # Import
9  from coopr.pyomo import *
10
11 # Data for Linear Optimization Problem
12 N = 3  # Number of variables
13 # Index list for decision variables x
14 xindx = range(1, N+1)
15
16 #Concrete Model
17 model = ConcreteModel(name="Relaxed Optimization Problem")
18
19 # Decision Variables
20 model.x = Var(xindx, within=NonNegativeIntegers)
21
22 # The objective function
23 model.obj = Objective(expr= 20*model.x[1] + 10*model.x[2]
       + 10*model.x[3], sense = maximize)
24 # Constraints
25 model.AConstraint1 = Constraint(expr=
26    2*model.x[1] + 20*model.x[2] + 4*model.x[3] <= 15)
27 model.AConstraint2 = Constraint(expr=
28    6*model.x[1] + 20*model.x[2] + 4*model.x[3] <= 20)
```

After running the model with Pyomo, this linear optimization problem produces the results that can be observed in the following listing. The result of solving this integer linear problem shows the objective function with the value 60, and the values of the integer variables are $x_1 = 2$, $x_2 = 0$, and $x_3 = 2$.

```
$ pyomo intprob.py --summary
[    0.00] Setting up Pyomo environment
[    0.00] Applying Pyomo preprocessing actions
Integer Optimization Model
[    0.00] Creating model
[    0.00] Applying solver
[    0.06] Processing results
    Number of solutions: 1
    Solution Information
      Gap: 0.0
      Status: optimal
      Function Value: 60.0
    Solver results file: results.json


==============================================================
Solution Summary
==============================================================

Model Integer Optimization Problem

  Variables:
    x : Size=3, Index=x_index, Domain=NonNegativeIntegers
        Key : Lower : Value : Upper : Initial : Fixed : Stale
          1 :     0 :   2.0 :  None :    None : False : False
          2 :     0 :   0.0 :  None :    None : False : False
          3 :     0 :   2.0 :  None :    None : False : False

  Objectives:
    obj : Size=1, Index=None, Active=True
        Key  : Active : Value
        None :   True :  60.0

  Constraints:
    AConstraint1 : Size=1
        Key  : Lower : Body : Upper
        None :  None : 12.0 :  15.0
    AConstraint2 : Size=1
        Key  : Lower : Body : Upper
        None :  None : 20.0 :  20.0

[    0.08] Applying Pyomo postprocessing actions
[    0.08] Pyomo Finished
```

Relaxing the constraints that the variables $x_i, i = 1 \ldots 3$ be integer variables by changing line 20 and declaring the variables as non-negative reals, results in a different numerical solution and can be observed in the following listing.

```
$ pyomo intprob.py --summary
[    0.00] Setting up Pyomo environment
[    0.00] Applying Pyomo preprocessing actions
Integer Optimization Model
[    0.02] Creating model
[    0.02] Applying solver
[    0.06] Processing results
    Number of solutions: 1
    Solution Information
      Gap: 0.0
      Status: feasible
      Function Value: 66.6666666667
    Solver results file: results.json


==============================================================
Solution Summary
==============================================================


Model Integer Optimization Problem

  Variables:
    x : Size=3, Index=x_index, Domain=NonNegativeReals
        Key : Lower : Value         : Upper : Initial : Fixed : Stale
          1 :     0 : 3.33333333333 :  None :    None : False : False
          2 :     0 :           0.0 :  None :    None : False : False
          3 :     0 :           0.0 :  None :    None : False : False

  Objectives:
    obj : Size=1, Index=None, Active=True
        Key  : Active : Value
        None :   True : 66.6666666667

  Constraints:
    AConstraint1 : Size=1
        Key  : Lower : Body          : Upper
        None :  None : 6.66666666667 :  15.0
    AConstraint2 : Size=1
        Key  : Lower : Body : Upper
```

```
     None :  None : 20.0 :   20.0

[    0.16] Applying Pyomo postprocessing actions
[    0.16] Pyomo Finished
```

25.3 APPLICATIONS OF INTEGER LINEAR OPTIMIZATION

A brief description of some typical problems that can be formulated as integer optimization problems follows.

- Knapsack Problem. Given a knapsack with fixed capacity and a collection of items, each with a weight and value, find the number of items to put in the knapsack that maximizes the total value carried subject to the requirement that that weight limitation not be exceeded.

- The Transportation Problem. Given a finite number of suppliers, each with fixed capacity, a finite number of demand centers, each with a given demand, and costs of transporting a unit from a supplier to a demand center, find the minimum cost method of meeting all of the demands without exceeding supplies.

- Assignment Problem. Given equal numbers of people and jobs and the value of assigning any given person to any given job, find the job assignment (each person is assigned to a different job) that maximizes the total value.

- Shortest Route Problem. Given a collection of locations and the distance between each pair of locations, find the cheapest way to get from one location to another.

- Maximum Flow Problem. Given a series of locations connected by pipelines of fixed capacity and two special locations (an initial location or source and a final location or sink), find the way to send the maximum amount from source to sink without violating capacity constraints.

The techniques used to solve integer linear problems are branch-and-bound and branch-and-cut algorithms. They are both implicit enumeration techniques, "implicit" meaning that (hopefully) many solutions will be skipped during enumeration as they are known to be non-optimal.

25.3.1 Branch and Bound

The most widely used method for solving integer linear optimization models is branch and bound. Subproblems are created by restricting the range of the integer variables. For binary variables, there are only two possible restrictions:

setting the variable to 0, or setting the variable to 1. More generally, a variable with lower bound l and upper bound u will be divided into two problems with ranges l to q and $q + 1$ to u, respectively. Lower bounds are provided by the linear optimization relaxation of the problem. If the optimal solution to a relaxed problem is (coincidentally) integral, it is an optimal solution to the subproblem, and the value can be used to terminate searches of subproblems whose lower bound is higher.

25.3.2   Branch and Cut

For branch and cut, the lower bound is again provided by the linear optimization relaxation of the integer program. The optimal solution to this linear program is at a corner of the feasible region (the set of all variable settings which satisfy the constraints). If the optimal solution to the problem is not integral, this algorithm searches for a constraint which is violated by this solution, but is not violated by any optimal integer solutions. This constraint is called a *cutting plane*.

When this constraint is added to the model, the old optimal solution is no longer valid, and so the new optimal will be different, potentially providing a better lower bound. Cutting planes are searched iteratively until either an integral solution is found or it becomes impossible or too expensive to find another cutting plane. In the latter case, a traditional branch operation is performed and the search for cutting planes continues on the subproblems.

Almost all the sample problems described in this chapter are formulated with the Pyomo and Pulp modelers. The underlying solver is GLPK, which is a linear (integer) optimization solver based on the revised simplex method and the branch-and-bound method for the integer variables.

25.4   INTEGER LINEAR OPTIMIZATION: CASE STUDY 1

In the knapsack problem, a hiker needs to take as many items as possible in his knapsack for the next hike. The knapsack has a capacity of 25 pounds. Each item has a priority from 1 to 10 that indicates the relative importance of the item, and a weight. This data is included in Table 25.1.

The decision variables, $x_i, i = 1, \ldots, 5$, for this integer linear problem can have only two possible values: (0, 1). $x_i = 1$ indicates that item i is put in the knapsack and $x_i = 0$ indicates that it is not put in the knapsack. The type of these variables are also known as *binary*.

The goal of this problem is to find the best way to pack the items in the knapsack by priority, given the constraint of the weight capacity of the knapsack. The objective function is to maximize the following expression:

$$z = 7x_1 + 5x_2 + 10x_3 6x_4 + 8x_5,$$

Table 25.1 Items for knapsack.

Item	Priority	Weight
1	7	5.5
2	5	9.5
3	10	13.5
4	6	6.5
5	8	6

subject to the following restrictions:

$$5.5x_1 + 9.5x_2 + 13.5x_3 + 6.5x_4 + 6x_5 \leq 25$$

$$x_i \in \{0, 1\}, i = 1, \ldots, 5.$$

25.4.1 Formulation of the Model Using Pyomo

The following listing shows the formulation of the model using the Pyomo modeler and is stored in file **knapsack1.py**. Note that the decision variables are declared with type *binary* in line 21. The version of the abstract model is stored in file **knapsack2.py** and needs the data in file **knapsack2.dat**.

```
1 """
2 Python Formulation using Pyomo Modeler
3 The knapsack problem. File: knapsack.py
4 J M Garrido, September  2014
5 usage: pyomo knapsack.py --summary
6 """
7 print "Knapsack Model"
8 # Import
9 from coopr.pyomo import *
10
11 # Data for Linear Optimization Problem
12 N = 5  # Number of variables
13 wlimit = 25  # weight limit
14 # Index list for decision variables x
15 xindx = range(1, N+1)
16
17 #Concrete Model
18 model = ConcreteModel(name="Knapsack Problem")
19
20 # Decision Variables
21 model.x = Var(xindx, within=Binary)
```

```
22
23 # The objective function
24 model.obj = Objective(expr= 7*model.x[1] + 5*model.x[2]
     + 10*model.x[3] + 6*model.x[4] + 8*model.x[5], sense = maximize)
25 # Weight Constraint (limit)
26 model.WConstraint1 = Constraint(expr=
27   5.5*model.x[1] + 9.5*model.x[2] + 13.5*model.x[3]
     + 6.5*model.x[4] + 6*model.x[5] <= wlimit)
```

The following listing is produced when running the model with Pyomo. Note that only items 1 3, and 5 are selected.

```
$ pyomo knapsack1.py --summary
[    0.00] Setting up Pyomo environment
[    0.00] Applying Pyomo preprocessing actions
Knapsack Model
[    0.00] Creating model
[    0.00] Applying solver
[    0.05] Processing results
    Number of solutions: 1
    Solution Information
      Gap: 0.0
      Status: optimal
      Function Value: 25.0
    Solver results file: results.json


============================================================
Solution Summary
============================================================

Model Knapsack Problem

  Variables:
    x : Size=5, Index=x_index, Domain=Binary
        Key : Lower : Value : Upper : Initial : Fixed : Stale
          1 :     0 :   1.0 :     1 :    None : False : False
          2 :     0 :   0.0 :     1 :    None : False : False
          3 :     0 :   1.0 :     1 :    None : False : False
          4 :     0 :   0.0 :     1 :    None : False : False
          5 :     0 :   1.0 :     1 :    None : False : False

  Objectives:
    obj : Size=1, Index=None, Active=True
        Key  : Active : Value
```

```
     None :   True :   25.0

 Constraints:
   WConstraint1 : Size=1
       Key  : Lower : Body : Upper
       None :  None : 25.0 :  25.0
```

```
[    0.05] Applying Pyomo postprocessing actions
[    0.05] Pyomo Finished
```

The results listing shows that only items 1, 3, and 5 are put in the knapsack because of the weight constraint.

25.4.2 Formulation of the Model Using Pulp

The following listing shows the formulation of the model using the Pulp modeler and is stored in file **knapsack1.py**. Note that the decision variables are declared with lower and upper bounds 0 and 1, respectively, and type *LpInteger* in line 21.

```
 1 """
 2 Python Formulation using the Pulp Modeler
 3 Knapsack problem. File: knapsack1.py
 4 J M Garrido, September  2014
 5 usage: python knapsack1.py
 6 """
 7 print "Knapsack Problem"
 8 # Import PuLP modeler functions
 9 from pulp import *
10
11 # Data for Linear Optimization Problem
12 N = 5        # Number of decision variables
13 wlimit = 25 # weight limit
14 # Index list for decision variables x
15 xindx = range(1, N+1)
16
17 # Create the model to contain the problem data
18 model = LpProblem("Knapsack",LpMaximize)
19
20 # Decision variables
21 x = LpVariable.dicts("X", xindx,0,1, LpInteger)
22
23 # The Pulp objective function
24 model += 7*x[1] + 5* x[2] + 10*x[3] + 6*x[4] + 8*x[5],
```

```
       "Maximum items to take"
25
26 # Weight Constraint
27 model += 5.5*x[1] + 9.5* x[2] + 13.5*x[3] + 6.5*x[4] + 6*x[5]
       <= wlimit,"Weight"
28
29
30 # Solve the optimization problem using the PuLP Solver
31 model.solve(GLPK())
32
33 # Print the status of the solution
34 print "Status:", LpStatus[model.status]
35
36 # Print each of the variables with it's optimum value
37 for v in model.variables():
38     print v.name, "=", v.varValue
39
40 # Print the optimized value of the objective function
41 print "Objective Function", value(model.objective)
```

25.5   INTEGER LINEAR OPTIMIZATION: CASE STUDY 2

A factory manufactures three types of automobile parts. The following table
has the data on the unit requirements of materials (pounds) and labor (hours),
as well as the unit sales price and unit variable cost. The total available labor
per week is 200 hours and the total of 170 pounds of material available per
week. There are three types of machines that need to be rented, one for each
type of part. The weekly costs for renting these machines are $150.00 for
machine 1, $100.00 for machine 2, and $85.00 for machine 3. The goal of the
problem is to maximize the profit of producing the three types of automobile
parts.

Part type	Material	Labor	Sales price	Var cost
1	7	5	34.00	18.00
2	5	4	22.00	12.00
3	7	12	40.00	22.00

In formulating this linear optimization model, let x_1 denote the number of
parts manufactured of type 1, x_2 denote the number of parts manufactured
of type 2, and x_3 denote the number of parts manufactured of type 3. These
integer variables are used to formulate the total sales, S, with the following
expression:

$$S = 34x_1 + 22x_2 + 40x_3.$$

The total costs variable, V, is formulated with the following expression:

$$V = 18x_1 + 12x_2 + 22x_3.$$

The cost of renting the machines (the fixed cost) depends on whether the parts of a specific type are produced. Let y_1 denote whether the parts of type 1 are produced, y_2 denote whether the parts of type 2 are produced, and y_3 denote whether the parts of type 3 are produced. These binary variables are used to formulate the total fixed cost, F, with the following expression:

$$F = 150y_1 + 100y_2 + 85y_3.$$

The weekly profit is the objective function of the problem and can be expressed as:

$$P = S - V - F.$$

The objective function can then be formulated with the following expression:

$$z = \begin{array}{lll} 34x_1 & + 22x_2 & + 40x_3 \\ -18x_1 & - 12x_2 & - 22x_3 \\ -150y_1 & - 100y_2 & - 85y_3. \end{array}$$

There are two types of problem constraints: The first type of constraint derives from the total available labor and material per week. The second type of constraint associates the type of part produced with the corresponding machine that needs to be rented.

$$\begin{array}{llll} 7x_1 & + 5x_2 & + 7x_3 & \leq 170 \quad \text{(Material available)} \\ 5x_1 & + 4x_2 & + 12x_3 & \leq 200 \quad \text{(Labor available)} \end{array}$$

In this problem, given the total available material and labor per week, the maximum possible number of parts of type 1 that can be produced is 24. In a similar manner, the maximum possible number of parts of type 2 that can be produced is 34, and the maximum possible number of parts of type 3 that can be produced is 16. These constraints are:

$$\begin{array}{ll} x_1 & \leq 24y_1 \\ x_2 & \leq 34y_2 \\ x_3 & \leq 16y_3. \end{array}$$

25.5.1   Formulation of the Model Using Pyomo

The following listing shows the formulation of the problem for using the Pyomo modeler and is stored in file `autoparts.py`.

```
 1 """
 2 Python Formulation using Pyomo Modeler
 3 Auto parts production problem. File: autoparts.py
 4 J M Garrido, September  2014
 5 usage: pyomo autoparts.py --summary
 6 """
 7 print "Auto Parts Production Problem"
 8 # Import
 9 from coopr.pyomo import *
10
11 # Data for Linear Optimization Problem
12 N = 3        # Number of auto parts, decision variables
13 # Index list for decision variables x
14 xindx = range(1, N+1)
15 yindx = range(1, N+1)
16
17 #Concrete Model
18 model = ConcreteModel(name="Auto Parts Production Problem")
19
20 # Decision Variables
21 model.x = Var(xindx, within=NonNegativeIntegers)
22 model.y = Var(yindx, within=Binary)
23 model.yy = Var(within=NonNegativeIntegers)
24
25 # The objective function
26 model.obj = Objective(expr= 16*model.x[1] + 10*model.x[2]
        + 18*model.x[3] - 150*model.y[1] - 100*model.y[2]
        - 85*model.y[3], sense = maximize)
27
28 # Material Constraint
29 model.MConstraint1 = Constraint(expr=
30   7*model.x[1] + 5*model.x[2] + 7*model.x[3] <= 170)
31
32 # Labor Constraint
33 model.LConstraint1 = Constraint(expr=
34   5*model.x[1] + 4*model.x[2] + 12*model.x[3] <= 200)
35
36 # General Constraints
37 model.GConstraint1 = Constraint(expr=
38   model.x[1] - 24*model.y[1] <= 0)
39 model.GConstraint2 = Constraint(expr=
```

```
40    model.x[2] - 34*model.y[2] <= 0)
41 model.GConstraint3 = Constraint(expr=
42    model.x[3] - 16*model.y[3] <= 0)
43 model.GConstraint4 = Constraint(expr=
44    model.x[1] - model.y[1] >= 0)
45 model.GConstraint5 = Constraint(expr=
46    model.x[2] - model.y[2] >= 0)
47 model.GConstraint6 = Constraint(expr=
48    model.x[3] - model.y[3] >= 0)
49
50 # Additional Constraints
51 # At least two machines must be rented
52 #model.AConstraint1 = Constraint(expr=
53 #  model.y[1] + model.y[2] + model.y[3] >= 2)
54
55 # If machine 1 is rented, machine 3 must also
56 #model.AConstraint2 = Constraint(expr=
57 #  model.y[1] - model.y[3] <= 1)
58
59 # Al three machines must be rented
60 #model.AConstraint3 = Constraint(expr=
61 #  model.y[1] + model.y[2] + model.y[3] - model.yy >= 3)
```

Running this linear optimization model with Pyomo produces the results that can be observed in the following listing.

```
$ pyomo autoparts.py --summary
[    0.00] Setting up Pyomo environment
[    0.00] Applying Pyomo preprocessing actions
Auto Parts Production Problem
[    0.02] Creating model
[    0.02] Applying solver
[    0.06] Processing results
    Number of solutions: 1
    Solution Information
      Gap: 0.0
      Status: optimal
      Function Value: 240.0
    Solver results file: results.json

============================================================
Solution Summary
============================================================
```

```
Model Auto Parts Production Problem

   Variables:
    x : Size=3, Index=x_index, Domain=NonNegativeIntegers
        Key : Lower : Value : Upper : Initial : Fixed : Stale
          1 :     0 :   0.0 :  None :    None : False : False
          2 :     0 :  34.0 :  None :    None : False : False
          3 :     0 :   0.0 :  None :    None : False : False
    y : Size=3, Index=y_index, Domain=Binary
        Key : Lower : Value : Upper : Initial : Fixed : Stale
          1 :     0 :   0.0 :     1 :    None : False : False
          2 :     0 :   1.0 :     1 :    None : False : False
          3 :     0 :   0.0 :     1 :    None : False : False
    yy : Size=1, Index=None, Domain=NonNegativeIntegers
        Key  : Lower : Value : Upper : Initial : Fixed : Stale
        None :     0 :  None :  None :    None : False :  True

   Objectives:
    obj : Size=1, Index=None, Active=True
        Key  : Active : Value
        None :   True : 240.0

   Constraints:
    MConstraint1 : Size=1
        Key  : Lower : Body  : Upper
        None :  None : 170.0 : 170.0
    LConstraint1 : Size=1
        Key  : Lower : Body  : Upper
        None :  None : 136.0 : 200.0
    GConstraint1 : Size=1
        Key  : Lower : Body : Upper
        None :  None :  0.0 :   0.0
    GConstraint2 : Size=1
        Key  : Lower : Body : Upper
        None :  None :  0.0 :   0.0
    GConstraint3 : Size=1
        Key  : Lower : Body : Upper
        None :  None :  0.0 :   0.0
    GConstraint4 : Size=1
        Key  : Lower : Body : Upper
        None :   0.0 :  0.0 :  None
    GConstraint5 : Size=1
        Key  : Lower : Body : Upper
        None :   0.0 : 33.0 :  None
    GConstraint6 : Size=1
```

```
Key  : Lower : Body : Upper
None :   0.0 :  0.0 : None
```

```
[    0.08] Applying Pyomo postprocessing actions
[    0.08] Pyomo Finished
```

The results show that the optimal value of profit is $240.00 and only 34 automobile parts of type 2 are to be produced.

An additional constraint on the problem is: at least two machines must be rented. This constraint is formulated with the following expression:

$$y_1 + y_2 + y_3 \geq 2.$$

Running the model with this additional constraint produces the following results.

```
Objective function: 195.00000000
Values of the variables:
x1                           0
x2                          20
x3                          10
y1                           0
y2                           1
y3                           1
```

The results show that the optimal value of profit is now $195.00 and 20 automobile parts of type 2 and 10 automobile parts of type 3 are to be produced.

Instead of the previous constraint, another constraint on the problem is: if machine 1 is rented, then machine 3 should also be rented. This constraint is formulated with the following expression:

$$y_1 - y_3 \leq 1.$$

Running the model with this constraint produces the results that can be observed in the following listing.

```
Objective function: 240.00000000
Values of the variables:
x1                           0
x2                          34
x3                           0
y1                           0
```

y2	1
y3	0

The results are the same as the first solution to this problem. The results show that the optimal value of profit is $240.00 and only 34 automobile parts of type 2 are to be produced.

Instead of the previous constraint, another constraint on the problem is: all three machines should be rented. This constraint is formulated with the following expression:

$$y_1 + y_2 + y_3 = 3.$$

Running the model with this constraint produces the results that can be observed in the following listing.

```
Objective function: 69.00000000

Values of the variables:
x1                      11
x2                       3
x3                      11
y1                       1
y2                       1
y3                       1
```

The results show that the optimal value of profit is $69.00 and 11 automobile parts of type 1, 3 parts of type 2, and 11 parts of type 3 are to be produced.

25.5.2 Formulation of the Model Using Pulp

The following listing shows the formulation of the model for using the Pulp modeler.

```
"""
Python Formulation using the Pulp Modeler
Auto parts production problem. File: autoparts.py
J M Garrido, September  2014
usage: python autoparts.py
"""

print "Automobile Parts"
# Import PuLP modeler functions
```

```
from pulp import *

# Data for Linear Optimization Problem
N = 3          # Number of auto parts, decision variables

# Index list for decision variables x
xindx = range(1, N+1)
yindx = range(1, N+1)

# Create the model to contain the problem data
model = LpProblem("Auto Parts",LpMaximize)

# Decision variables
x = LpVariable.dicts("X", xindx,0,None, LpInteger)
y = LpVariable.dicts("Y", yindx,0,1, LpInteger)
# Auxiliary variable
yy = LpVariable("YY", 0, None, LpInteger)

# The Pulp objective function
model += 16*x[1] + 10* x[2] + 18*x[3] - 150*y[1] - 100*y[2] - 85*y[3],
    "Maximum  auto parts"

# Material Constraint
model += 7*x[1] + 5*x[2] + 7*x[3]  <= 170,"Material"

# Labor Constraint
model += 5*x[1] + 4*x[2] + 12*x[3] <= 200, "Labor"

# Parts Constraints
model += x[1] - 24*y[1] <= 0, "Max production type 1"
model += x[2] - 34*y[2] <= 0, "Max production part 2"
model += x[3] - 16*y[3] <= 0, "max production part 3"

# General Constraints
model += x[1] - y[1] >= 0, "Part type 1"
model += x[2] - y[2] >= 0, "Part type 2"
model += x[3] - y[3] >= 0, "Part type 3"

# Machine Constraints
#model += y[1] + y[2] + y[3] >= 2, "At least 2 machines"
#model += y[1] - y[3] <= 1, "If machine 1 then also machine 3"
#model += y[1] + y[2] + y[3] + yy <= 3, "All 3 machines must be rented"

# Solve the optimization problem using the specified PuLP Solver
model.solve(GLPK())
```

```
# Print the status of the solution
print "Status:", LpStatus[model.status]

# Print each of the variables with it's resolved optimum value
for v in model.variables():
    print v.name, "=", v.varValue

# Print the optimized value of the objective function
print "Objective Function", value(model.objective)
```

25.6 SUMMARY

Integer linear optimization modeling can be used to study and calculate various problems that are represented by variables that have only integer values. The additional restrictions on decision variables, is that the type of these variables are to be binary. If not all variables have integer values, then it is a mixed integer linear problem. Typical applications of problems that can be solved with integer linear optimization are knapsack problems, transportation problems, assignment problems, shortest route problems, and maximum flow problems.

Key Terms

integer variables	binary variables	integer relaxation
shortest route	knapsack	transportation
assignment	maximum flow	branch & bound
branch & cut	mixed integer	

25.7 EXERCISES

25.1 A company manufactures two types of duffel traveling bags. Three units of material are used to manufacture bags of type 1, and six units of material to manufacture bags of type 2. The total units of available material are 150. The initial setup cost to manufacture bags is $15 for type 1 and $30 for type 2. The company can sell bags of type 1 with a profit of $2 per bag and type 2 with a profit of $5 per bag. Compute the number of bags of each type to produce and the maximum profits possible.

25.2 A computer distributer has four warehouses (W1, W2, W3, and W4) and each can ship 120 units per week. The operational costs of the warehouses are $385 for W1, $480 for W2, $290 for W3, and $145 for W4. There are three demand points (D1, D2, and D3) that have the following weekly demands: 75 for D1, 68 for D2, and 35 for D3. The distributer has the following operational restrictions: at most, two warehouses can be operating; if warehouse W1 is operating, then W2 must also be operating; and either warehouse W4 or warehouse W2 must be operating. Compute the minimum weekly cost while meeting demand. The following table includes data on the cost of transporting a unit from a warehouse to a destination point.

From	D1	D2	D3
W1	$18	$36	$47
W2	$45	$12	$22
W3	$22	$34	$15
W4	$20	$46	$32

25.3 Traffic engineers are studying traffic patterns in part of a city. The immediate problem is to find the maximum flow of vehicles from a source point, S, to a destination point, D. Figure 25.1 and the following table show the flow capacity of the various roads (between nodes). Note that the direction of the traffic is important. Formulate and solve a linear optimization problem that computes the maximum traffic flow from point S to point D.

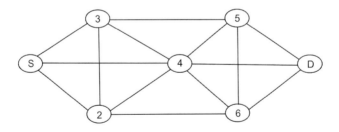

Figure 25.1 Graph of road capacity in city.

Arc	Capacity	Arc	Capacity
S-3	6	4-5	5
3-S	1	5-4	1
S-4	7	4-6	7
4-S	1	6-4	1
S-2	5	4-D	2
2-S	0	D-4	0
2-4	1	5-D	8
4-2	4	D-5	0
2-3	2	5-6	2
3-2	1	6-5	2
3-4	2	6-D	8
4-3	1	D-6	2
3-5	4	5-3	3

Bibliography

[1] Hossein Arsham. *Deterministic Modeling: Linear Optimization with Applications*. Web site (University of Baltimore): `http://home.ubalt.edu/ntsbarsh/opre640a/partIII.htm`.

[2] Joyce Farrell. *Programming Logic and Design*. Second edition. Course Technology (Thompson) 2002.

[3] Behrouz A. Forouzan. *Foundations of Computer Science: From Data manipulation to Theory of Computation*. Brooks/Cole (Thompson), 2003.

[4] G. Fulford, P. Forrester, and A. Jones. *Modeling with Differential and Difference Equations*. Cambridge University Press, 1997.

[5] Garrido, J. M. *Introduction to Computational Modeling Using C and Open-Source Tools*. Boca Raton: CRC Press/Taylor and Francis, 2014.

[6] Guttag, John V. *Introduction to Computation and Programming Using Python*. Cambridge: MIT Press, 2013.

[7] Rod Haggarty. *Discrete Mathematics for Computing*. Addison Wesley (Pearson), Harlow, UK, 2002.

[8] Dan Kalman. *Elementary Mathematical Models*. The Mathematical Association of America, 1997.

[9] Roland E. Larsen, Robert P. Hostteller, and Bruce H. Edwards. *Brief Calculus with Applications*. Alternate third ed., D. C. Heath and Company, 1991.

[10] Charles F. Van Loan and K.-Y. Daisy Fan. *Insight through Computing: A MATLAB Introduction to Computational Science and Engineering*. SIAM-Society for Industrial and Applied Mathematics, 2009.

[11] E. B. Magrab, S. Azarm, B. Balachandran, J. H. Duncan, K. E. Herold, and G. C. Walsh. *An Engineer's Guide to MATLAB: With Applications from Mechanical, Aerospace, Electrical, Civil, and Biological Systems Engineering*. Third Ed. Prentice Hall, Pearson, Upper Saddle River, NJ, 2011.

[12] Douglas Mooney and Randall Swift. *A Course in Mathematical Modeling*. The Mathematical Association of America, 1999.

[13] Holly Moore. *MATLAB for Engineers*. Sec Ed. Prentice Hall, Pearson, Upper Saddle River, NJ, 2009.

[14] E. Part-Enander, A. Sjoberg, B. Melin, and P. Isaksson. *The MATLAB Handbook*. Addison-Wesley Longman, Harlow, UK, 1996.

[15] D. M. Etter. *Engineering Problem Solving with C*. Pearson/Prentice Hall, 2005.

[16] Harold J. Rood. *Logic and Structured Design for Computer Programmers*. Third edition. Brooks/Cole (Thompson), 2001.

[17] Rama N. Reddy and Carol A. Ziegler. *C Programming for Scientists and Engineers*. Jones and Bartlett Pub. Sudbury, Massachusetts, 2010.

[18] Angela B. Shiflet and George W. Shiflet. *Introduction to Computational Science: Modeling and Simulation for the Sciences*. Princeton University Press, Princeton, NJ, 2006.

[19] L. F. Sampine, R. C. Allen, and S. Prues. *Fundamentals of Numerical Computing*. John Wiley and Sons, New York, 1997.

[20] David M. Smith. *Engineering Computation with MATLAB*. Addison-Wesley, Pearson Education, Boston MA, 2010.

[21] Robert E. White. *Computational Mathematics: Models, Methods, and Analysis with MATLAB and MPI*. Chapman and Hall/CRC. September 17, 2003.

[22] Wayne L. Winston. *Operations Research: Applications and Algorithms*. PWS-Kent Pub. Co., 1987.

[23] Wing, Jeannette M. "Computational Thinking." *Communications of the ACM*. March 2006. Vol. 49, No. 3.

Index

abstract model, 342
abstraction, 4, 14, 58, 135, 189
access mode, 134
accessor methods, 140
accumulator, 87
accuracy, 20
activation record, 173
activity, 422
adjacent solution, 329
algorithm, 4, 57, 87
algorithm notation, 4, 58
alternation, 72
analytical methods, 178
approximation error, 21
approximations, 21
arc, 407
area, 298
argument, 53
arguments, 52
arithmetic, 27
array elements, 257
artificial arc, 408
assignment operator, 32
assignment problem, 397
assignment statement, 32
association, 137
assumption, 178
assumptions, 189
attributes, 133
auxiliary methods, 140
average, 182

back of list, 158
backward difference, 294
balanced problem, 372
base case, 167
base class, 143, 144, 148
basic variables, 329
binding constraint, 327

bits, 19
bytecode, 17

capacity, 415
Cartesian plane, 37
central difference, 294
characteristics, 143
class, 132
class hierarchy, 143
class relationship, 137
class variables, 139
coefficients vector, 214, 218
cofactor, 260
collections, 132
column vector, 241, 243, 258
commands, 57
communication diagram, 134
compilation, 17
complex numbers, 27
complex roots, 219
components, 131
compound condition, 71
computational model, 4
computational science, 5
computational thinking, 5
computations, 29
computer implementation, 4, 58
computer tools, 5
conceptual model, 8
concrete model, 342
condition, 83, 86
conditional expressions, 69
conditions, 69
conjugate matrix, 260
conjugate transpose, 261
console, 184
constraints, 332
constructor, 145
constructor methods, 140